METHODS IN MOLECULAR BIOLOGY

Series Editor
John M. Walker
School of Life and Medical Sciences
University of Hertfordshire
Hatfield, Hertfordshire, UK

For further volumes:
http://www.springer.com/series/7651

For over 35 years, biological scientists have come to rely on the research protocols and methodologies in the critically acclaimed *Methods in Molecular Biology* series. The series was the first to introduce the step-by-step protocols approach that has become the standard in all biomedical protocol publishing. Each protocol is provided in readily-reproducible step-by-step fashion, opening with an introductory overview, a list of the materials and reagents needed to complete the experiment, and followed by a detailed procedure that is supported with a helpful notes section offering tips and tricks of the trade as well as troubleshooting advice. These hallmark features were introduced by series editor Dr. John Walker and constitute the key ingredient in each and every volume of the *Methods in Molecular Biology* series. Tested and trusted, comprehensive and reliable, all protocols from the series are indexed in PubMed.

High Throughput Gene Screening

Methods and Protocols

Edited by

Valerie J. Carabetta and Olaitan Akintunde

Department of Biomedical Sciences, Cooper Medical School of Rowan University, Camden, NJ, USA

Humana Press

Editors
Valerie J. Carabetta
Department of Biomedical Sciences
Cooper Medical School of Rowan University
Camden, NJ, USA

Olaitan Akintunde
Department of Biomedical Sciences
Cooper Medical School of Rowan University
Camden, NJ, USA

ISSN 1064-3745 ISSN 1940-6029 (electronic)
Methods in Molecular Biology
ISBN 978-1-0716-4194-1 ISBN 978-1-0716-4192-7 (eBook)
https://doi.org/10.1007/978-1-0716-4192-7

This Humana imprint is published by the registered company Springer Science+Business Media, LLC, part of Springer Nature.
The registered company address is: 1 New York Plaza, New York, NY 10004, U.S.A.

Preface

Within the last 20 years, there have been enormous strides in our capacity to study entire genomes, global gene expression, microbial populations, and the underlying genetics of disease states. This is largely due to significant advances in high-throughput sequencing (HTS) technologies, also called next-generation sequencing (NGS), resulting in a shift to massive parallel sequencing platforms instead of the traditional Sanger sequencing approach. This technology is applicable in all fields of biology and led to exciting breakthroughs in our understanding of complex biological processes and human diseases. The basic workflow of most HTS approaches is nucleic acid isolation, library construction, amplification, sequencing, and data analysis. The modern HTS approaches can generate over 100 gigabytes of data, so the downstream computational analyses have concordantly evolved. As the technology advances with improvements in the error rate and read length, the number of applications of HTS continues to grow. With decreasing costs per reaction and increasing accuracy and reproducibility, the use of this technology for clinical applications is steadily increasing.

The assembly of this collection was only possible thanks to the valuable contributions of experts who routinely perform HTS as part of their research. The collection begins with a chapter discussing the evolution of HTS technologies (Chap. 1), from the early years to where we are today. Next, Chaps. 2, 3 and 4 focus on genome-based sequencing technologies, including whole genome sequencing, targeted sequencing, and metagenomic analysis. Chapters 5, 6 and 7 cover epigenome-based technologies, including EM-seq, MNase-seq, and ATAC-seq. Chapters 8, 9, 10, 11 and 12 describe transcriptome-based sequencing technologies, which encompass RNA-seq, scRNA-seq, miRNA-seq, RIL-seq, and performing HTS techniques on challenging formalin-fixed and paraffin-embedded (FFPE) tissue. Chapters 13 and 14 explore HTS modalities for structural studies of DNA, which include Hi-C and DamID-seq. Finally, Chaps. 15 and 16 discuss potential applications of HTS for clinical use, including cancer research and treatment, and personalized medicine.

We would like to personally thank all the contributors to this collection who shared their knowledge and expertise with the community. We have full confidence that this collection will become a valuable resource for all those interested in incorporating HTS into their research. We are excited to see the next generation of discoveries that result from this technology.

Camden, NJ, USA

Valerie J. Carabetta
Olaitan Akintunde

Contents

Preface .... *v*
Contributors .... *ix*

PART I GENOME-BASED SEQUENCING TECHNOLOGIES

1 The Evolution of Next-Generation Sequencing Technologies 3
Olaitan Akintunde, Trichina Tucker, and Valerie J. Carabetta

2 A Step-by-Step Guide to Sequencing and Assembly of Complete Bacterial Genomes Using the Oxford Nanopore MinION 31
Anil Kumar, Max M. Häggblom, and Lee J. Kerkhof

3 Targeted Next-Generation Sequencing in Rare Diseases 45
Aleša Kristan and Nataša Debeljak

4 Collection and Processing of Samples for Next-Generation Sequencing to Study the Gut Microbiome 59
Maxwell Bokor, Emily Chiacchiaro, and Sangita Phadtare

PART II EPIGENOME-BASED SEQUENCING TECHNOLOGIES

5 Whole Genome Methylation Sequencing via Enzymatic Conversion (EM-seq): Protocol, Data Processing, and Analysis 73
Nelly N. Olova and Simon Andrews

6 MNase-seq to Identify Genome-Wide DNA-Protein Interactions 99
Benjamin R Carone

7 ATAC-seq for Characterizing Host and Pathogen Genome Accessibility During Virus Infection 111
Dawei Liu, Timothy R. Howard, and Ileana M. Cristea

PART III TRANSCRIPTOME-BASED SEQUENCING TECHNOLOGIES

8 RNA Sequencing Protocols for Short-Read Sequencing 125
Laura Vasquez-Velez, Veera D'Mello, and Patricia Soteropoulos

9 Dissection of Gene Expression at the Single-Cell Level: scRNA-seq 159
Samar Nasser Chehimi

10 The Study of Small RNA Sequencing from Biological Samples (miRNA-seq) 175
Weam Othman Elbezanti, Kishore B. Challagundla, and Manoj K. Pandey

11 Biological Insights from RNA–RNA Interactomes in Bacteria, as Revealed by RIL-seq 189
Aviezer Silverman and Sahar Melamed

12 Extraction, Purification, and Next-Generation Sequencing (NGS) Analysis of DNA and RNA from Formalin-Fixed and Paraffin-Embedded (FFPE) Tissue 207
Mainul Hoque, Veera D'Mello, Seema Husain, and Patricia Soteropoulos

PART IV SEQUENCING TECHNOLOGIES FOR DNA STRUCTURE

13 Using Chromosome Conformation Capture Combined with Deep Sequencing (Hi-C) to Study Genome Organization in Bacteria........ 231
Qin Liao and Xindan Wang

14 An Optimized Adaptation of DamID for NGS Applications........ 245
Karen L. Reddy and Xianrong Wong

PART V APPLICATIONS OF NGS TECHNOLOGY

15 How Have Massively Parallel Sequencing Technologies Furthered Our Understanding of Oncogenesis and Cancer Progression? 265
Diogo Ann Onuselogu, Saskia Benz, and Suparna Mitra

16 The Use of Next-Generation Sequencing in Personalized Medicine........ 287
Liya Popova and Valerie J. Carabetta

Correction to: The Study of Small RNA Sequencing from Biological Samples (miRNA-seq) C1
Weam Othman Elbezanti, Kishore B. Challagundla, and Manoj K. Pandey

Index *317*

Contributors

OLAITAN AKINTUNDE • *Department of Biomedical Sciences, Cooper Medical School of Rowan University, Camden, NJ, USA*
SIMON ANDREWS • *Bioinformatics Group, Babraham Institute, Cambridge, UK*
SASKIA BENZ • *Faculty of Medicine and Health, University of Leeds, Leeds, UK*
MAXWELL BOKOR • *Department of Biomedical Sciences, Cooper Medical School of Rowan University, Camden, NJ, USA*
VALERIE J. CARABETTA • *Department of Biomedical Sciences, Cooper Medical School of Rowan University, Camden, NJ, USA*
BENJAMIN R CARONE • *Department of Biological & Biomedical Sciences, Rowan University, Glassboro, NJ, USA*
KISHORE B. CHALLAGUNDLA • *Department of Biochemistry and Molecular Biology, The Fred and Pamela Buffett Cancer Center, The Child Health Research Institute, University of Nebraska Medical Center, Omaha, NE, USA*
SAMAR NASSER CHEHIMI • *Department of Psychiatry, University of Pennsylvania, Philadelphia, PA, USA*
EMILY CHIACCHIARO • *Department of Biomedical Sciences, Cooper Medical School of Rowan University, Camden, NJ, USA*
ILEANA M. CRISTEA • *Department of Molecular Biology, Princeton University, Princeton, NJ, USA*
VEERA D'MELLO • *Genomics Center, Rutgers New Jersey Medical School, Newark, NJ, USA*
NATAŠA DEBELJAK • *Faculty of Medicine, Medical Centre for Molecular Biology, University of Ljubljana, Institute of Biochemistry and Molecular Genetics, Ljubljana, Slovenia*
WEAM OTHMAN ELBEZANTI • *Department of Surgery, Cooper University Hospital, Camden, NJ, USA*
MAX M. HÄGGBLOM • *Department of Biochemistry and Microbiology, Rutgers University, New Brunswick, NJ, USA*
MAINUL HOQUE • *Genomics Center, Rutgers New Jersey Medical School, Newark, NJ, USA*
TIMOTHY R. HOWARD • *Department of Molecular Biology, Princeton University, Princeton, NJ, USA*
SEEMA HUSAIN • *Genomics Center, Rutgers New Jersey Medical School, Newark, NJ, USA*
LEE J. KERKHOF • *Department of Marine and Coastal Sciences, Rutgers University, New Brunswick, NJ, USA*
ALEŠA KRISTAN • *Faculty of Medicine, Medical Centre for Molecular Biology, University of Ljubljana, Institute of Biochemistry and Molecular Genetics, Ljubljana, Slovenia*
ANIL KUMAR • *Department of Biochemistry and Microbiology, Rutgers University, New Brunswick, NJ, USA*
QIN LIAO • *Department of Biology, Indiana University, Bloomington, IN, USA*
DAWEI LIU • *Department of Molecular Biology, Princeton University, Princeton, NJ, USA*
SAHAR MELAMED • *Department of Microbiology and Molecular Genetics, Institute for Medical Research Israel-Canada, Faculty of Medicine, The Hebrew University of Jerusalem, Jerusalem, Israel*
SUPARNA MITRA • *Faculty of Medicine and Health, University of Leeds, Leeds, UK*

Nelly N. Olova • *MRC Human Genetics Unit, MRC Institute of Genetics and Cancer, University of Edinburgh, Edinburgh, UK; School of Biodiversity, One Health and Veterinary Medicine, University of Glasgow, Glasgow, UK*
Diogo Ann Onuselogu • *Faculty of Medicine and Health, University of Leeds, Leeds, UK*
Manoj K. Pandey • *Department of Biomedical Sciences, Cooper Medical School of Rowan University, Camden, NJ, USA*
Sangita Phadtare • *Department of Biomedical Sciences, Cooper Medical School of Rowan University, Camden, NJ, USA*
Liya Popova • *Department of Biomedical Sciences, Cooper Medical School of Rowan University, Camden, NJ, USA*
Karen L. Reddy • *Department of Biological Chemistry and Center for Epigenetics, Johns Hopkins University of Medicine, Baltimore, MD, USA*
Aviezer Silverman • *Department of Microbiology and Molecular Genetics, Institute for Medical Research Israel-Canada, Faculty of Medicine, The Hebrew University of Jerusalem, Jerusalem, Israel*
Patricia Soteropoulos • *The Genomics Center, Department of Microbiology, Biochemistry, and Molecular Genetics, Rutgers New Jersey Medical School, Newark, NJ, USA; Genomics Center, Rutgers New Jersey Medical School, Newark, NJ, USA*
Trichina Tucker • *Department of Biomedical Sciences, Cooper Medical School of Rowan University, Camden, NJ, USA*
Laura Vasquez-Velez • *Genomics Center, Rutgers New Jersey Medical School, Newark, NJ, USA*
Xindan Wang • *Department of Biology, Indiana University, Bloomington, IN, USA*
Xianrong Wong • *Sidney Kimmel Cancer Institute, Johns Hopkins University School of Medicine, Baltimore, MD, USA; A*STAR Skin Research Labs (A*SRL), Agency for Science, Technology and Research (A*STAR), Singapore, Republic of Singapore*

Part I

Genome-Based Sequencing Technologies

Chapter 1

The Evolution of Next-Generation Sequencing Technologies

Olaitan Akintunde, Trichina Tucker, and Valerie J. Carabetta

Abstract

The genetic information that dictates the structure and function of all life forms is encoded in the DNA. In 1953, Watson and Crick first presented the double helical structure of a DNA molecule. Their findings unearthed the desire to elucidate the exact composition and sequence of DNA molecules. Discoveries and the subsequent development and optimization of techniques that allowed for deciphering the DNA sequence has opened new doors in research, biotech, and healthcare. The application of high-throughput sequencing technologies in these industries has positively impacted and will continue to contribute to the betterment of humanity and the global economy. Improvements, such as the use of radioactive molecules for DNA sequencing to the use of florescent dyes and the implementation of polymerase chain reaction (PCR) for amplification, led to sequencing a few hundred base pairs in days, to automation, where sequencing of thousands of base pairs in hours became possible. Significant advances have been made, but there is still room for improvement. Here, we look at the history and the technology of the currently available next-generation sequencing platforms and the possible applications of such technologies to biomedical research and beyond.

Key words Next-generation sequencing, RNA-seq, DNA-seq, High-throughput sequencing, Single-molecule sequencing

1 A Brief History of DNA Sequencing

Deoxyribonucleic acid (DNA) is made up of four nitrogenous bases, which encodes information that provides the blueprint for the cell and directs physiology. Proper cell functioning requires proper DNA transcription and eventual translation into proteins. Any mutation or alteration in the DNA sequence in the genome could lead to possible cellular defects or disease. The early attempts to determine the identity and correct order of bases present in a DNA fragment, referred to as first-generation sequencing, involve the use of chemicals to cleave bases within a DNA molecule or the use of chain-terminating nucleotides, followed by the manual separation of fragments generated via electrophoresis [1, 2]. Maxam and Gilbert sequencing, also called the partial chemical degradation

Valerie J. Carabetta and Olaitan Akintunde (eds.), *High Throughput Gene Screening: Methods and Protocols*, Methods in Molecular Biology, vol. 2866, https://doi.org/10.1007/978-1-0716-4192-7_1,

method, utilizes chemicals that target specific, individual purines or pyrimidines to cleave the radiolabeled DNA backbone into fragments. The DNA is labeled with radioactive ^{32}P at the 5′ end. Next, specific chemicals are used to modify the nucleotides. For example, hydrazine removes the nitrogenous base from cytosine and thymine, but in the presence of high-salt concentrations, it preferentially cleaves cytosine. Formic acid is used to methylate adenine and guanine is methylated by dimethyl sulfate. Following base modification, piperidine is used to cleave the sugar-phosphate backbone, producing fragments that are analyzed using polyacrylamide gel electrophoresis [3].

Later, Sanger sequencing was introduced, which uses dideoxynucleotides (ddNTPs). ddNTPs are chemically modified dNTPs that lack hydroxyl groups at both the 2′ and 3′ positions and terminate extension once incorporated. This methodology requires a single-stranded DNA (ssDNA) template and a short oligonucleotide primer. Each ddNTP is added to the reaction at a fraction of the concentration of regular dNTPs, in the presence of the Klenow fragment for synthesis of the complementary strand. When a ddNTP is incorporated by the polymerase, chain termination occurs (*see* Fig. 1). Four parallel reactions are carried out yielding fragments of varying lengths, each ending with the specific ddNTP in use. This technique was originally performed with radiolabeled nucleotides, but the development of fluorescence labeling now allows for fluorometric detection [1, 4]. The read length using this methodology is typically less than 1 kilobase (kb), which allows for low-throughput sequencing.

To sequence larger fragments or entire genomes, the DNA must be fragmented, cloned, individually sequenced, and then computationally assembled into genomes. Automated Sanger sequencing was first introduced by Applied Biosystems (ABI) in 1986, following the unveiling of the ABI Prism 310, a commercial DNA sequencing machine that employed automated capillary electrophoresis and an advanced imaging system. Four separate reactions, each with one of the four ddNTPs, were combined on a single gel and fluorescence was measured. A computer program was designed to interpret and compile the sequences, removing the need for manual sequence interpretation, reducing human error [5]. This automation allowed for early error detection and higher resolution [6]. Craig Venter and colleagues established the first, fully equipped sequencing facility, with six automated sequencers at the National Institutes of Health (NIH). In the 1990s, larger facilities were established, like the Institute for Genomic Research (TIGR), which had 30 sequencers and the Welcome Trust Sanger Institute. The availability of automated sequencing led to the discovery of 337 new and 48 homologous human genes via the expressed sequence tag (EST) method. ESTs are short DNA sequences (<500 bp) obtained from the 5′ and 3′ regions of

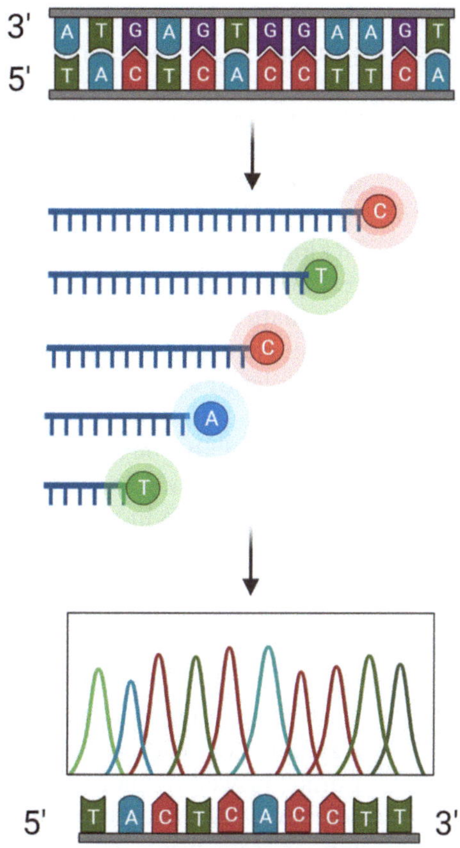

Fig. 1 Sanger sequencing workflow. The reverse strand of linear DNA is synthesized using DNA polymerase and four fluorescently labeled dideoxyribonucleotides (ddNTPs), which when incorporated will terminate chain elongation and emit a fluorescence signal unique to each ddNTP. The resulting fragments are analyzed via capillary gel electrophoresis. Created with BioRender.com

cDNA. They are used as an efficient tool for the rapid identification of coding sequences in a genome by sequence matching, which reduces time and costs compared to other methods. This approach was widely used in the search for genes linked to human diseases, such as Huntington's disease. The National Center for Biotechnology Information (NCBI) has an EST database, a collection of annotated ESTs that serve as a gene discovery tool [7].

Later, next-generation sequencing (NGS), also referred to as high-throughput sequencing (HTS) or second generation sequencing, was developed. NGS technology started with the development of pyrosequencing [8] and was first commercially available in 2005 as the 454/Roche platform [9]. NGS utilizes a massive parallel sequencing approach, with the ability to process large DNA samples at a significantly reduced cost and time. One downside of Sanger sequencing is that it is relatively low throughput and costly. For example, the first human genome sequence cost approximately 1 billion dollars [10]. With NGS, millions or even billions of reads can be produced in a single run within a few hours or days, making it more efficient than Sanger sequencing [11, 12]. The development of reversible dye terminator technology was a large contributing factor to the success of NGS technology [13]. Unlike sanger sequencing, which uses ddNTPs lacking the 3'-OH group to terminate extension irreversibly, this technology utilizes modified nucleotides attached to a reversible termination group. This group includes a fluorescent dye attached to a nucleotide via cleavable linker and can be photocleaved using ultraviolet irradiation after the nucleotide is incorporated and identified. This allows for the continuation of DNA extension by incorporation of the next nucleotide [14]. Photocleavage of the tags eliminates the need to use chemical reagents in the reaction, yielding a cleaner product without additional purification steps.

The human genome (~3 billion bp) sequencing using the 454 Genome Sequencer FLX was completed in 2 months, with a cost of about $100,000. In contrast, it took 15 years with Sanger sequencing, required multiple collaborations across the world, and was costly [15, 16]. A shortcoming of NGS is the inability to read the entire length of the genome sequence, as millions of short fragments or reads are generated. Therefore, it takes a lot of computing power to piece these short reads together into one large sequence [12]. Recently, the development of new technologies for longer reads was developed, which can sequence double-stranded DNA one strand at a time and yield read lengths of >10 kb. These technologies eliminate the need for amplification and make downstream genome assembly easier. While there will always remain some imperfections, notable improvements have been made and eventually will be a great option for disease diagnosis, by detecting structural variations linked to pathology [2]. In this review, we explore the popular NGS platforms currently being used. We discuss the diverse applications of NGS, including whole-genome sequencing and transcriptomics. We highlight the strengths and limitations of each approach and discuss some clinical applications.

2 Popular Next-Generation Sequencing Technologies

DNA sample preparation, immobilization, and sequencing are the three critical steps of NGS technology. Sample preparation for sequencing usually involves attachment of specifically defined adapter sequences to the ends of random DNA fragments. The resulting products of this DNA preparation is commonly referred to as the "sequencing library." The addition of adapters is necessary to secure the DNA fragments of the sequencing library to a solid platform and specify the site where the sequencing reaction is initiated. Some NGS platforms are largely based on pyrosequencing, like the Roche 454 series. Pyrosequencing is a technique where a DNA fragment is attached to a bead in an emulsion and following emulsion PCR, multiple copies of the same DNA fragment surround the bead. Luminescence is used to detect pyrophosphate (PP_i) release as a nucleotide is incorporated. ATP sulfurylase in the reaction mixture converts PP_i into ATP, which is used by the enzyme luciferase to produce light. Advantages of this technique are that it uses natural nucleotides instead of modified ones for chain termination and allows for observation in real time. A modified version of this technique, known as ion torrent, measures the pH change for sequence determination [10, 15].

Other NGS platforms, such as Illumina, require the use of a reversible dye terminator that allows for the addition of one nucleotide at a time during the DNA extension phase [1, 16]. Illumina uses bridge amplification, in which adapters ligated on both ends of a fragment will bind oligonucleotides on a flow cell, which bend and form a loop or bridge structure. This loop serves as a primer for reverse strand synthesis (*see* Fig. 2). Multiple rounds of extension yield millions of fragments and the original sequence is determined by the reversible dye terminator method [13, 17]. Another popular platform is SOLiD (Small Oligonucleotide Ligation and Detection

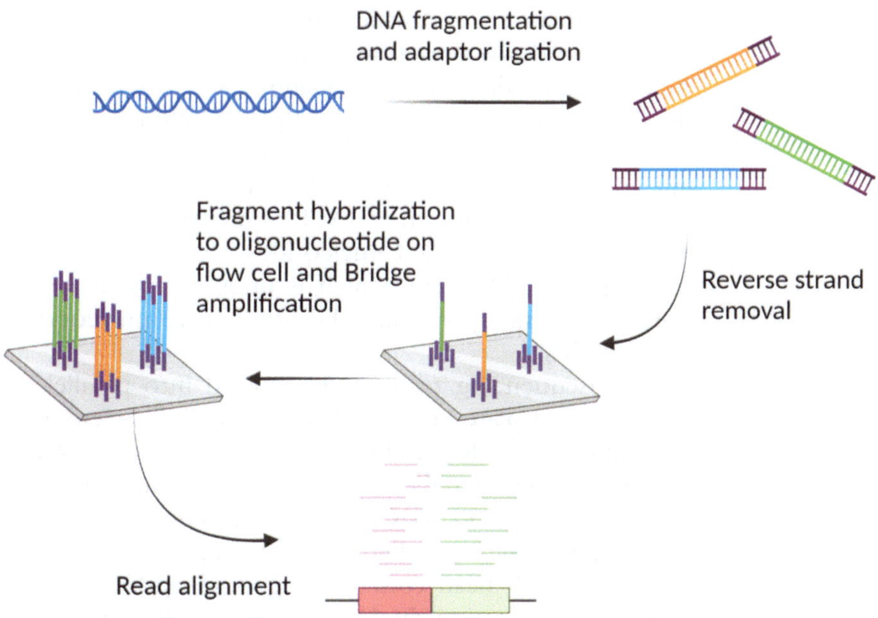

Fig. 2 Illumina sequencing platform. DNA is fragmented, linearized, and ligated to adapters that will bind surface-tethered oligos via complementarity on a flow cell. The DNA fragments bend over to form bridges, followed by bridge amplification and cluster formation. Sequences are determined following incorporation of reversible dye ddNTPs. The fragments in the library are then assembled using various bioinformatic pipelines. Created with BioRender.com

system) sequencing that sequences oligonucleotides with fixed lengths by ligation of fluorescent molecules. DNA ligases are used to add fluorescent oligonucleotides that anneal to DNA templates in a sequential fashion [15, 18]. Multiple ligation cycles to a fluorescently labeled probe containing dinucleotides allows for sequence determination two nucleotides at a time [19]. In the next sections, we will describe four commonly used NGS platforms. For a recent, detailed review of the performance of the specific benchtop instruments developed by each company, *see* [20].

2.1 454 GS FLX Titanium (Roche)

The 454 GS FLX Titanium sequencer utilizes the pyrosequencing method, which was first introduced in 1996, by Nyren and Ronaghi at the Royal Institute of Technology, Stockholm [8]. Both Sanger's dideoxy and this pyrosequencing method are sequence-by-synthesis (SBS) techniques, as DNA polymerase is required to generate the desired output. Pyrosequencing is based on generating a complementary strand of ssDNA, while concurrently releasing a signal from the incorporated nucleotides. With pyrosequencing, one nucleotide at a time is passed over multiple copies of the template DNA to be identified, leading to nucleotide incorporation by polymerases via complementary base pairing. Following the synthesis of the longest possible complementary strand by the polymerase, base incorporation will end. One PP_i molecule is released per nucleotide incorporated, followed by ATP sulfurylase-driven conversion of pyrophosphate to ATP, which emits light in the presence of luciferase. This works by a reaction of ATP with the substrate luciferin, to yield luciferyl-adenylate, which later reacts with oxygen to produce the light emitting oxyluciferin [21]. The light intensity is measured after base incorporation, followed by removal of unincorporated nucleotides, after which, the next nucleotide is introduced [8]. To protect the consumption of dATP by luciferase during the sequencing reaction, a modified nucleotide, deoxyadenosine-5′-(α-thio)-triphosphate (dATPαS), which is unusable by luciferase, is used during base incorporation. The other three nucleotides are standard dNTPs.

Pyrosequencing technology evolved into parallel sequencing on a picotiter plate in 2005, first by 454 Life Sciences, which was later acquired by Roche Diagnostics (*see* Table 1, [9]). This plate harbors about two million wells, each of which can hold one 28-μm diameter microbead. DNA from a sample is fragmented and each ssDNA fragment is attached to a single microbead via adapters. The surface of each bead contains oligonucleotides that bind to the adapter end, securing the DNA. Emulsion PCR (emPCR) of each bead produces a DNA library with millions of copies of that fragment. Clonal amplification occurs during emPCR, as each unique template molecule is physically separated from all others, with daughter molecules remaining bound to the microbeads. The plate is then incubated and washed with each dNTP, applied sequentially, in the presence of ATP sulfurylase and luciferase. PP_i

Table 1
Comparing the properties of different NGS technologies

Platform	Mechanism/Chemistry	Year	Read length (bp)	Output/ day (Mb)	Common error type	Cost/ Mb
Automated Sanger	Reversible dye terminator	2002	800	6	NA	$500.00
454 (Roche)	EmPCR, PPi release, luminescence	2005	100–500	700	InDel	$20.00
Ion torrent (life technologies)	EmPCR, proton release	2011	100–400	8000	InDel	$0.50
Illumina	Bridge amplification PCR, dye terminator	2011	100–300	50,000	Substitution	$0.50
SOLiD (applied biosystems)	PCR, Octomer ligation	2011	25–75	5000	Mismatch	$0.50
SMRT (Pacific biosciences)	Zero-mode wavelength, SMS	2011	14,000-60,000	1000	InDel/ mismatch	$1.00
Nanopore (Oxford)	Membrane nanopores, SMS	2015	150,000	> 90	InDel/ mismatch	$2.00

bp base pairs, *Mb* megabases, *EmPCR* emulsion PCR, PP_i pyrophosphate, *SMS* single-molecule sequencing, *InDel* insertion–deletion mutation
Data used in this table is from references [10, 11, 16, 31]

release is detected from light emission that is captured during nucleotide incorporation using a high-resolution, charge-coupled device (CCD) camera under the well [9, 10]. The read length for 454/Roche is 400–500 bp. The average substitution error rate, excluding insertions–deletions (InDels), is in the range of 10^{-3} to 10^{-4}. While this rate is higher than that of Sanger sequencing, it is among the lowest average substitution error rate of the newer sequencing technologies [16].

2.2 Ion-Torrent (Life Technologies)

This platform detects and measures hydrogen ion release or pH change following nucleotide incorporation instead of fluorescence. Like 454 sequencing, beads covered with multiple copies of a DNA fragment produced via emPCR are added to a picotiter plate. Next, DNA polymerase and one nucleotide at a time are added, with nucleotide incorporation measured via pH change driven by proton release, in place of PP_i release (*see* Fig. 3). This technology is driven by complementary metal-oxide-semiconductor (CMOS) technology used in the microprocessor chips production. This is a low-cost semiconductor that allows for non-optical DNA sequencing by using sensors to detect ions produced during polymerase-driven sequencing reactions, which enables rapid sequencing during the actual detection phase. The read lengths of the ion-torrent is 200–600 bp, depending on the specific instrument available (*see* Table 1, [1, 22]).

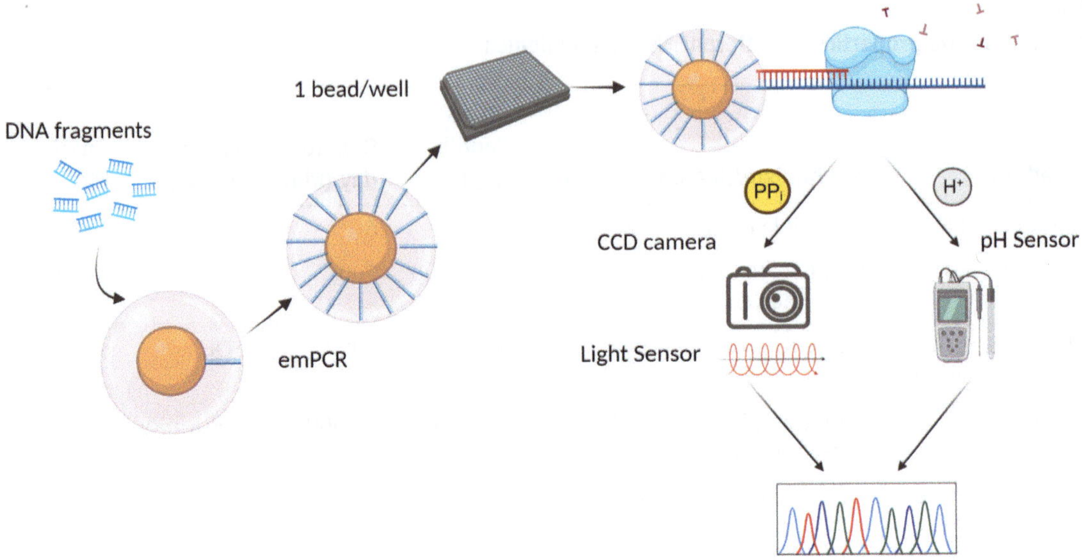

Fig. 3 454/Roche and Ion Torrent sequencing. DNA is fragmented and attached to a bead via ligated adapter sequences. Amplification of the fragment attached to the beads occurs by emulsion PCR (emPCR). Beads (clones) are collected in wells of picotiter plates followed by the addition of DNA polymerase and dNTPs. For 454/Roche sequencing, the sequence is determined by measuring the light signal, which is unique to each dNTP, emitted using a charged couple device (CCD) camera, following pyrophosphate (PP_i) release during base incorporation and extension. Ion Torrent sequencing follows a similar workflow, but sequence determination is made by measuring H^+ ion release during incorporation by a pH sensor. Created with BioRender.com

2.3 Illumina Sequencing

The Illumina sequencing process involves adapter-ligated DNA fragments immobilized on a glass surface that are subjected to clonal amplification [23]. The reversible dye terminator technique is the foundation for the Illumina Genome Analyzer (GA). This platform utilizes a sequencing-by-synthesis approach. Like the 454, the protocol requires conversion of DNA into an adapter-ligated library and immobilization onto a surface for sequencing [16, 23, 24]. Sodium hydroxide is used in the denaturation of the dsDNA library. The resulting ssDNA molecules, at a low concentration, are pushed through the channels of a flow cell with fixed oligonucleotides on a glass surface that will anneal to the two different adapters attached to the 5′ and 3′ ends, based on complementarity. This arrangement allows for bridge amplification, in which the synthesis of the reverse strand starts from the annealed portion and the newly synthesized strands are covalently attached to the flow cell (*see* Fig. 2, [25]). The tethered DNA molecule then bends and binds to another oligonucleotide complementary to the second adapter on the open end of the strand, generating a second covalently attached reverse strand cell (*see* Fig. 2). Several bridge amplifications create clusters of both forward and reverse strands, consisting of thousands of copies of the original sequence, closely packed on the flow cell [10, 16, 23, 24]. After denaturation, each cluster will have only single stranded, identically oriented copies of the original DNA molecule. Visualization of the clusters produced by bridge amplification occurs by detection of fluorescent reversible-terminator nucleotides. In this approach, fluorescently labeled 3′-O-azidomethyl-dNTPs are used to pause the polymerization reaction, enabling the removal of unincorporated bases and fluorescent imaging to determine the identity of the added nucleotide [26]. Following scanning of the flow cell with a CCD camera, the fluorescent moiety and the 3′ block are removed and the process is repeated. Here, the read lengths are shorter than for Sanger sequencing, but it has increased sequencing speed and output. For examples, using this technology, millions of reads are obtained within hours. It is also more cost efficient and eliminates the need for gel electrophoresis (*see* Table 1). Illumina is among the most widely used and arguably the most successful sequencing technology [1].

2.4 SOLiD (Applied Biosystems)

Pioneered by Harvard Medical School and the Howard Hughes Medical Institute in 2005 and unlike other methods where the DNA extension reaction is executed by polymerases, SOLiD utilizes ligases (*see* Table 1). The ligases attach fluorescently labeled primers in a sequence-specific manner to the template strand [18, 19]. Once the initial universal sequencing primer has been annealed to a single-stranded template of DNA molecules in a library, the primer is extended by introducing a mixture of octamer probes. Each probe contains four unique 5′ fluorescent labels, with one of the 16 possible dinucleotide combinations (e.g., AT, AG, AC, etc.) at the 3′ end. The octamers compete to be ligated to the sequencing primer based on complementarity to the template sequence. The last two complementary nucleotides at the 3′ end of the probe are read. On the 5′ end of the probe, three bases and the dye present are removed, exposing a free 5′ phosphate for continued extension. After 10 cycles of ligations or the desired length is reached, the extended product is denatured and a new sequencing primer, shifted by one nucleotide, is annealed to the template to repeat the process. This same process is repeated three more times with three additional sequencing primers, each shifted by one nucleotide compared to the previous. Performing five rounds of ligations with five different, shifted primers allows for dual measurement of each base in the sequence and therefore, increases sequencing accuracy [16, 18].

3 Single-Molecule Sequencing

Single-molecule sequencing (SMS), also known as third-generation sequencing, is executed without fragment amplification, so both DNA strands can be sequenced, providing more information and increased accuracy [27]. Advantages of SMS are that it reduces errors that occur during DNA amplification or the library preparation step and is effective to analyze low DNA concentrations. In addition, this technique can identify non-standard nucleotides, like modifications such as methylation, that are omitted during the amplification step of the other platforms [16]. Another benefit is the generation of longer reads compared to other NGS platforms (*see* Table 1). Longer reads reduce the computing requirements of fragment assembly during applications such as whole-genome sequencing. A final advantage of this technology is that it can also detect epigenetic markers without the need for chromatin immunopurification. The downside is that SMS techniques have a higher error rate in comparison to other NGS technologies. Next, we will discuss the two common SMS platforms that are readily in use.

3.1 Single-Molecule Real-Time (SMRT) Sequencing (Pacific Biosciences)

In SMRT sequencing, nucleotide detection occurs in tiny chambers called zero-mode waveguides (ZMWs). DNA polymerase is fixed to the bottom of each ZMW, to which target DNA and nucleotides tagged with different colored fluorophores are added [28]. Single DNA strand extension can be monitored in real time, as only incorporated nucleotides emit fluorescence. This eliminates interference from other unincorporated labeled dNTPs in solution. The ZMW is an optical waveguide that guides laser energy through pores with diameter narrower than the wavelength. Therefore, the electromagnetic energy from the excitation beam decays as it penetrates the nanoscale aperture. This allows a small detection volume, on the order of 100 zeptoliters, that helps reduce background noise. In other words, a single incorporation event is easily captured against a background of other nucleotides [10, 11, 28]. For this technique, read lengths are significantly increased, and typically will be 14–60 kb (*see* Table 1).

3.2 Nanopore DNA Sequencing (Oxford)

For this SMS platform, ssDNA is electrophoresed by a molecular motor protein, such as DNA helicase. The DNA helicase binds to ssDNA and pushes it through a biological nanopore embedded in a synthetic membrane, across which a voltage is applied (*see* Fig. 4). As the ssDNA passes through the nanopore, individual bases disrupt current flow in a unique manner, allowing the sequence of the molecule to be inferred by monitoring the changes in current. In other words, the nucleotide movement through the pores impedes the flow of ions, reducing current for a time proportional to the DNA length and composition of nucleotides. This allows for the

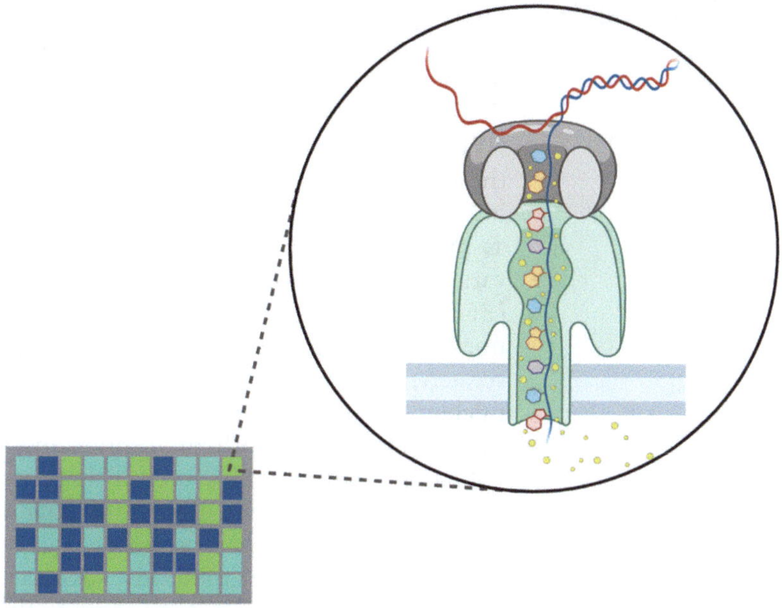

Fig. 4 Oxford nanopore sequencing. A ssDNA fragment is pushed through a membrane nanopore composed of a motor enzyme. Each nucleotide will alter the current through the pore by a different magnitude as it moves through the pore, allowing for sequence determination. Created with BioRender.com

sequencing of both strands of a DNA molecule [10, 29, 30]. An advantage of the nanopore is the long read lengths achievable, with different modes capable of detecting 10–300 kb (*see* Table 1).

4 Applications of NGS Technologies

To better understand the underlying pathophysiology of a disease, there may be a need to study mutations in genes, but also the expressed RNAs and proteins to provide clues to its management and treatment [23]. Whole-genome sequencing allows for the detection of mutations or genetic variants that may be responsible for a particular disease. Transcriptomic analysis will aid in understanding exactly which genes are expressed at different times and can also be used to monitor disease progression. Advances in sequencing technology did not just reduce the cost and streamline the process; they have also paved the way for new applications to elucidate molecular mechanisms of genomic structure and cellular functions [15, 31]. Next, we discuss some traditional and newer applications of NGS technologies. We highlight the methodology, clinical or research applications, and discuss any noted limitations of each approach.

4.1 DNA Sequencing Techniques

Detecting DNA alterations that affect human health is now possible because of NGS technologies. Some techniques that can be applied to the study of disease are DNA sequencing (DNA-seq [32]), single-cell DNA-seq (scDNA-seq), whole exome sequencing (WES, [33]), and targeted sequencing (TS, [34]). Sequencing of the entire genome provides global information about exons and introns, which can reveal the regulatory components of genes, such as promoters, enhancers, and intronic regulators, and structural variants, like copy number variants, inversions, and translocations [35, 36]. The read depth, meaning the number of times each nucleotide is detected in a sequencing reaction, is crucial for sequencing accuracy. To observe small changes, such as single nucleotide substitutions (SNPs) or point mutations linked to diseases, a high read depth and sequencing of multiple affected individuals are required. However, if there are large structural changes relative to a reference genome, information is achievable even with a low read depth. In most cases, increasing read depth can be expensive and cheaper methods, such as customized hybridization chips, may be preferred. One issue is that most sequencing runs yield a limited number of fragments reads. For larger genomes, a more cost-effective approach is WES, in which only the DNA sequence that is transcribed into mRNA (exons) are analyzed. WES is selective and enriches for sections of genomic DNA captured using commercially available capture arrays. These arrays harbor unique baits to capture target exons. Frequently, streptavidin or magnetic beads coated with oligonucleotides that bind and capture exome sequences are used. A benefit of WES for analyzing the human genome, is that the price can be significantly reduced since only about 1% of the ~20,000 human genes are coding sequences. The downside is that the bait oligos can sometimes miss unknown coding regions or regions with key mutations that could be important for an underlying disease state [31, 36].

TS is another popular technique, which allows for a gene or region of a genome to be assessed for variations that may be linked to a phenotype or specific disease. It is faster, more accurate, less expensive, and requires less computing power than whole-genome sequencing. Sequencing a small number of target genes typically requires PCR amplification, followed by NGS. For many diseases, multiple genetic panels are predesigned or customized to isolate the desired genomic regions by hybridization probes. The high read depth of TS, considered ultra-deep sequencing, makes it possible to identify rare mutations present in only a small subpopulation of cells, which is highly advantageous for the study of malignancies. TS has become a mainstay in cancer research and treatment [37–39]. Identification of genetic variations linked to a disease or response to drug treatment is essential for optimizing drug design or individual pharmacotherapy [37, 40]. Another popular technique used for the study of tumor development and progression is scDNA-seq [41, 42]. This technique involves isolation of single cells from a population utilizing either low throughput methods, like serial dilution, laser capture microdissection or fluorescence-activated cell sorting (FACS), or high throughput methods, such as microfluidic chips or combinatorial indexing [43]. The cells are then lysed to recover genetic material. A sequencing library is prepared by fragmenting the DNA and ligating adapters for NGS. While DNA-seq provides the sequence of the average genomic sequence of a population of cells, single-cell sequencing shines light on the genome of a single cell within a population. scDNA-seq is becoming increasingly popular and is well suited for the study of mutational processes, mosaicism, lineage tracing, germ-line mutations, and cancer progression [44].

As with any technology, DNA sequencing has its limitations. One common problem encountered is the underrepresentation of high GC- or AT-rich regions. There is often an inherent difficulty in differentiating sequencing errors from actual variants within a population, especially if they are rare. Mapping errors frequently occur with short read alignments around repetitive chromosomal regions. In addition, errors occur when arranging short sequencing reads into larger assemblages and there are problems in identifying large structural variants, like inversions. Despite these limitations, DNA sequencing technology has revolutionized our understanding of cellular physiology in health and disease.

4.2 RNA Sequencing Techniques

Early attempts to understand and quantify transcript levels on the global scale relied on microarray technology. Microarray technology is an approach where a desired sample is bound to a surface or chip containing thousands to millions of immobilized nucleic acid fragments. mRNA is recovered from different samples or individuals and reverse transcribed to cDNA, which are then exposed to the microarray chip to determine which genes are expressed. One downslide to microarrays is that they require a sequenced reference genome and transcriptome for its design and interpretation. NGS techniques, such as RNA-seq, can reveal the expression profile of organisms with un-sequenced genomes [45]. RNA-seq is more sensitive, provides better accuracy, and is not influenced by chip sequence biases [15]. It relies on high-throughput sequencing of cDNA fragments produced from RNA or RNA fragments, which results in an accurate mapping of transcripts to unique regions of the genome and eliminates most of the background noise. As transcripts can be precisely quantified, it is possible to analyze transcript isoforms within a 5000-fold dynamic range [15, 46]. RNA-seq has led to the discovery of novel genes and RNAs that were not detected by microarray technology, such as novel gene fusions [47]. It can also reveal possible differences between exomes and the transcriptome caused by RNA-editing [48].

For this technique, mRNA is reversed transcribed to cDNA, following careful and rigorous isolation techniques that yield quality mRNA and often deplete rRNA. As rRNA represents the majority of the RNA species in a cell, depletion can lead to the identification and quantification of lower abundance RNAs. Different RNA preparation methodologies have emerged over time, but all are based on purifying desired mRNA, non-coding RNAs, or miRNAs from other RNA species and contaminants. A typical workflow involves the selection of polyadenylated mRNA, followed by rRNA depletion, and reverse transcription. However, this approach is biased for the 3′ end, works better with high-quality mRNA, and yields poor data recovery from non-coding portions of RNA [49, 50]. RNA-seq with prokaryotic RNAs is possible, even though they lack poly-A tails and cannot be enriched or selected with poly T primers. This requires the use of random oligonucleotides, often times hexamers consisting of six random nucleotides [51]. Data analysis and interpretation occur via freely available bioinformatic platforms. Sequencing reads are aligned to a reference genome and assembled into transcripts. Different software packages have been specifically designed for RNA-seq analyses [52–54].

RNA-seq has been used for many applications in medicine, including the study of disease progression overtime, such as heart failure [45]. RNA-seq is commonly used to study differential gene expression [55], which can reveal functions of new genes or identify new cancer biomarkers [54, 56, 57]. An extension of RNA-seq is microRNA (miRNA)-seq. miRNAs are a class of RNAs, typically ~22 nucleotides in length, that regulate multiple physiological and pathological processes. Their main function is to regulate translation and stability of their target mRNAs. The study of miRNA expression patterns in cells and tissues has been used to study development and disease pathology, especially in cancer cells [58, 59]. As with DNA sequencing, single-cell RNA-seq (scRNA-seq) is also being increasingly performed to study how gene expression varies among individual cells within a population [60]. Traditional RNA-seq measures gene expression across a population and does not address population heterogeneity. This has been especially powerful in the study of the underlying pathogenesis of cancer cells. Tumor cell heterogeneity is believed to be a significant cause of treatment failure, so this understanding is necessary to optimize treatment regimens [61].

4.3 NGS for the Study of Epigenetics

NGS technology has been developed to study epigenetics, which are reversible changes to nucleosomes or nucleic acids that produce phenotypic variation without corresponding alterations to the genome. In general, eukaryotic DNA is organized into actively transcribed regions called euchromatin and transcriptionally silent regions called heterochromatin. These loosely or tightly packed regions depend on the combination of modifications present on the N-terminal tails of histone proteins, which make up the nucleosomes, and other silencing factors [23, 62]. A technique to identify regions of euchromatin is Assay for Transposase-Accessible Chromatin (ATAC)-seq [63], which utilizes a hyperactive Tn5 transposase mutant to introduce sequencing adapters into the actively transcribed, open regions of the native chromatin. The tagged sequences are purified, amplified, and sequenced using NGS technologies. Sequencing reads are used to identify regions with increased accessibility and are used to map binding sites of histones. ATAC-seq can determine the location and interactions between transcription factors, nucleosomes, and open chromatin, and shed light onto how chromatin compaction is regulated [63].

Another technique to study epigenetic modifications is whole-genome bisulfite sequencing (WGBS). WGBS detects the presence of 5-methylcytosine, which is an abundant epigenetic DNA modification that is typically associated with gene silencing [64]. For this technique, extracted genomic DNA is treated with sodium bisulfite to convert unmethylated cytosines into uracil, while methylated cytosine will remain unchanged. During the PCR step, the uracil bases are changed to thymine, which means that the sequencing results will be composed of mostly thymine, adenine, and guanine. The presence of cytosine indicates the methylation sites [65]. WGBS has been successfully used to study stem cell differentiation, development, disease diagnosis, and forensic science [66–68].

4.4 NGS for the Study of DNA-Binding Proteins

Chromatin immunoprecipitation sequencing (ChIP-seq) is a widely used technique for mapping protein–DNA interactions. For this methodology, DNA-binding proteins are crosslinked to their target DNA binding sites often using chemical crosslinkers, like formaldehyde, and the DNA is typically sheared by sonication or enzymatic treatment with micrococcal nuclease (MNase). The immunoprecipitation step requires a specific antibody directed against the DNA-binding protein of interest. The use of a high-quality, efficient antibody or ChIP-certified antibodies is necessary to reduce the signal to noise ratio to ensure quality data. The crosslink is reversed, often through extensive heat treatment, proteins and RNA degraded by treatment with appropriate enzymes, and target DNA isolated. A sequencing library with appropriate adapters is then produced [65]. Peak detection software maps sequencing reads to a reference genome and detects regions with enriched frequency (different software tools reviewed in [69]). Applications of ChIP-seq include mapping transcription factor binding sites, studying chromatin structure, and epigenetics [70]. In fact, ChIP-seq is often used for annotating the state of chromatin under specific conditions and to study histone modifications [71]. By performing the immunopurification with antibodies specific for different histone modifications, more information on which chromosomal regions are enriched with specific modified species is available. Modified histones are promising as biomarkers for various diseases and ChIP-seq is an effective technique for their study [72].

4.5 NGS for the Analysis of Chromosome Structure and Positioning

Tyramide signal amplification (TSA)-seq has been described as a cytological ruler that measures intracellular distance between the chromosome and intranuclear structures, enabling the spatial understanding of the three-dimensional genome organization. It can reveal distances between genetic loci and nuclear compartments and provide information on how the chromosome spans or travels across those different compartments. The proximity of a gene to the center of the nucleus can influence its activity and may change as the cell grows. This method involves the use of primary antibodies directed against specific proteins of interest combined with secondary antibodies that are coupled with the enzyme horseradish-peroxidase (HRP). Following cross-linking with formaldehyde, cells are labeled with biotin-tyramide and HRP catalyzes the formation of biotin-tyramide free radicals, which creates a concentration gradient with a radius of 1 μm. TSA labels both proteins and DNA. DNA fragments in close proximity will be labeled and isolated by immunopurification of biotin with streptavidin beads. The signal intensity increases as the distance between the protein of interest and the chromosome decreases [73].

Another technique used to study long-range DNA–DNA interactions and chromosome architecture is chromosome conformation capture (3C) sequencing (3C-seq). This technique is used to identify the functional interactions among unique chromosomal segments [74, 75]. In other words, 3C-seq, also called Hi-C, allows for the study of chromosomal loci that may interact when a genome is folded in three-dimensional space versus when it is linear. For this methodology, chromatin is cross-linked with formaldehyde, which creates covalent bonds between segments of DNA that are bridged together by proteins in 3D space. The DNA is digested with restriction enzymes, re-ligated, and crosslinks reversed. This creates 3C templates which are used for NGS [76]. TSA-seq and 3C-seq have led to an unparalleled understanding of the 3D structure and organization of genome in living cells.

4.6 NGS Applications for Microbial Population Studies

Transposons are mobile genetic elements that can be excised and insert in different loci throughout a genome. The insertion of a transposon will most likely disrupt gene function and thus transposons became a powerful tool to study gene functions. Transposon sequencing (Tn-seq) is a technique which requires the generation and use of a transposon library in a microorganism. Transposons can only be inserted into genes that are non-essential for viability, but the fitness of the resulting mutant strains under specific growth conditions may vary greatly. Underrepresentation or overrepresentation of specific mutants in a population provides clues about their necessity in each specific condition. Mutant strains with a growth deficiency will be underrepresented in a population and contain an insertion in a gene that is conditionally essential. Mutants that are overrepresented in a population will represent insertions in genes that when deleted, confer a selective growth advantage. To identify the disrupted gene and determine the position and frequency of insertion in the population, the downstream genomic region from the 3′ end of the insertion site is amplified. As with other techniques, adapters are ligated to create a library, which is then sequenced using NGS technologies. Typically, sequencing primers designed internal to the 3′ end of the transposon are used [77, 78]. Tn-seq is one of the few techniques that can directly link genotype to phenotype in microbiology.

Transposon-directed insertion site sequencing (TraDIS) is an extension of Tn-seq, whereby amplification and sequencing of transposon-rich fragment libraries are used to determine the rate of insertion. TraDIS uses random, mechanical shearing of the DNA by sonication, which improves mapping accuracy due to longer sequencing reads. This is different from Tn-seq, which uses restriction digests to produce uniform fragments during library preparation [79]. In the last decade, there have been even more advances in Tn-seq technologies, with novel combinations of techniques emerging. These include motility TIS, density-TraDISort, TraDISort, and dTn-seq [80]. The first three methodologies involve separating cells based on phenotypes other than growth, like motility, density (capsule production), and FACS, respectively. Droplet Tn-seq (dTn-seq) utilizes microfluidics to sort single cells, in which individual mutant cells are encased in an oil droplet. The cells will grow in their own droplet, without the influence of other mutants in the population. As dTn-seq is compatible with FACS or microscopy, droplets can be re-encapsulated with other droplets containing different cells to allow for the signals to diffuse between the two. Sometimes, the true fitness of a mutant cannot be ascertained when it is in the presence of other mutants, which is when dTn-seq may be desired. In addition, the effects of secreted products or signals can be evaluated [81]. Another useful extension of Tn-seq is the TraDIS-Xpress technique. A limitation of traditional Tn-seq is that insertion of transposons in essential genes is lethal, so the study of their function is not possible. TraDIS-Xpress uses a Tn5 transposon where an inducible promoter is outward facing. Gene disruptions of non-essential genes will still occur, but insertion of the transposon upstream of an essential gene will lead to overexpression. This means that all genes can be analyzed using this technique [82].

5 Challenges and Limitations of NGS Technologies

As with any new methodology, there are great benefits, but there will always be limitations. Each sequencing technology has its own shortcomings and the capacity to detect variants differs [83]. One such limitation, especially in the DNA of higher eukaryotic organisms, is the high error rate when sequencing homopolymers, which are stretches of DNA containing multiple repeats of the same nucleotide. Homopolymers, also called mononucleotide microsatellites, are more difficult to sequence because of the absence of a blocking moiety in the nucleotides used, since the repeating nucleotide will completely incorporate during a single cycle, which leads to uncertainty in base calling. The sequencing of homopolymers is a challenge when using every NGS platform, especially for those which utilize sequencing-by-synthesis, such as the Ion Torrent and 454 [84]. Therefore, homopolymers may end up significantly shortened or expanded in the final genome assemblage. Homopolymers are not the only DNA sequence that is challenging. The accuracy of GC-rich regions remains a challenge for all sequencing platforms, including traditional Sanger sequencing. Furthermore, the short-read lengths of most platforms undermines the ability to correctly sequence large repeat regions and other large-scale variations that may be present in a genome [10]. Insertions, deletions, and larger structural variants are typically difficult to identify. Assembling of a de novo genome from short read assemblages remains a challenge and often leads to incomplete assembled sequences with gaps. Increasing read length and accuracy will significantly improve the coverage and accuracy of de novo genomes, which also will enable precise variant mapping between individuals [15].

As the number of platforms and applications increases, the need for standardization between researchers also increases. The National Institute of Standards and Technology (NIST) has been working with other agencies towards regulating NGS protocols by developing standards and metrics that are widely acceptable [85]. One area of focus is on the analysis human genomic DNA from multiple cell lines, which will serve as benchmarks and will be characterized for uniformity, stability, and quality using the available sequencing platforms and protocols. This will yield high confidence datasets that are reproducible and can be used by both clinical and research laboratories to better evaluate sequencing and bioinformatics techniques. Similarly, Brownstein et al. proposed the CLARITY challenge, to help compare the efficiency of various sequencing technologies. The CLARITY challenge was designed to standardize methods for diagnosing genetic disease using genomic sequencing data. In general, the participating researchers used similar workflows and downstream bioinformatic analyses. The study also identified important needs, like quality publicly available reference genomes and the importance of well-documented clinical reports, including reference accession numbers and other bioinformatic statistics. The hope is to be able to set standards for diagnosing genetic diseases using both clinical case history and appropriately processed and analyzed genomic data [86].

6 Conclusions

Since the 1990s, considerable efforts to improve whole-genome sequencing has led to significant advancements in sequencing capacity and output (*see* Tables 1 and 2, [87]). Important advancements include: 1. the development of reversible dye terminators and mutant T7 DNA polymerase that allowed for easy incorporation of such terminators [88], 2. amplification of DNA by PCR, which reduced the need for large quantities of template DNA and enabled small sample processing [89], 3. DNA fragment isolation and purification using magnetic beads enabled higher quality and purer starting material [90], 4. double-stranded DNA sequencing technologies, which allowed for cloned fragment and paired-end sequencing, increasing the read depth, 5. capillary electrophoresis, which eliminated manual gel electrophoresis and promoted fluorescence interpretation [91], and 6. automation of techniques,

Table 2
Notable improvements in NGS technology

Improvement [87]	Consequence
Use of dye labeled terminators instead of labeled primers	Allowed for one sequencing reaction per sample instead of multiple reactions.
Use of mutant form of T7 DNA polymerase	Effectively incorporated dye-labelled terminators.
Use of PCR amplification	Reduced required template concentration and aided miniaturization.
Use of magnetic beads for DNA purification	Simplified the automation of pre-sequencing steps.
The abiltity to sequence double-stranded DNA	Enabled the use of plasmid clones and allowed for paired end sequencing.
Use of capillary electrophoresis	Eliminated use of agarose gels, and simplified the extraction and interpretation of the fluorescent signals.
Comercialization of platforms	Improved efficiencies and reduced error rates.

which led to a reduction in errors and improved efficiency. Like the NGS technologies, great strides have been made towards developing and improving software platforms to identify variants, quantify, and assemble large amounts of sequencing data. Software such as Phred eliminated the need for the manual editing of reads, improved read quality, and helped analyze sequences with repeats [87]. Now, there are many freely available to software programs to analyze all the different types of data sets that are generated (reviewed in [92, 93]).

With these tools, a small research group can generate large amounts of sequence data faster and cheaper than with conventional Sanger sequencing, with costs reduced to 0.1–4% and time shortened by a factor of 100–1000 based on daily throughput. One challenge facing smaller research groups is the cost of acquiring these instruments. Recently, some smaller-scale, more cost-effective versions of the larger instruments have become available, like the Illumina GA IIe or 454/Roche GS Junior, but they have a lower sequencing capacity. However, the overall costs can still be rather high, as the cost of sequencing per base is increased compared to the larger instruments and still requires knowledge and familiarity with protocols. Sequencing technology choice depends on the experimental requirements and sometimes it may be necessary to combine multiple platforms. This scenario may create competition, allowing more companies to provide sequencing services as needed or required, which will ultimately save individual laboratories the time and expertise required for sequence generation and data analysis [16].

NGS technologies will continue to play a leading role in identifying new disease biomarkers and make importation contributions to our fundamental understanding of how genetic abnormalities contribute to different phenotypes [26]. Continuous efforts towards the improvement of these NGS platforms will ensure better accuracy and reproducibility, will reduce error rates, and the overall cost of sequencing [2]. As the number of potential applications and uses for NGS expands, these technological improvements will be necessary to keep up with the increased demands for faster turnaround times, higher accuracy, and a reasonable cost.

Acknowledgements

This work was supported by grant GM138303 from the National Institute of General Medical Sciences, awarded to V.J.C.

References

1. Heather JM, Chain B (2016) The sequence of sequencers: the history of sequencing DNA. Genomics 107(1):1–8. https://doi.org/10.1016/j.ygeno.2015.11.003
2. Hu T, Chitnis N, Monos D, Dinh A (2021) Next-generation sequencing technologies: an overview. Hum Immunol 82(11):801–811. https://doi.org/10.1016/j.humimm.2021.02.012
3. Gilbert W, Maxam A (1973) The nucleotide sequence of the *lac* operator. Proc Natl Acad Sci USA 70(12):3581–3584. https://doi.org/10.1073/pnas.70.12.3581
4. Sanger F, Nicklen S, Coulson AR (1977) DNA sequencing with chain-terminating inhibitors. Proc Natl Acad Sci 74(12):5463–5467. https://doi.org/10.1073/pnas.74.12.5463
5. Smith LM, Sanders JZ, Kaiser RJ, Hughes P, Dodd C, Connell CR, Heiner C, Kent SB, Hood LE (1986) Fluorescence detection in automated DNA sequence analysis. Nature 321(6071):674–679. https://doi.org/10.1038/321674a0
6. Hutchison CA 3rd (2007) DNA sequencing: bench to bedside and beyond. Nucleic Acids Res 35(18):6227–6237. https://doi.org/10.1093/nar/gkm688
7. Adams MD, Kelley JM, Gocayne JD, Dubnick M, Polymeropoulos MH, Xiao H, Merril CR, Wu A, Olde B, Moreno RF et al (1991) Complementary DNA sequencing: expressed sequence tags and human genome project. Science 252(5013):1651–1656. https://doi.org/10.1126/science.2047873
8. Ronaghi M, Karamohamed S, Pettersson B, Uhlén M, Nyrén P (1996) Real-time DNA sequencing using detection of pyrophosphate release. Anal Biochem 242(1):84–89. https://doi.org/10.1006/abio.1996.0432
9. Margulies M, Egholm M, Altman WE, Attiya S, Bader JS, Bemben LA, Berka J, Braverman MS, Chen Y-J, Chen Z, Dewell SB, Du L, Fierro JM, Gomes XV, Godwin BC, He W, Helgesen S, Ho CH, Irzyk GP, Jando SC, Alenquer MLI, Jarvie TP, Jirage KB, Kim J-B, Knight JR, Lanza JR, Leamon JH, Lefkowitz SM, Lei M, Li J, Lohman KL, Lu H, Makhijani VB, McDade KE, McKenna MP, Myers EW, Nickerson E, Nobile JR, Plant R, Puc BP, Ronan MT, Roth GT, Sarkis GJ, Simons JF, Simpson JW, Srinivasan M, Tartaro KR, Tomasz A, Vogt KA, Volkmer GA, Wang SH, Wang Y, Weiner MP, Yu P, Begley RF, Rothberg JM (2005) Genome sequencing in microfabricated high-density picolitre reactors. Nature 437(7057):376–380. https://doi.org/10.1038/nature03959
10. Reuter Jason A, Spacek DV, Snyder Michael P (2015) High-throughput sequencing technologies. Mol Cell 58(4):586–597. https://doi.org/10.1016/j.molcel.2015.05.004
11. Kchouk M, Gibrat J-F, Elloumi M (2017) Generations of sequencing technologies: from first to next generation. Biol Med 9:1–8
12. Ruparel H, Bi L, Li Z, Bai X, Kim DH, Turro NJ, Ju J (2005) Design and synthesis of a 3-*O*-allyl photocleavable fluorescent nucleotide as a reversible terminator for DNA sequencing by synthesis. Proc Natl Acad Sci 102(17):5932–5937. https://doi.org/10.1073/pnas.0501962102
13. Li Z, Bai X, Ruparel H, Kim S, Turro NJ, Ju J (2003) A photocleavable fluorescent nucleotide for DNA sequencing and analysis. Proc Natl Acad Sci 100(2):414–419. https://doi.org/10.1073/pnas.242729199
14. Turcatti G, Romieu A, Fedurco M, Tairi AP (2008) A new class of cleavable fluorescent nucleotides: synthesis and optimization as reversible terminators for DNA sequencing by synthesis. Nucleic Acids Res 36(4):e25. https://doi.org/10.1093/nar/gkn021
15. Soon WW, Hariharan M, Snyder MP (2013) High-throughput sequencing for biology and medicine. Mol Syst Biol 9(1):640. https://doi.org/10.1038/msb.2012.61
16. Kircher M, Kelso J (2010) High-throughput DNA sequencing – concepts and limitations. BioEssays 32(6):524–536. https://doi.org/10.1002/bies.200900181
17. Chen F, Dong M, Ge M, Zhu L, Ren L, Liu G, Mu R (2013) The history and advances of reversible terminators used in new generations of sequencing technology. Genomics Proteomics Bioinformatics 11(1):34–40. https://doi.org/10.1016/j.gpb.2013.01.003
18. Shendure J, Porreca GJ, Reppas NB, Lin X, McCutcheon JP, Rosenbaum AM, Wang MD, Zhang K, Mitra RD, Church GM (2005) Accurate multiplex polony sequencing of an evolved bacterial genome. Science 309(5741):1728–1732. https://doi.org/10.1126/science.1117389
19. Voelkerding KV, Dames SA, Durtschi JD (2009) Next-generation sequencing: from basic research to diagnostics. Clin Chem 55(4):641–658. https://doi.org/10.1373/clinchem.2008.112789

20. Pervez MT, Hasnain MJU, Abbas SH, Moustafa MF, Aslam N, Shah SSM (2022) A comprehensive review of performance of next-generation sequencing platforms. Biomed Res Int 2022:3457806. https://doi.org/10.1155/2022/3457806
21. Morciano G, Sarti AC, Marchi S, Missiroli S, Falzoni S, Raffaghello L, Pistoia V, Giorgi C, Di Virgilio F, Pinton P (2017) Use of luciferase probes to measure ATP in living cells and animals. Nat Protoc 12(8):1542–1562. https://doi.org/10.1038/nprot.2017.052
22. Rothberg JM, Hinz W, Rearick TM, Schultz J, Mileski W, Davey M, Leamon JH, Johnson K, Milgrew MJ, Edwards M, Hoon J, Simons JF, Marran D, Myers JW, Davidson JF, Branting A, Nobile JR, Puc BP, Light D, Clark TA, Huber M, Branciforte JT, Stoner IB, Cawley SE, Lyons M, Fu Y, Homer N, Sedova M, Miao X, Reed B, Sabina J, Feierstein E, Schorn M, Alanjary M, Dimalanta E, Dressman D, Kasinskas R, Sokolsky T, Fidanza JA, Namsaraev E, McKernan KJ, Williams A, Roth GT, Bustillo J (2011) An integrated semiconductor device enabling non-optical genome sequencing. Nature 475(7356):348–352. https://doi.org/10.1038/nature10242
23. Bentley DR, Balasubramanian S, Swerdlow HP, Smith GP, Milton J, Brown CG, Hall KP, Evers DJ, Barnes CL, Bignell HR, Boutell JM, Bryant J, Carter RJ, Keira Cheetham R, Cox AJ, Ellis DJ, Flatbush MR, Gormley NA, Humphray SJ, Irving LJ, Karbelashvili MS, Kirk SM, Li H, Liu X, Maisinger KS, Murray LJ, Obradovic B, Ost T, Parkinson ML, Pratt MR, Rasolonjatovo IM, Reed MT, Rigatti R, Rodighiero C, Ross MT, Sabot A, Sankar SV, Scally A, Schroth GP, Smith ME, Smith VP, Spiridou A, Torrance PE, Tzonev SS, Vermaas EH, Walter K, Wu X, Zhang L, Alam MD, Anastasi C, Aniebo IC, Bailey DM, Bancarz IR, Banerjee S, Barbour SG, Baybayan PA, Benoit VA, Benson KF, Bevis C, Black PJ, Boodhun A, Brennan JS, Bridgham JA, Brown RC, Brown AA, Buermann DH, Bundu AA, Burrows JC, Carter NP, Castillo N, Chiara ECM, Chang S, Neil Cooley R, Crake NR, Dada OO, Diakoumakos KD, Dominguez-Fernandez B, Earnshaw DJ, Egbujor UC, Elmore DW, Etchin SS, Ewan MR, Fedurco M, Fraser LJ, Fuentes Fajardo KV, Scott Furey W, George D, Gietzen KJ, Goddard CP, Golda GS, Granieri PA, Green DE, Gustafson DL, Hansen NF, Harnish K, Haudenschild CD, Heyer NI, Hims MM, Ho JT, Horgan AM, Hoschler K, Hurwitz S, Ivanov DV, Johnson MQ, James T, Huw Jones TA, Kang GD, Kerelska TH, Kersey AD, Khrebtukova I, Kindwall AP, Kingsbury Z, Kokko-Gonzales PI, Kumar A, Laurent MA, Lawley CT, Lee SE, Lee X, Liao AK, Loch JA, Lok M, Luo S, Mammen RM, Martin JW, McCauley PG, McNitt P, Mehta P, Moon KW, Mullens JW, Newington T, Ning Z, Ling Ng B, Novo SM, O'Neill MJ, Osborne MA, Osnowski A, Ostadan O, Paraschos LL, Pickering L, Pike AC, Pike AC, Chris Pinkard D, Pliskin DP, Podhasky J, Quijano VJ, Raczy C, Rae VH, Rawlings SR, Chiva Rodriguez A, Roe PM, Rogers J, Rogert Bacigalupo MC, Romanov N, Romieu A, Roth RK, Rourke NJ, Ruediger ST, Rusman E, Sanches-Kuiper RM, Schenker MR, Seoane JM, Shaw RJ, Shiver MK, Short SW, Sizto NL, Sluis JP, Smith MA, Ernest Sohna Sohna J, Spence EJ, Stevens K, Sutton N, Szajkowski L, Tregidgo CL, Turcatti G, Vandevondele S, Verhovsky Y, Virk SM, Wakelin S, Walcott GC, Wang J, Worsley GJ, Yan J, Yau L, Zuerlein M, Rogers J, Mullikin JC, Hurles ME, McCooke NJ, West JS, Oaks FL, Lundberg PL, Klenerman D, Durbin R, Smith AJ (2008) Accurate whole human genome sequencing using reversible terminator chemistry. Nature 456(7218):53–59. https://doi.org/10.1038/nature07517
24. Fedurco M, Romieu A, Williams S, Lawrence I, Turcatti G (2006) BTA, a novel reagent for DNA attachment on glass and efficient generation of solid-phase amplified DNA colonies. Nucleic Acids Res 34(3):e22. https://doi.org/10.1093/nar/gnj023
25. Bentley DR, Balasubramanian S, Swerdlow HP, Smith GP, Milton J, Brown CG, Hall KP, Evers DJ, Barnes CL, Bignell HR, Boutell JM, Bryant J, Carter RJ, Keira Cheetham R, Cox AJ, Ellis DJ, Flatbush MR, Gormley NA, Humphray SJ, Irving LJ, Karbelashvili MS, Kirk SM, Li H, Liu X, Maisinger KS, Murray LJ, Obradovic B, Ost T, Parkinson ML, Pratt MR, Rasolonjatovo IMJ, Reed MT, Rigatti R, Rodighiero C, Ross MT, Sabot A, Sankar SV, Scally A, Schroth GP, Smith ME, Smith VP, Spiridou A, Torrance PE, Tzonev SS, Vermaas EH, Walter K, Wu X, Zhang L, Alam MD, Anastasi C, Aniebo IC, Bailey DMD, Bancarz IR, Banerjee S, Barbour SG, Baybayan PA, Benoit VA, Benson KF, Bevis C, Black PJ, Boodhun A, Brennan JS, Bridgham JA, Brown RC, Brown AA, Buermann DH, Bundu AA, Burrows JC, Carter NP, Castillo N, Chiara E, Catenazzi M, Chang S, Neil Cooley R, Crake NR, Dada OO, Diakoumakos KD, Dominguez-Fernandez B, Earnshaw DJ, Egbujor UC, Elmore DW, Etchin SS, Ewan MR, Fedurco M, Fraser LJ, Fuentes Fajardo KV, Scott Furey W, George D, Gietzen KJ, Goddard CP, Golda GS, Granieri PA,

Green DE, Gustafson DL, Hansen NF, Harnish K, Haudenschild CD, Heyer NI, Hims MM, Ho JT, Horgan AM, Hoschler K, Hurwitz S, Ivanov DV, Johnson MQ, James T, Huw Jones TA, Kang G-D, Kerelska TH, Kersey AD, Khrebtukova I, Kindwall AP, Kingsbury Z, Kokko-Gonzales PI, Kumar A, Laurent MA, Lawley CT, Lee SE, Lee X, Liao AK, Loch JA, Lok M, Luo S, Mammen RM, Martin JW, McCauley PG, McNitt P, Mehta P, Moon KW, Mullens JW, Newington T, Ning Z, Ling Ng B, Novo SM, O'Neill MJ, Osborne MA, Osnowski A, Ostadan O, Paraschos LL, Pickering L, Pike AC, Pike AC, Chris Pinkard D, Pliskin DP, Podhasky J, Quijano VJ, Raczy C, Rae VH, Rawlings SR, Chiva Rodriguez A, Roe PM, Rogers J, Rogert Bacigalupo MC, Romanov N, Romieu A, Roth RK, Rourke NJ, Ruediger ST, Rusman E, Sanches-Kuiper RM, Schenker MR, Seoane JM, Shaw RJ, Shiver MK, Short SW, Sizto NL, Sluis JP, Smith MA, Ernest Sohna Sohna J, Spence EJ, Stevens K, Sutton N, Szajkowski L, Tregidgo CL, Turcatti G, van de Vondele S, Verhovsky Y, Virk SM, Wakelin S, Walcott GC, Wang J, Worsley GJ, Yan J, Yau L, Zuerlein M, Rogers J, Mullikin JC, Hurles ME, McCooke NJ, West JS, Oaks FL, Lundberg PL, Klenerman D, Durbin R, Smith AJ (2008) Accurate whole human genome sequencing using reversible terminator chemistry. Nature 456(7218):53–59. https://doi.org/10.1038/nature07517

26. Guo J, Xu N, Li Z, Zhang S, Wu J, Kim DH, Sano Marma M, Meng Q, Cao H, Li X, Shi S, Yu L, Kalachikov S, Russo JJ, Turro NJ, Ju J (2008) Four-color DNA sequencing with *O*-modified nucleotide reversible terminators and chemically cleavable fluorescent dideoxynucleotides. Proc Natl Acad Sci 105(27): 9145–9150. https://doi.org/10.1073/pnas.0804023105

27. Simpson JT, Workman RE, Zuzarte PC, David M, Dursi LJ, Timp W (2017) Detecting DNA cytosine methylation using nanopore sequencing. Nat Methods 14(4):407–410. https://doi.org/10.1038/nmeth.4184

28. Levene MJ, Korlach J, Turner SW, Foquet M, Craighead HG, Webb WW (2003) Zero-mode waveguides for single-molecule analysis at high concentrations. Science 299(5607):682–686. https://doi.org/10.1126/science.1079700

29. Haque F, Li J, Wu H-C, Liang X-J, Guo P (2013) Solid-state and biological nanopore for real-time sensing of single chemical and sequencing of DNA. Nano Today 8(1): 56–74. https://doi.org/10.1016/j.nantod.2012.12.008

30. Wang Y, Yang Q, Wang Z (2015) The evolution of nanopore sequencing. Front Genet 5. https://doi.org/10.3389/fgene.2014.00449

31. Churko JM, Mantalas GL, Snyder MP, Wu JC (2013) Overview of high throughput sequencing technologies to elucidate molecular pathways in cardiovascular diseases. Circ Res 112(12):1613–1623. https://doi.org/10.1161/CIRCRESAHA.113.300939

32. Gautam SS, Kc R, Leong KW, Mac Aogáin M, O'Toole RF (2019) A step-by-step beginner's protocol for whole genome sequencing of human bacterial pathogens. J Biol Methods 6(1):e110. https://doi.org/10.14440/jbm.2019.276

33. Mahajan MC, McLellan AS (2020) Whole-exome sequencing (WES) for Illumina short read sequencers using solution-based capture. Methods Mol Biol 2076:85–108. https://doi.org/10.1007/978-1-4939-9882-1_5

34. Dilliott AA, Farhan SMK, Ghani M, Sato C, Liang E, Zhang M, McIntyre AD, Cao H, Racacho L, Robinson JF, Strong MJ, Masellis M, Bulman DE, Rogaeva E, Lang A, Tartaglia C, Finger E, Zinman L, Turnbull J, Freedman M, Swartz R, Black SE, Hegele RA (2018) Targeted next-generation dequencing and bioinformatics pipeline to evaluate genetic determinants of constitutional disease. J Vis Exp 134. https://doi.org/10.3791/57266

35. Wheeler DA, Srinivasan M, Egholm M, Shen Y, Chen L, McGuire A, He W, Chen Y-J, Makhijani V, Roth GT, Gomes X, Tartaro K, Niazi F, Turcotte CL, Irzyk GP, Lupski JR, Chinault C, Song X-z, Liu Y, Yuan Y, Nazareth L, Qin X, Muzny DM, Margulies M, Weinstock GM, Gibbs RA, Rothberg JM (2008) The complete genome of an individual by massively parallel DNA sequencing. Nature 452(7189):872–876. https://doi.org/10.1038/nature06884

36. Chou J, Ohsumi TK, Geha RS (2012) Use of whole exome and genome sequencing in the identification of genetic causes of primary immunodeficiencies. Curr Opin Allergy Clin Immunol 12(6):623–628. https://doi.org/10.1097/ACI.0b013e3283588ca6

37. Bewicke-Copley F, Arjun Kumar E, Palladino G, Korfi K, Wang J (2019) Applications and analysis of targeted genomic sequencing in cancer studies. Comput Struct Biotechnol J 17:1348–1359. https://doi.org/10.1016/j.csbj.2019.10.004

38. Han S-W, Kim H-P, Shin J-Y, Jeong E-G, Lee W-C, Lee K-H, Won J-K, Kim T-Y, Oh D-Y, Im S-A, Bang Y-J, Jeong S-Y, Park KJ, Park

J-G, Kang GH, Seo J-S, Kim J-I, Kim T-Y (2013) Targeted sequencing of cancer-related genes in colorectal cancer using next-generation sequencing. PLoS One 8(5): e64271. https://doi.org/10.1371/journal.pone.0064271

39. Dongre HN, Haave H, Fromreide S, Erland FA, Moe SEE, Dhayalan SM, Riis RK, Sapkota D, Costea DE, Aarstad HJ, Vintermyr OK (2021) Targeted next-generation sequencing of cancer-related genes in a Norwegian patient cohort with head and neck squamous cell carcinoma reveals novel actionable mutations and correlations with pathological parameters. Front Oncol 11. https://doi.org/10.3389/fonc.2021.734134
40. Gulilat M, Lamb T, Teft WA, Wang J, Dron JS, Robinson JF, Tirona RG, Hegele RA, Kim RB, Schwarz UI (2019) Targeted next generation sequencing as a tool for precision medicine. BMC Med Genet 12(1):81. https://doi.org/10.1186/s12920-019-0527-2
41. Lim B, Lin Y, Navin N (2020) Advancing cancer research and medicine with single-cell genomics. Cancer Cell 37(4):456–470. https://doi.org/10.1016/j.ccell.2020.03.008
42. Evrony GD, Hinch AG, Luo C (2021) Applications of single-cell DNA sequencing. Annu Rev Genomics Hum Genet 22(1):171–197. https://doi.org/10.1146/annurev-genom-111320-090436
43. Hu P, Zhang W, Xin H, Deng G (2016) Single cell isolation and analysis. Front Cell Dev Biol 4. https://doi.org/10.3389/fcell.2016.00116
44. Evrony GD, Hinch AG, Luo C (2021) Applications of single-cell DNA sequencing. Annu Rev Genomics Hum Genet 22:171–197. https://doi.org/10.1146/annurev-genom-111320-090436
45. Lee J-H, Gao C, Peng G, Greer C, Ren S, Wang Y, Xiao X (2011) Analysis of transcriptome complexity through RNA sequencing in normal and failing murine hearts. Circ Res 109(12):1332–1341. https://doi.org/10.1161/CIRCRESAHA.111.249433
46. Wang Z, Gerstein M, Snyder M (2009) RNA-Seq: a revolutionary tool for transcriptomics. Nat Rev Genet 10(1):57–63. https://doi.org/10.1038/nrg2484
47. Edgren H, Murumagi A, Kangaspeska S, Nicorici D, Hongisto V, Kleivi K, Rye IH, Nyberg S, Wolf M, Borresen-Dale A-L, Kallioniemi O (2011) Identification of fusion genes in breast cancer by paired-end RNA-sequencing. Genome Biol 12(1):R6. https://doi.org/10.1186/gb-2011-12-1-r6
48. Łabaj PP, Leparc GG, Linggi BE, Markillie LM, Wiley HS, Kreil DP (2011) Characterization and improvement of RNA-Seq precision in quantitative transcript expression profiling. Bioinformatics 27(13):i383–i391. https://doi.org/10.1093/bioinformatics/btr247
49. Sánchez-Pla A, Reverter F, Ruíz de Villa MC, Comabella M (2012) Transcriptomics: mRNA and alternative splicing. J Neuroimmunol 248(1–2):23–31. https://doi.org/10.1016/j.jneuroim.2012.04.008
50. Hoeijmakers WA, Bártfai R, Stunnenberg HG (2013) Transcriptome analysis using RNA-Seq. Methods Mol Biol 923:221–239. https://doi.org/10.1007/978-1-62703-026-7_15
51. Rosenow C, Saxena RM, Durst M, Gingeras TR (2001) Prokaryotic RNA preparation methods useful for high density array analysis: comparison of two approaches. Nucleic Acids Res 29(22):E112. https://doi.org/10.1093/nar/29.22.e112
52. Corchete LA, Rojas EA, Alonso-López D, De Las Rivas J, Gutiérrez NC, Burguillo FJ (2020) Systematic comparison and assessment of RNA-seq procedures for gene expression quantitative analysis. Sci Rep 10(1):19737. https://doi.org/10.1038/s41598-020-76881-x
53. Conesa A, Madrigal P, Tarazona S, Gomez-Cabrero D, Cervera A, McPherson A, Szcześniak MW, Gaffney DJ, Elo LL, Zhang X, Mortazavi A (2016) A survey of best practices for RNA-seq data analysis. Genome Biol 17(1):13. https://doi.org/10.1186/s13059-016-0881-8
54. Hong M, Tao S, Zhang L, Diao L-T, Huang X, Huang S, Xie S-J, Xiao Z-D, Zhang H (2020) RNA sequencing: new technologies and applications in cancer research. J Hematol Oncol 13(1):166. https://doi.org/10.1186/s13045-020-01005-x
55. Oshlack A, Robinson MD, Young MD (2010) From RNA-seq reads to differential expression results. Genome Biol 11(12):220. https://doi.org/10.1186/gb-2010-11-12-220
56. Govindarajan M, Wohlmuth C, Waas M, Bernardini MQ, Kislinger T (2020) High-throughput approaches for precision medicine in high-grade serous ovarian cancer. J Hematol Oncol 13(1):134. https://doi.org/10.1186/s13045-020-00971-6
57. Wu H, Li X, Li H (2019) Gene fusions and chimeric RNAs, and their implications in cancer. Genes Dis 6(4):385–390. https://doi.org/10.1016/j.gendis.2019.08.002
58. Wang N, Zheng J, Chen Z, Liu Y, Dura B, Kwak M, Xavier-Ferrucio J, Lu Y-C, Zhang M, Roden C, Cheng J, Krause DS,

Ding Y, Fan R, Lu J (2019) Single-cell microRNA-mRNA co-sequencing reveals non-genetic heterogeneity and mechanisms of microRNA regulation. Nat Commun 10(1): 95. https://doi.org/10.1038/s41467-018-07981-6
59. Pritchard CC, Cheng HH, Tewari M (2012) MicroRNA profiling: approaches and considerations. Nat Rev Genet 13(5):358–369. https://doi.org/10.1038/nrg3198
60. Mannarapu M, Dariya B, Bandapalli OR (2021) Application of single-cell sequencing technologies in pancreatic cancer. Mol Cell Biochem 476(6):2429–2437. https://doi.org/10.1007/s11010-021-04095-4
61. Li L, Xiong F, Wang Y, Zhang S, Gong Z, Li X, He Y, Shi L, Wang F, Liao Q, Xiang B, Zhou M, Li X, Li Y, Li G, Zeng Z, Xiong W, Guo C (2021) What are the applications of single-cell RNA sequencing in cancer research: a systematic review. J Exp Clin Cancer Res 40(1):163. https://doi.org/10.1186/s13046-021-01955-1
62. Lamond AI, Earnshaw WC (1998) Structure and function in the nucleus. Science 280(5363):547–553. https://doi.org/10.1126/science.280.5363.547
63. Buenrostro JD, Giresi PG, Zaba LC, Chang HY, Greenleaf WJ (2013) Transposition of native chromatin for fast and sensitive epigenomic profiling of open chromatin, DNA-binding proteins and nucleosome position. Nat Methods 10(12):1213–1218. https://doi.org/10.1038/nmeth.2688
64. Breiling A, Lyko F (2015) Epigenetic regulatory functions of DNA modifications: 5-methylcytosine and beyond. Epigenetics Chromatin 8(1):24. https://doi.org/10.1186/s13072-015-0016-6
65. Kawakatsu T (2020) Whole-genome bisulfite sequencing and epigenetic variation in cereal methylomes. Methods Mol Biol 2072:119–128. https://doi.org/10.1007/978-1-4939-9865-4_10
66. Lister R, Pelizzola M, Kida YS, Hawkins RD, Nery JR, Hon G, Antosiewicz-Bourget J, O'Malley R, Castanon R, Klugman S, Downes M, Yu R, Stewart R, Ren B, Thomson JA, Evans RM, Ecker JR (2011) Hotspots of aberrant epigenomic reprogramming in human induced pluripotent stem cells. Nature 471(7336):68–73. https://doi.org/10.1038/nature09798
67. Yan JG, Fu HY, Shen JZ, Zhou HR, Zhang YY, Huang JL, Chen CJ, Huang SH (2016) Application of bisulfite sequencing PCR in detecting the abnormal methylation of suppressor gene of Wnt signaling pathway in acute promyelocytic leukemia. Zhongguo Shi Yan Xue Ye Xue Za Zhi 24(5):1299–1304. https://doi.org/10.7534/j.issn.1009-2137.2016.05.003
68. Xu H, Zhao Y, Liu Z, Zhu W, Zhou Y, Zhao Z (2012) Bisulfite genomic sequencing of DNA from dried blood spot microvolume samples. Forensic Sci Int Genet 6(3):306–309. https://doi.org/10.1016/j.fsigen.2011.06.007
69. Eder T, Grebien F (2022) Comprehensive assessment of differential ChIP-seq tools guides optimal algorithm selection. Genome Biol 23(1):119. https://doi.org/10.1186/s13059-022-02686-y
70. Mundade R, Ozer HG, Wei H, Prabhu L, Lu T (2014) Role of ChIP-seq in the discovery of transcription factor binding sites, differential gene regulation mechanism, epigenetic marks and beyond. Cell Cycle 13(18):2847–2852. https://doi.org/10.4161/15384101.2014.949201
71. Nakato R, Sakata T (2021) Methods for ChIP-seq analysis: a practical workflow and advanced applications. Methods 187:44–53. https://doi.org/10.1016/j.ymeth.2020.03.005
72. Dirks RA, Stunnenberg HG, Marks H (2016) Genome-wide epigenomic profiling for biomarker discovery. Clin Epigenetics 8:122. https://doi.org/10.1186/s13148-016-0284-4
73. Chen Y, Zhang Y, Wang Y, Zhang L, Brinkman EK, Adam SA, Goldman R, van Steensel B, Ma J, Belmont AS (2018) Mapping 3D genome organization relative to nuclear compartments using TSA-Seq as a cytological ruler. J Cell Biol 217(11):4025–4048. https://doi.org/10.1083/jcb.201807108
74. Tanizawa H, Noma K (2012) Unravelling global genome organization by 3C-seq. Semin Cell Dev Biol 23(2):213–221. https://doi.org/10.1016/j.semcdb.2011.11.003
75. Dekker J, Rippe K, Dekker M, Kleckner N (2002) Capturing chromosome conformation. Science 295(5558):1306–1311. https://doi.org/10.1126/science.1067799
76. Rebouissou C, Sallis S, Forné T (2022) Quantitative chromosome conformation capture (3C-qPCR). Methods Mol Biol 2532:3–13. https://doi.org/10.1007/978-1-0716-2497-5_1
77. van Opijnen T, Bodi KL, Camilli A (2009) Tn-seq: high-throughput parallel sequencing for fitness and genetic interaction studies in microorganisms. Nat Methods 6(10): 767–772. https://doi.org/10.1038/nmeth.1377

78. Carabetta VJ, Esquilin-Lebron K, Zelzion E, Boyd JM (2021) Genetic approaches to uncover gene products involved in iron-sulfur protein maturation: high-throughput genomic screening using transposon sequencing. In: Dos Santos PC (ed) Fe-S proteins: methods and protocols. Springer, New York, pp 51–68. https://doi.org/10.1007/978-1-0716-1605-5_3

79. Barquist L, Mayho M, Cummins C, Cain AK, Boinett CJ, Page AJ, Langridge GC, Quail MA, Keane JA, Parkhill J (2016) The TraDIS toolkit: sequencing and analysis for dense transposon mutant libraries. Bioinformatics 32(7):1109–1111. https://doi.org/10.1093/bioinformatics/btw022

80. Cain AK, Barquist L, Goodman AL, Paulsen IT, Parkhill J, van Opijnen T (2020) A decade of advances in transposon-insertion sequencing. Nat Rev Genet 21(9):526–540. https://doi.org/10.1038/s41576-020-0244-x

81. Thibault D, Jensen PA, Wood S, Qabar C, Clark S, Shainheit MG, Isberg RR, van Opijnen T (2019) Droplet Tn-Seq combines microfluidics with Tn-Seq for identifying complex single-cell phenotypes. Nat Commun 10(1): 5729. https://doi.org/10.1038/s41467-019-13719-9

82. Yasir M, Turner AK, Bastkowski S, Baker D, Page AJ, Telatin A, Phan MD, Monahan L, Savva GM, Darling A, Webber MA, Charles IG (2020) TraDIS-Xpress: a high-resolution whole-genome assay identifies novel mechanisms of triclosan action and resistance. Genome Res 30(2):239–249. https://doi.org/10.1101/gr.254391.119

83. Lam HYK, Clark MJ, Chen R, Chen R, Natsoulis G, O'Huallachain M, Dewey FE, Habegger L, Ashley EA, Gerstein MB, Butte AJ, Ji HP, Snyder M (2012) Performance comparison of whole-genome sequencing platforms. Nat Biotechnol 30(1):78–82. https://doi.org/10.1038/nbt.2065

84. Ivády G, Madar L, Dzsudzsák E, Koczok K, Kappelmayer J, Krulisova V, Macek M Jr, Horváth A, Balogh I (2018) Analytical parameters and validation of homopolymer detection in a pyrosequencing-based next generation sequencing system. BMC Genomics 19(1):158. https://doi.org/10.1186/s12864-018-4544-x

85. Zook JM, Salit M (2011) Genomes in a bottle: creating standard reference materials for genomic variation – why, what and how? Genome Biol 12(1):P31. https://doi.org/10.1186/gb-2011-12-s1-p31

86. Brownstein CA, Beggs AH, Homer N, Merriman B, Yu TW, Flannery KC, DeChene ET, Towne MC, Savage SK, Price EN, Holm IA, Luquette LJ, Lyon E, Majzoub J, Neupert P, McCallie D Jr, Szolovits P, Willard HF, Mendelsohn NJ, Temme R, Finkel RS, Yum SW, Medne L, Sunyaev SR, Adzhubey I, Cassa CA, de Bakker PI, Duzkale H, Dworzyński P, Fairbrother W, Francioli L, Funke BH, Giovanni MA, Handsaker RE, Lage K, Lebo MS, Lek M, Leshchiner I, MacArthur DG, McLaughlin HM, Murray MF, Pers TH, Polak PP, Raychaudhuri S, Rehm HL, Soemedi R, Stitziel NO, Vestecka S, Supper J, Gugenmus C, Klocke B, Hahn A, Schubach M, Menzel M, Biskup S, Freisinger P, Deng M, Braun M, Perner S, Smith RJ, Andorf JL, Huang J, Ryckman K, Sheffield VC, Stone EM, Bair T, Black-Ziegelbein EA, Braun TA, Darbro B, DeLuca AP, Kolbe DL, Scheetz TE, Shearer AE, Sompallae R, Wang K, Bassuk AG, Edens E, Mathews K, Moore SA, Shchelochkov OA, Trapane P, Bossler A, Campbell CA, Heusel JW, Kwitek A, Maga T, Panzer K, Wassink T, Van Daele D, Azaiez H, Booth K, Meyer N, Segal MM, Williams MS, Tromp G, White P, Corsmeier D, Fitzgerald-Butt S, Herman G, Lamb-Thrush D, McBride KL, Newsom D, Pierson CR, Rakowsky AT, Maver A, Lovrečić L, Palandačić A, Peterlin B, Torkamani A, Wedell A, Huss M, Alexeyenko A, Lindvall JM, Magnusson M, Nilsson D, Stranneheim H, Taylan F, Gilissen C, Hoischen A, van Bon B, Yntema H, Nelen M, Zhang W, Sager J, Zhang L, Blair K, Kural D, Cariaso M, Lennon GG, Javed A, Agrawal S, Ng PC, Sandhu KS, Krishna S, Veeramachaneni V, Isakov O, Halperin E, Friedman E, Shomron N, Glusman G, Roach JC, Caballero J, Cox HC, Mauldin D, Ament SA, Rowen L, Richards DR, San Lucas FA, Gonzalez-Garay ML, Caskey CT, Bai Y, Huang Y, Fang F, Zhang Y, Wang Z, Barrera J, Garcia-Lobo JM, González-Lamuño D, Llorca J, Rodriguez MC, Varela I, Reese MG, De La Vega FM, Kiruluta E, Cargill M, Hart RK, Sorenson JM, Lyon GJ, Stevenson DA, Bray BE, Moore BM, Eilbeck K, Yandell M, Zhao H, Hou L, Chen X, Yan X, Chen M, Li C, Yang C, Gunel M, Li P, Kong Y, Alexander AC, Albertyn ZI, Boycott KM, Bulman DE, Gordon PM, Innes AM, Knoppers BM, Majewski J, Marshall CR, Parboosingh JS, Sawyer SL, Samuels ME, Schwartzentruber J, Kohane IS, Margulies DM (2014) An international effort towards developing standards for best practices in analysis,

interpretation and reporting of clinical genome sequencing results in the CLARITY challenge. Genome Biol 15(3):R53. https://doi.org/10.1186/gb-2014-15-3-r53
87. Shendure J, Balasubramanian S, Church GM, Gilbert W, Rogers J, Schloss JA, Waterston RH (2017) DNA sequencing at 40: past, present and future. Nature 550(7676):345–353. https://doi.org/10.1038/nature24286
88. Prober JM, Trainor GL, Dam RJ, Hobbs FW, Robertson CW, Zagursky RJ, Cocuzza AJ, Jensen MA, Baumeister K (1987) A system for rapid DNA sequencing with fluorescent chain-terminating dideoxynucleotides. Science 238(4825):336–341. https://doi.org/10.1126/science.2443975
89. Craxton M (1991) Linear amplification sequencing, a powerful method for sequencing DNA. Methods 3(1):20–26. https://doi.org/10.1016/S1046-2023(05)80159-8
90. DeAngelis MM, Wang DG, Hawkins TL (1995) Solid-phase reversible immobilization for the isolation of PCR products. Nucleic Acids Res 23(22):4742–4743. https://doi.org/10.1093/nar/23.22.4742
91. Zhang J, Fang Y, Hou JY, Ren HJ, Jiang R, Roos P, Dovichi NJ (1995) Use of non-cross-linked polyacrylamide for four-color DNA sequencing by capillary electrophoresis separation of fragments up to 640 bases in length in two hours. Anal Chem 67(24):4589–4593. https://doi.org/10.1021/ac00120a026
92. Pereira R, Oliveira J, Sousa M (2020) Bioinformatics and computational tools for next-generation sequencing analysis in clinical genetics. J Clin Med 9(1). https://doi.org/10.3390/jcm9010132
93. Kanzi AM, San JE, Chimukangara B, Wilkinson E, Fish M, Ramsuran V, de Oliveira T (2020) Next generation sequencing and bioinformatics analysis of family genetic inheritance. Front Genet 11:544162. https://doi.org/10.3389/fgene.2020.544162

Chapter 2

A Step-by-Step Guide to Sequencing and Assembly of Complete Bacterial Genomes Using the Oxford Nanopore MinION

Anil Kumar, Max M. Häggblom, and Lee J. Kerkhof

Abstract

The Oxford Nanopore (ONT) MinION enables sequencing of longer DNA/RNA fragments compared to other sequencers, such as Illumina, etc. This nanopore method provides distinct advantages for generating complete genome assemblies from microorganisms. Specifically, the R9.4 flow cells used for MinION sequencing have much lower error rates compared with earlier versions of the ONT platform. Coupled with base calling using Dorado software, higher-quality long reads can now be generated for complete bacterial genome assembly. In this chapter, we describe a detailed MinION method to assemble a complete genome from a microorganism, polish the final assembly, and evaluate the genome quality using various software tools. Because of the low cost for MinION sequencing, this platform could be an asset for virtually any laboratory interested in generating complete genomes from microorganisms.

Key words Oxford Nanopore, Bacterial genome, Long-read assemblers

1 Introduction

The genome represents the complete genetic material of an organism that stores information needed for its functioning, including growth and metabolism. Determining the base-pair order of the genome is vital in many research fields including microbiology, biotechnology, forensic sciences, and genetics [1]. Furthermore, DNA sequencing of short- and long-read amplicons can also provide information regarding the composition of the microbial community in a particular environment [2, 3]. This evaluation of microbial community composition using amplicons requires databases of curated and reference sequences/genomes [2, 4]. Having high-quality genomes of representative strains also provides foundational information for elucidating the functioning of the microbes in their habitats via subsequent metagenome and metatranscriptome sequencing.

Valerie J. Carabetta and Olaitan Akintunde (eds.), *High Throughput Gene Screening: Methods and Protocols*, Methods in Molecular Biology, vol. 2866, https://doi.org/10.1007/978-1-0716-4192-7_2,

DNA sequencing has improved significantly with time since Sanger methods were introduced [5]. Currently, most nucleic acid sequencing is done by different next-generation sequencing (NGS) platforms [6]. The Illumina sequencer is probably the most widely used second-generation NGS platform because of its high throughput and low error rates. These second-generation sequencers have been followed by third-generation sequencers, such as PacBio and Oxford Nanopore, which can provide longer sequence length, albeit with higher error rates. However, sequencing longer segments of the genomes enables reconstruction of a complete gapless genome assembly [7]. In this chapter, we provide a detailed methodology for generating a complete, closed genome sequence from a microorganism using the Oxford Nanopore sequencer. Additionally, software has been developed to polish the long-read assembly by combining Nanopore and Illumina raw reads [8, 9]. We have found that hybrid assembly methods using both long and short reads is often not necessary with the increased sequence accuracy of the R9.4 chemistry on the MinION platform.

2 Materials

2.1 DNA Extraction, Assembly Preparation, and Sequencing

1. High molecular-weight DNA extraction kit.
2. Rapid sequencing or ligation sequencing kit (Oxford Nanopore Technologies, UK).
3. Thermocycler.
4. R-9.4 flow-cells.
5. ONT MK1C or MK1B nanopore sequencer.

2.2 Software Tools

1. Dorado basecaller (https://github.com/nanoporetech/dorado).
2. Flye [10].
3. Canu [11].
4. Miniasm [12].
5. Trycycler [8].
6. Raven assembler [13].
7. Racon assemblers [14].
8. Polypolish [15].
9. Medaka (https://github.com/nanoporetech/medaka).
10. Minimap2 [16].
11. Filtlong (https://github.com/rrwick/Filtlong).
12. bwa [17].

13. POLCA [18].
14. CheckM [19].
15. BUSCO [20].
16. QUAST [21].

3 Methods

3.1 DNA Extraction, Assembly Preparation, and Sequencing

The first crucial step for generating genomes using the Nanopore sequencer is obtaining high molecular-weight (HMW) genomic DNA. Certain commercially available DNA extraction kits, such as the Monarch HMW DNA Extraction Kit (New England Biolabs, Ipswich, USA) and Wizard HMW DNA Extraction Kit (Promega Corporation, Madison, USA) can provide high molecular-weight nucleic acids for genome generation (*see* **Note 1**). Additional, manual methods such as hexadecyl-trimethylammonium bromide (CTAB)-phenol-chloroform extraction [22] or phenol/chloroform extraction [23] can also generate HMW DNA. These manual methods generally require additional clean-up steps, such as cesium chloride gradient ultracentrifugation or Ampure bead (Beckmann Coulter, Brea, USA) purification prior to sequencing.

Nanopore sequence library construction is most easily accomplished using the R9.4 rapid sequencing kit (RAD-004), following the manufacturer's instructions (*see* **Note 2**). This is a transposase-based system which randomly inserts ONT sequencing adaptors into HMW DNA at 4–8 kb intervals. The procedure takes approximately 5 min and is best accomplished in a PCR thermocycler. During the library preparation, it is also possible to prime the nanopore flow cell using the flush tether and flush buffer for sequencing. Detailed methods for preparing the sequence library and running on the sequencer are given on the Oxford Nanopore website: https://nanoporetech.com/.

3.2 Basecalling

The nanopore sequencer delivers the raw reads in a fast5 or pod5 format. These raw reads must be converted into standard fastq formats using an ONT basecaller. Dorado is our current choice for basecalling and this program also performs other functions, such as removing barcode sequences and demultiplexing. The basecaller can be downloaded from (https://github.com/nanoporetech/dorado). Before running the software, the user needs to download the appropriate model based on the chemistry and the type of flow cell. For example, the recommended model for DNA sequencing using an R9.4 flow cell is dna_r9.4.1_e8_sup@v3.6. The standard code for running Dorado is given as:

- dorado basecaller dna_r9.4.1_e8_sup@v3.6 [pod5/fast5-file-path] > -r --emit-fastq >[output-file.fastq]

3.3 Assembly Generation and Polishing

Three long-read assemblers for whole genome assembly generation are used to compare the final assemblies generated from the assemblers (*see* **Note 3**).

3.3.1 Canu Assembler

Canu is a long-read assembler designed for polishing and assembling PacBio and Oxford Nanopore reads [11]. The assembler can be downloaded from (https://github.com/marbl/canu), and a detailed tutorial can be accessed at (https://canu.readthedocs.io/en/latest/tutorial.html). The Linux command to run the assembler is given as:

- canu -p [assembly-prefix] -d [assembly-directory] genomeSize = [number[g|m|k] -Nanopore [*fastq]

The command will correct, trim, and assemble the whole genome from the fastq reads. The polishing of the Canu-generated assembly is done in several steps to improve the final assembly.

1. First, map the Canu-generated assembly to the raw Nanopore reads using minimap2. The command used for mapping of the reads is as follows:
   - minimap2 -ax map-ont [canu-assembly.fasta] [nanopore-reads.fastq] > [canu.sam]
2. Then, polish the Canu assembly using two rounds of Racon assembler (round 1)
   - racon [nanopore-reads.fastq] [canu.sam] [canu-assembly.fasta] -t [number] > [r1-racon-polished-canu.fasta]
3. After that, map the racon polished assembly to the Nanopore raw reads using the command:
   - minimap2 -ax map-ont [r1-racon-polished-canu.fasta] [nanopore-reads.fastq] > [r2-canu.sam]
4. Then, perform the second round of Racon polishing using the command:
   - racon [canu-assembly.fasta] [r2-canu.sam] [r1-racon-polished-canu.fasta] -t [number] > [r2_racon-polished-canu.fasta]
5. Lastly, polish the final assembly using the Medaka tool (https://github.com/nanoporetech/medaka) as follows:
   - medaka_consensus -d [r2-racon-polished-canu.fasta] -i [nanopore-reads.fastq] -t [number] -o [dir-path]

3.3.2 Flye Assembler

Flye is a genome assembler for PacBio and Nanopore long reads [10]. The assembler can be downloaded from (https://github.com/fenderglass/Flye) and the user manual can be assessed from (https://github.com/fenderglass/Flye/blob/flye/docs/USAGE.md). The Linux command to run the assembler is:

- flye –nano-raw [nanopore-reads.fastq] -g [number[g|m|k] -o [dir-path] -t [number]

The polishing of the Flye-generated assembly is done just like the Canu assembly polishing as follows.

1. First, map the Flye-generated assembly to the raw Nanopore reads using minimap2. The command used for mapping of the reads is as follows:
   - minimap2 -ax map-ont [flye-assembly.fasta] [nanopore-reads.fastq] > [Flye.sam]
2. Second, polish the Flye assembly using two rounds of Racon assembler (round 1)
   - racon [nanopore-reads.fastq] [Flye.sam] [flye-assembly.fasta] -t [number] > [r1-racon-polished-flye.fasta]
3. After that, map the Racon-polished assembly to the Nanopore raw reads using the command:
   - minimap2 -ax map-ont [r1-racon-polished-flye.fasta] [nanopore-reads.fastq] > [r2-flye.sam]
4. Then, perform a second round of Racon polishing using the command:
   - racon [flye-assembly.fasta] [r2-flye.sam] [r1-racon-polished-flye.fasta] -t [number] > [r2-racon-polished-flye.fasta]
5. Lastly, polish the final assembly using the medaka tool as follows:
   - medaka_consensus -d [r2-racon-polished-flye.fasta] -i [nanopore-reads.fastq] -t [number] -o [dir-path]

3.3.3 Trycycler Assembler

Trycycler is a long-read assembler effective in generating complete genome assemblies [8]. The assembler can be downloaded from (https://github.com/rrwick/Trycycler/wiki). To run Trycycler, the user should have at least 60x read coverage. Trycycler uses different assemblers such as Flye [10], Raven [13], and Miniasm [12] for genome assembly. All the commands are provided on the Trycycler homepage. We have also indicated the commands used to run the assembler below (*see* **Note 4**).

1. Reads QC:
   - Filtlong –min_length 1000 –keep_percent 95 [input_reads.fastq.gz] > [reads.fastq]
2. Reads subsampling:
   - trycycler subsample –reads [reads.fastq] –out_dir [read_subsets]

The command will subsample the reads into 12 default reads sets.

3. Assemblies: the assembly will be performed using three different assemblers on 12 reads sets as follows:
   - Flye –nano-hq [read_subsets/sample_01.fastq] –threads [number] –out-dir [assembly_01] && cp [assembly_01/assembly.fasta] [assemblies/assembly_01.fasta] && cp [assembly_01/assembly_graph.gfa] [assemblies/assembly_01.gfa] && rm -r [assembly_01]
   - miniasm_and_minipolish.sh [read_subsets/sample_02.fastq] [number] > [assemblies/assembly_02.gfa] && any2-fasta [assemblies/assembly_02.gfa] > [assemblies/assembly_02.fasta]
   - raven --threads [number] --disable-checkpoints --graphical-fragment-assembly [assemblies/assembly_03.gfa] [read_-subsets/sample_03.fastq] > [assemblies/assembly_03.fasta]

   Rerun all the three assemblers on the other reads set, i.e., Flye for the fourth, seventh, and tenth set, Miniasm for the fifth, eighth, and 11th set, and Raven for the sixth, ninth, and 12th set.

4. Clustering contigs: the command will cluster the assemblies into pre-replicon groups.
   - trycycler cluster --assemblies [assemblies/*.fasta] --reads [reads.fastq] --out_dir [Trycycler]

5. Reconciling contigs: The command should be run separately for all clusters that are sufficiently similar to each other as follows:
   - trycycler reconcile --reads [reads.fastq] --cluster_dir [trycycler/cluster_number]

6. Multiple sequence alignment: this command will take the reconciled contigs and run the multiple sequence alignment. The command should be run on each of the good clusters separately.
   - trycycler msa –cluster_dir [trycycler/cluster_number]

7. Partitioning reads: the command is used to partition the reads between clusters, i.e., each read is assigned to the clusters to which it aligned best.
   - trycycler partition --reads [reads.fastq] --cluster_dirs [trycycler/cluster_*]

8. Generating a consensus: this command generates the consensus contig for every cluster. The command should be run for each of the good clusters separately.

- trycycler consensus --cluster_dir [trycycler/cluster_number]

if the user has multiple clusters, combine the consensus sequences into one fasta file using the command:

- cat [trycycler/cluster_*/7_final_consensus.fasta] > [trycycler/consensus.fasta],

9. Polishing using long reads: if the user only has the long reads, then polishing of the final assembly can be done using Medaka polisher using the command:
   - for c in trycycler/cluster_*; do

     medaka_consensus -i "$c"/4_reads.fastq -d "$c"/7_final_consensus.fasta -o "$c"/medaka -m r941_min_sup_g507 -t 12

     mv "$c"/medaka/consensus.fasta "$c"/8_medaka.fasta

     rm -r "$c"/medaka "$c"/*.fai "$c"/*.mmi # clean up

     done
10. Polishing using short reads: if the user has short reads of the sequenced genome, then the short reads can also be used for polishing using Polypolish and POLCA as:

    I.
    Using Polypolish [15]:
    - bwa index [consensus.fasta]
    - bwa mem -t [number] -a [consensus.fasta 1.fastq.gz] > [alignments_1.sam]
    - bwa mem -t [number] -a [consensus.fasta 2.fastq.gz] > [alignments_2.sam]
    - polypolish [consensus.fasta] [alignments_1.sam] [alignments_2.sam] > [polypolish.fasta]

    II.
    Using POLCA [18]:
    - polca.sh -a [polypolish.fasta] -r [1.fastq.gz 2.fastq.gz] -t [number] -m 1G
    - mv [*.PolcaCorrected.fa] [polypolish_polca.fasta]

    The final assembly will be stored as polypolish_polca.fasta.

3.4 Checking the Quality of the Assembly

Different genome assemblers generate distinct assemblies of the same genome, so an assessment of the genome assembly's quality is necessary to determine the best assembly. Three different tools are used routinely to assess the quality of the genome assembly.

3.4.1 CheckM

CheckM is a standard tool for assessing the quality of a prokaryotic genome assembly constructed from isolates, single cells, or metagenomes [19]. The tool uses universal, single-copy genes of the phylogenetic lineage and provides the completeness and contamination levels of the genome assembly. The tool can be run using the command:

- checkm lineage_wf -t [number] -x [assembly-extension] [bin--folder] [output-folder]

3.4.2 BUSCO

BUSCO is a tool to assess the quality of a prokaryotic and eukaryotic genome and metagenome assemblies [20]. The tool uses universal single-copy orthologs to assess the assembly quality and provides results in the form of completeness and contamination percentages. To run the tool, the following command is used:

- busco -i [sequence-file] -l [lineage] -o [output-name] -m [genome/transcriptome/proteins]

3.4.3 Quast

The Quast tool is used to evaluate the genome assembly by computing various matrices such as the number of contigs, N50-value, and total genome size [21]. The tool can be used with or without a reference genome of the closest strain or species sequenced previously. The tool can be run using the following command:

- quast.py [contigs-1.fasta] [contigs-2.fasta] -r [reference.fasta.gz] -o [quast-test-output]

3.5 Comparison of the Assembly Programs

To test the quality of various assemblers, the bacterial genome of *Mucilaginibacter* sp. strain E4BP6 was sequenced and assembled using the methods given above. A DNA library was prepared using the rapid sequencing kit, sequenced using a R9.4 flow cell, and basecalled using Dorado. Genome assembly utilized either the Canu, Flye, or Trycycler assemblers (*see* **Note 5**). The assemblies generated from the Canu and Flye assembler were polished using the Nanopore-only fastq reads, while the assembly generated using the Trycycler assembler was polished using both Nanopore and Illumina fastq reads. The unpolished and polished assemblies were screened and compared for completion and quality using QUAST (Table 1), CheckM (Table 2), and BUSCO (Table 3) tools. Comparison of the assembly quality indicated that the Trycycler-generated assembly was the best, indicating hybrid approaches using short and long reads outperform other methods. However, it is worth noting that the Flye assembler was also capable of generating a complete genome sequence (one contig; 97% completeness; ~1.2% contamination) with CheckM. These results are comparable to the Trycycler-generated assembly. The BUSCO results suggested the Flye-generated assembly was 94.4% complete while Trycycler assembly was 98.4% complete. CheckM and

Table 1
QUAST report on the unpolished and polished assemblies of *Mucilaginibacter* sp. strain E4BP6 generated from Canu, Flye, and Trycycler assemblers

Statistics without reference	Canu-unpolished	Canu-polished	Fly-unpolished	Flye-polished	Trycycler-unpolished	Trycycler-polished
# contigs	4	4	1	1	1	1
# contigs (>= 0 bp)	4	4	1	1	1	1
# contigs (>= 1000 bp)	4	4	1	1	1	1
# contigs (>= 5000 bp)	4	4	1	1	1	1
# contigs (>= 10000 bp)	4	4	1	1	1	1
# contigs (>= 25000 bp)	4	4	1	1	1	1
# contigs (>= 50000 bp)	2	2	1	1	1	1
Largest contig	4890185	4890116	4864724	4864673	4864759	4864841
Total length	5016475	5016275	4864724	4864673	4864759	4864841
Total length (>= 0 bp)	5016475	5016275	4864724	4864673	4864759	4864841
Total length (>= 1000 bp)	5016475	5016275	4864724	4864673	4864759	4864841
Total length (>= 5000 bp)	5016475	5016275	4864724	4864673	4864759	4864841
Total length (>= 10000 bp)	5016475	5016275	4864724	4864673	4864759	4864841
Total length (>= 25000 bp)	5016475	5016275	4864724	4864673	4864759	4864841
Total length (>= 50000 bp)	4951474	4951315	4864724	4864673	4864759	4864841
N50	4890185	4890116	4864724	4864673	4864759	4864841
N90	4890185	4890116	4864724	4864673	4864759	4864841
auN	4768245	4768298	4864724	4864673	4864759	4864841
L50	1	1	1	1	1	1
L90	1	1	1	1	1	1
GC (%)	40.02	40.02	39.99	39.99	39.99	39.99
Mismatches						
# N's per 100 kbp	0	0	0	0	0	0
# N's	0	0	0	0	0	0

Worst Median Best

BUSCO results indicate that the Trycycler generated assembly was better compared to other assemblers, but the Flye assembler also produced a good quality assembly. The Canu assembler was unable to generate the complete closed genome sequence using the tested bacterial strain.

Table 2
CheckM report on the unpolished and polished assemblies of *Mucilaginibacter* sp. strain E4BP6 generated from Canu, Flye and Trycycler assemblers. The table shows the completeness and contamination percentage of each assembly

Bin Id	Marker lineage	Genomes	Markers	Marker sets	0	1	2	3	4	5+	Completeness	Contamination	Strain heterogeneity
Trycycler unpolished	p__Bacteroidetes (UID2605)	350	316	210	5	308	3	0	0	0	97.62	1.19	0
Trycycler polished	p__Bacteroidetes (UID2605)	350	316	210	5	308	3	0	0	0	97.62	1.19	0
Flye unpolished	p__Bacteroidetes (UID2605)	350	316	210	7	306	3	0	0	0	96.9	1.19	0
Flye polished	p__Bacteroidetes (UID2605)	350	316	210	5	308	3	0	0	0	97.62	1.19	0
Canu unpolished	p__Bacteroidetes (UID2605)	350	316	210	6	304	5	1	0	0	97.38	3.1	62.5
Canu polished	p__Bacteroidetes (UID2605)	350	316	210	5	305	5	1	0	0	97.62	3.1	62.5

Table 3
BUSCO report on the unpolished and polished assemblies of *Mucilaginibacter* sp. strain E4BP6 generated from Canu, Flye, and Trycycler assemblers. The table shows the completeness and contamination percentage of each assembly

Input file	Dataset	Complete	Single	Duplicated	Fragmented	Missing	n_markers	Scaffold N50	Contigs N50	Percent gaps	Number of scaffolds
Trycycler unpolished	bacteria_odb10	96.8	96.8	0	3.2	0	124	4,864,759	4,864,759	0.00%	1
Trycycler polished	bacteria_odb10	98.4	98.4	0	1.6	0	124	4,864,841	4,864,841	0.00%	1
Fly unpolished	bacteria_odb10	94.4	94.4	0	4.8	0.8	124	4,864,724	4,864,724	0.00%	1
Flye polished	bacteria_odb10	94.4	94.4	0	4.8	0.8	124	4,864,673	4,864,673	0.00%	1
Canu unpolished	bacteria_odb10	94.4	94.4	0	5.6	0	124	4,890,185	4,890,185	0.00%	4
Canu polished	bacteria_odb10	94.4	94.4	0	4.8	0.8	124	4,890,116	4,890,116	0.00%	4

4 Notes

1. The rapid sequencing kit works well for library preparation and takes only 5–10 mins compared to the ligation sequencing kit.
2. Use the R9.4.1 and R10.4.1 flow cells as they have low error rates.
3. The final assembly generated from the assemblers should be polished using Nanopore raw fastq reads.
4. If the user has Illumina reads, those reads can also be used for assembly polishing using the Trycycler assembler, which generally gives a better assembly.
5. Generally, Flye and Trycycler assemblers give better assembly quality with a complete, closed genome. Both methods should both be tried, and the user needs to evaluate the final assembly quality to determine the better genome assembly.

Acknowledgements

This work was funded in part by the U.S. National Science Foundation (Award Number 2129351) and the USDA National Institute of Food and Agriculture Hatch project accession number 1012785 through the New Jersey Agricultural Experiment Station (Hatch Project NJ01160).

References

1. Giani AM, Gallo GR, Gianfranceschi L, Formenti G (2020) Long walk to genomics: history and current approaches to genome sequencing and assembly. Comput Struct Biotechnol J 18:9–19
2. Kerkhof LJ, Roth PA, Deshpande SV, Bernhards RC, Liem AT, Hill JM, Häggblom MM, Webster NS, Ibironke O, Mirzoyan S, Polashock JJ, Sullivan RF (2022) A ribosomal operon database and MegaBLAST settings for strain-level resolution of microbiomes. FEMS Microbes 3. https://doi.org/10.1093/femsmc/xtac002
3. Bolyen E, Rideout JR, Dillon MR, Bokulich NA, Abnet CC, Al-Ghalith GA, Alexander H, Alm EJ, Arumugam M, Asnicar F, Bai Y, Bisanz JE, Bittinger K, Brejnrod A, Brislawn CJ, Brown CT, Callahan BJ, Caraballo-Rodríguez AM, Chase J, Cope EK, Da Silva R, Diener C, Dorrestein PC, Douglas GM, Durall DM, Duvallet C, Edwardson CF, Ernst M, Estaki M, Fouquier J, Gauglitz JM, Gibbons SM, Gibson DL, Gonzalez A, Gorlick K, Guo J, Hillmann B, Holmes S, Holste H, Huttenhower C, Huttley GA, Janssen S, Jarmusch AK, Jiang L, Kaehler BD, Bin KK, Keefe CR, Keim P, Kelley ST, Knights D, Koester I, Kosciolek T, Kreps J, Langille MGI, Lee J, Ley R, Liu YX, Loftfield E, Lozupone C, Maher M, Marotz C, Martin BD, McDonald D, McIver LJ, Melnik AV, Metcalf JL, Morgan SC, Morton JT, Naimey AT, Navas-Molina JA, Nothias LF, Orchanian SB, Pearson T, Peoples SL, Petras D, Preuss ML, Pruesse E, Rasmussen LB, Rivers A, Robeson MS, Rosenthal P, Segata N, Shaffer M, Shiffer A, Sinha R, Song SJ, Spear JR, Swafford AD, Thompson LR, Torres PJ, Trinh P, Tripathi A, Turnbaugh PJ, Ul-Hasan S, van der Hooft JJJ, Vargas F, Vázquez-Baeza Y, Vogtmann E, von Hippel M, Walters W, Wan Y, Wang M, Warren J, Weber KC, Williamson CHD, Willis AD, Xu ZZ, Zaneveld JR, Zhang Y, Zhu Q, Knight R, Caporaso JG (2019) Reproducible, interactive, scalable and

extensible microbiome data science using QIIME 2. Nat Biotechnol 37:852–857

4. Quast C, Pruesse E, Gerken J, Schweer T, Yilmaz P, Peplies J, Glöckner FO (2015) SILVA Databases. In: Encyclopedia of metagenomics. Springer, Boston, pp 626–635
5. Sanger F, Nicklen S, Coulson AR (1977) DNA sequencing with chain-terminating inhibitors. Proc Natl Acad Sci USA 74:5463–5467. https://doi.org/10.1073/pnas.74.12.5463
6. Heather JM, Chain B (2016) The sequence of sequencers: the history of sequencing DNA. Genomics 107:1–8
7. Mukhia S, Kumar A, Kumar R (2023) Antioxidant prodigiosin-producing cold-adapted Janthinobacterium sp. ERMR3:09 from a glacier moraine: genomic elucidation of cold adaptation and pigment biosynthesis. Gene 857:147178. https://doi.org/10.1016/j.gene.2023.147178
8. Wick RR, Judd LM, Cerdeira LT, Hawkey J, Méric G, Vezina B, Wyres KL, Holt KE (2021) Trycycler: consensus long-read assemblies for bacterial genomes. Genome Biol 22:266. https://doi.org/10.1186/s13059-021-02483-z
9. Wick RR, Judd LM, Gorrie CL, Holt KE (2017) Unicycler: resolving bacterial genome assemblies from short and long sequencing reads. PLoS Comput Biol 13:e1005595. https://doi.org/10.1371/journal.pcbi.1005595
10. Kolmogorov M, Yuan J, Lin Y, Pevzner PA (2019) Assembly of long, error-prone reads using repeat graphs. Nat Biotechnol 37:540–546. https://doi.org/10.1038/s41587-019-0072-8
11. Koren S, Walenz BP, Berlin K, Miller JR, Bergman NH, Phillippy AM (2017) Canu: scalable and accurate long-read assembly via adaptive κ-mer weighting and repeat separation. Genome Res 27:722–736. https://doi.org/10.1101/gr.215087.116
12. Li H (2016) Minimap and miniasm: fast mapping and de novo assembly for noisy long sequences. Bioinformatics 32:2103–2110. https://doi.org/10.1093/bioinformatics/btw152
13. Vaser R, Šikić M (2021) Time- and memory-efficient genome assembly with raven. Nat Comput Sci 1:332–336. https://doi.org/10.1038/s43588-021-00073-4
14. Vaser R, Sović I, Nagarajan N, Šikić M (2017) Fast and accurate de novo genome assembly from long uncorrected reads. Genome Res 27:737–746. https://doi.org/10.1101/gr.214270.116
15. Wick RR, Holt KE (2022) Polypolish: short-read polishing of long-read bacterial genome assemblies. PLoS Comput Biol 18:e1009802. https://doi.org/10.1371/journal.pcbi.1009802
16. Li H (2018) Minimap2: pairwise alignment for nucleotide sequences. Bioinformatics 34:3094–3100. https://doi.org/10.1093/bioinformatics/bty191
17. Li H, Durbin R (2010) Fast and accurate long-read alignment with burrows-wheeler transform. Bioinformatics 26:589–595. https://doi.org/10.1093/bioinformatics/btp698
18. Zimin AV, Salzberg SL (2020) The genome polishing tool POLCA makes fast and accurate corrections in genome assemblies. PLoS Comput Biol 16. https://doi.org/10.1371/journal.pcbi.1007981
19. Parks DH, Imelfort M, Skennerton CT, Hugenholtz P, Tyson GW (2015) CheckM: assessing the quality of microbial genomes recovered from isolates, single cells, and metagenomes. Genome Res 25:1043–1055. https://doi.org/10.1101/gr.186072.114
20. Simão FA, Waterhouse RM, Ioannidis P, Kriventseva EV, Zdobnov EM (2015) BUSCO: assessing genome assembly and annotation completeness with single-copy orthologs. Bioinformatics 31:3210–3212. https://doi.org/10.1093/bioinformatics/btv351
21. Gurevich A, Saveliev V, Vyahhi N, Tesler G (2013) QUAST: quality assessment tool for genome assemblies. Bioinformatics 29:1072–1075. https://doi.org/10.1093/BIOINFORMATICS/BTT086
22. Gadkari PS, Mcguinness LR, Männistö MK, Kerkhof LJ, Häggblom MM (2020) Arctic tundra soil bacterial communities active at subzero temperatures detected by stable isotope probing. FEMS Microbiol Ecol 96. https://doi.org/10.1093/femsec/fiz192
23. Trigodet F, Lolans K, Fogarty E, Shaiber A, Morrison HG, Barreiro L, Jabri B, Eren AM (2022) High molecular weight DNA extraction strategies for long-read sequencing of complex metagenomes. Mol Ecol Resour 22:1786–1802. https://doi.org/10.1111/1755-0998.13588

Chapter 3

Targeted Next-Generation Sequencing in Rare Diseases

Aleša Kristan and Nataša Debeljak

Abstract

Targeted next-generation sequencing (NGS) in rare disease focuses on genetic analysis of specific regions in genome that are linked to a rare disease. In addition to library preparation, sequencing, and data analysis, targeted NGS includes an additional step of target enrichment of selected genes and regions. It allows for more sensitive and profound sequencing, as it is a fast and cost-effective approach with less data burden and is therefore often a method of choice for identifying rare variants in known genes, especially in diagnostics of rare diseases. Several in silico tools address the pathogenicity predictions of rare variants of unknown significance (VUS) and can therefore facilitate clinical interpretation.

Key words Library preparation, Targeted enrichment, Disease-specific DNA probes, Hybridization capture, Sequencing, Data analysis, Pathogenicity prediction

1 Introduction

There are three main types of NGS sequencing, namely whole genome sequencing (WGS), whole exome sequencing (WES), and targeted NGS. WGS provides the most comprehensive genome coverage, which is more suitable for novel gene discovery and research applications. On the other hand, WES and targeted NGS are target-based approaches that cover all exons (WES) or specific genes or regions of interest (targeted NGS) and are more often being used in clinical testing [1]. In the field of rare diseases, targeted NGS is focused on specific genes and pathways, known to be linked to a development of a particular rare disease [2–4]. Because panels cover only few genomic regions, targeted NGS enables deeper coverage than alternative sequencing methods. This allows for more sensitive and profound sequencing and is often a method of choice for identifying rare variants in known genes with high confidence (*see* **Note 1**). Compared to WES and WGS, targeted NGS has a faster turnaround time and is more cost-effective, which is essential to many clinical applications [1, 5, 6]. Additional advantages of targeted NGS are scalability (more samples per

Valerie J. Carabetta and Olaitan Akintunde (eds.), *High Throughput Gene Screening: Methods and Protocols*, Methods in Molecular Biology, vol. 2866, https://doi.org/10.1007/978-1-0716-4192-7_3,

sequencing run), faster data interpretation (smaller datasets leading to quicker computational analysis and fewer data burden), and suitability for development of commercial diagnostic tests (cost effectiveness and speed).

After sample isolation, library preparation is the first step in the NGS protocol. The workflow for targeted sequencing involves target enrichment of specific genes and regions associated with a disease as an additional step. Two main methods for target enrichment may be used, hybridization capture and amplicon sequencing [1]. In hybridization capture, biotinylated oligonucleotide probes are used to capture specific regions of interest. After library preparation, samples can be multiplexed (pooled) before target enrichment. In amplicon sequencing, PCR with specific primers containing a barcode and adapter sequences is used to amplify only the regions of interest before the library preparation. While preparation of libraries with amplicons is cheaper and involves fewer steps, it is limited to a maximum of 10,000 amplicons per panel, shows less uniform coverage, and produces more PCR duplicates than sequencing after hybridization capture. Also, certain regions in genome make primer design for amplicon enrichment difficult [5, 6]. Sequencing is followed by data analysis, that may be the most time-consuming part of the analysis. New variants of unknown significance (VUS) are often found, making clinical interpretation challenging. Tools enabling pathogenicity prediction of variants in coding and non-coding regions can facilitate clinical interpretation and are described in detail.

2 Materials

For targeted NGS using the hybridization approach, below are listed examples of commercially available kits and equipment that can be used. For an example of targeted NGS in rare disease setup, see diagnosis of rare erythrocytosis [3, 7].

2.1 Library Preparation

1. Reagent kit for library preparation (DNA Prep for Enrichment, Illumina, or similar) [8].
2. Sample indexes (Unique Dual (UD) indexes, Illumina, or similar).
3. Magnetic beads for DNA purification (MagSi-NGSPREP Plus, Magtivio, or similar).
4. Fluorometric assay (Qubit™ dsDNA Quantification Assay, Invitrogen, or similar).
5. Reagent kit for fragment analysis (Agilent High Sensitivity DNA Kit, Agilent Technologies, or similar) [9].

6. Fragment bioanalyzer (Agilent 2100, Agilent Technologies, or similar).
7. 1.5 mL DNA low-bind microcentrifuge tubes.
8. Pipette tips, filtered.
9. 96-well plates or 8-tube RNase/DNase free strips appropriate for PCR.
10. Ethanol (80%). Prepare 80% ethanol from absolute ethanol (96%) with nuclease-free water.
11. Nuclease-free water.
12. Single and multi-channel pipettes.
13. Magnetic stand for 1.5 ml tubes and PCR plates/8-tube strips.
14. Microcentrifuge.
15. Vortex mixer.
16. Thermal cycler.
17. Spectrophotometer for measuring UV absorbance.

2.2 Targeted Enrichment

1. Custom gene panel (xGen Custom Hyb Panels, Integrated DNA Technologies (IDT), or similar) [10].
2. Reagent kit for enrichment (DNA Prep for Enrichment, Illumina, or similar) [8].
3. Fluorometric assay (Qubit™ dsDNA Quantification Assay, Invitrogen, or similar).
4. Magnetic beads (MagSi-NGSPREP Plus, Magtivio, or similar).
5. 1.5 mL DNA low-bind microcentrifuge tubes.
6. Pipette tips, filtered.
7. 96-well plates or 8-tube RNase/DNase free strips appropriate for PCR.
8. Ethanol (80%). Prepare 80% ethanol from absolute ethanol (96%) with nuclease-free water.
9. Nuclease-free water.
10. Single and multi-channel pipettes.
11. Magnetic stand for 1.5 mL tubes and PCR plates/strips.
12. Microcentrifuge.
13. Thermal cycler.
14. Micro heating system.

2.3 Sequencing

1. Sequencing kit (Miniseq sequencing kit, Illumina, or similar) [11].
2. Sequencer (MiniSeq sequencer, Illumina, or similar).

3 Methods

3.1 Quality Control (QC) of Genomic DNA

Before library preparation, genomic DNA (gDNA) must be isolated from samples using DNA purification kit (recommended, QIAamp DNA Mini kit, Qiagen). It is recommended to assess the purity of DNA, measuring UV absorbance at 260, 280, and 230 nm, as a common method. The ratio of 260/280 is assessed for the main indication of purity, and ratio 260/230 as a secondary indication of sample purity. It is advised that genomic DNA samples have low amounts of EDTA and organic contaminants, such as phenol and ethanol, that could interfere with the next steps. It is recommended to quantify input DNA, especially if you have samples with low concentrations or highly degraded (formalin-fixed paraffin-embedded (FFPE) samples). The concentration measurements could be performed with a fluorometric assay, such as the Qubit dsDNA assay (Invitrogen), that is highly selective for double-stranded DNA (dsDNA) over other nucleic acids.

3.2 Library Preparation

Various reagent kits and protocols are applicable for library preparation and enrichment for targeted sequencing and are constantly modifying and improving. Here, we briefly describe the main principles for library preparation and hybridization-based enrichment, using one of the commercial reagent kits (DNA Prep for Enrichment, Illumina) [8], which we also used for the diagnostics of hereditary erythrocytosis [3, 7]. For a detailed protocol, see the manufacturer's instructions [8]. This protocol is suitable for sequencing of DNA isolated from different biological materials, like extracted cells, blood, FFPE tissue, or saliva. The DNA input can vary from 10–1000 ng; however, the recommended minimal DNA input is 50 ng. Library preparation is performed in 4 steps: (a) fragmentation of gDNA; (b) clean-up of fragmented DNA; (c) amplification of fragmented DNA; and (d) clean-up of amplified libraries.

3.2.1 Fragmentation of Genomic DNA and Post-fragmentation Clean-Up

For shearing of gDNA, different physical (sonication, nebulization, hydrodynamic shearing), chemical (heating with divalent metal cations), and enzymatic (transposase or endonuclease) methods exist [12]. Fragmentation using Illumina DNA Prep for Enrichment is performed with an enzymatic approach, using an enzyme transposase and free transposon (together called transposome) included in the reagent kit. In that way, fragmentation and tagging of genomic DNA with adapter sequences is performed in the same step (called also tagmentation) [13]. After fragmentation using transposases, the enzymatic reaction is stopped, and DNA is cleaned before PCR amplification.

3.2.2 Amplification of Fragmented DNA and Clean-Up of Amplified Libraries

In this step, the tagmented DNA is amplified in a PCR reaction and index adapters as oligonucleotides are added to the sample (*see* **Note 2**). Index adapters consist of the following sequences: a sequence that is complementary to the oligonucleotide on the flow-cell and it is important for binding on the flow cell; a sequence that is unique for the individual sample and is important as a sample label; and a sequence complementary to the oligonucleotide in sequencing reaction. Clean-up of amplified libraries is performed to isolate only fragments of a certain size. The most suitable way to do this is with magnetic beads (for instance, MagSi-NGSPREP Plus, Magtivio, or similar). Since larger fragments do not cluster well on the flow cell, it is useful to perform double-size selection using magnetic beads, to remove larger and smaller fragments and retain the average size of fragments between 300–500 bp [14].

3.2.3 Qualification of Pre-enriched Libraries

Before enrichment, the qualification of pre-enriched libraries can be performed on fragment analyzer, for instance Bioanalyzer Agilent 2100 (Agilent Technologies), to determine the size and distribution of fragments in a sample. A compatible reagent kit, for instance Agilent High Sensitivity DNA Kit (Agilent Technologies), is used for analysis on the fragment analyzer, following the manufacturer's instructions [9].

3.3 Targeted Enrichment

For capture enrichment, different methodologies could be used, and here we briefly describe the principles behind one of the commercial reagent kits (DNA Prep for Enrichment, Illumina) [8], which we also used for the diagnostics of hereditary erythrocytosis [3, 7]. For detailed protocol, see manufacturer's instructions [8]. Before enrichment, DNA libraries with unique indexes are combined into one pool of up to 12 libraries. Pooling is performed by combining libraries at the same volume (pooling by volume) or by combining libraries at equal mass (pooling by mass). In many cases, especially when initial DNA samples had low concentration or quality, pooling by mass helps achieve optimal sample balancing. Before pooling by mass, quantification of pre-enriched libraries must be performed. For accurate quantification, it is recommended to use a fluorometric assay, such as Qubit dsDNA assay, that are highly selective for double-stranded DNA (dsDNA) over other nucleic acids. The process of enriching target regions can be divided into the following steps: (a) probe design, (b) hybridization, (c) capture hybridized probes, (d) amplification of enriched libraries, and (e) clean-up of amplified enriched libraries, which will be discussed next.

3.3.1 Probe Design

Disease-specific genes and regions linked to a particular disease can be searched in the literature and public databases like Reactome [15], String [16], UniProt [17], GeneCards [18], Human Protein Atlas [19], National Center for Biotechnology Information

(NCBI) [20], and HUGO Gene Nomenclature Committee (HGNC) [21]. For an example, see the selection of genes and regions applied in targeted NGS for diagnosis of rare erythrocytosis [3, 7]. A custom gene panel could be designed and created at one of the commercial providers, for instance IDT offers xGen Custom Hyb Panels (*see* **Note 3**, [10]).

3.3.2 Hybridization, Capture, Amplification, and Clean-up

In the hybridization step, biotinylated probes bind to the target regions in the genome (*see* **Note 4**). After hybridization, the reaction probes are captured using magnetic beads, coated with streptavidin that bind to biotin. Captured target regions on the magnetic beads are subsequently washed and further eluted from the beads. The enriched library is next amplified using PCR, and the resulting libraries are purified with magnetic beads to remove unwanted products.

3.4 Illumina Sequencing

Enriched libraries are sequenced on one of the sequencers suitable for short-read sequencing, like a MiniSeq sequencer (Illumina), using appropriate reagent kit, following manufacturer's instructions [11]. Single-read or paired-end sequencing technology could be used, and we typically use paired-end sequencing with 2×150 cycles (*see* **Note 5**).

3.5 Data Analysis

3.5.1 Sequence Alignment, Variant Identification, and Annotation

Bioinformatic analysis of sequences can be performed with either the built-in bioinformatics tool within the sequencer, like Local Run Manager, Illumina, or custom-designed pipelines [22]. Two major steps in sequence data analysis are alignment (mapping) of reads to a reference genome and identification (calling) of variants (*see* **Note 6**). A variety tools for aligning sequence reads to a reference genome and variant calling exist, with different algorithms, and it is user preference for which suits their needs best [23]. A frequently used read aligner is Burrows-Wheeler Alignment (BWA) tool with algorithm BWA-MEM [24], and a popular variant caller is the Genome Analysis Toolkit (GATK) [25]. Both are included in the Illumina Local Run Manager.

The annotation of genetic variants can be performed within one of the available software tools for variant annotation, like Ensembl Variant Effect Predictor (VEP) [26] that has a user-friendly web interface, or some of the command line tools, like ANNOtate VARiation (ANNOVAR) [27] and snpEff [28]. There are also several commercially available analysis tools that offer variant annotation, and further interpretation and filtering of genetic variants, like Clinical Insight (QCI) Interpret (Qiagen). In the analysis of targeted NGS data for hereditary erythrocytosis, we used ANNOVAR and snpEff algorithms, with pathogenicity predictions from the dbNSFPv2 database [7, 22]. Only variants with a high-quality score (30) and adequate minor allele depth (> 5) should be considered for downstream analysis of rare germline variants.

3.5.2 Filtration of Rare Variants

In the case of rare diseases, the focus is on rare variants in the population. Rare variants (minor allele frequency (MAF) <0.01) are filtered based on the population frequency from different databases, like 1000 genomes [29], Exome Sequencing Project v.6500 (ESP6500) [30], and the Genome Aggregation (GnomAD) databases [31].

3.6 Clinical Evaluation of Variants

The clinical significance of variants is evaluated based on several criteria: (a) the pathogenicity predictions and conservation score from different in silico tools; (b) population frequency among GnomAD controls from the GnomAD database [31]; and (c) relevant data from the literature and clinical databases (listed below). Pathogenicity predictions of single-nucleotide variants in exonic regions can be examined with several in silico tools, including the Combined Annotation Dependent Depletion (CADD) score [32], Polymorphism Phenotyping ver-2.0 (PolyPhen-2) [33], Sorting Intolerant from Tolerant (SIFT) [34], Mutation Prediction (MutPred) [35], SNP&GO [36], Protein Analysis Through Evolutionary Relationships (PANTHER) [37], Predictor of Human Deleterious SNP (PhD-SNP) [38], Protein Variant Effect Analyzer (PROVEAN) [39], Mutation taster 2 [40], or other prediction tools from the VarSome database [41]. Pathogenicity predictions of variants in intronic regions and their impact on splicing can be assessed with in silico tools, such as the CADD score [32], the RegSNP-intron [42], Human Splicing Finder (HSF) [43], IntSplice [44], and a recent deep learning-based tool SpliceAI [45]. With the aim to determine the degree of variant position conservation during evolution, conservation analysis can be performed with the ConSurf server [46]. In order to perform conservation analysis of nucleotide sequences, a manual search for homologous sequences can be performed using the Basic Local Alignment Search Tool (BLAST) program BLASTN [47] and multiple sequence alignment with the ClustalW program [48].

Clinically relevant data on variants can be reviewed in databases like ClinVar [49], LOVD [50], COSMIC [51], and Clinical Interpretation of Variants in Cancer (CIViC) [52], and by performing a comprehensive review of the literature. The clinical interpretation of sequence variants is based on the American College of Medical Genetics and Genomics (ACMG) and Association for Molecular Pathology (AMP) standards and guidelines [53], that can be modified in accordance with the Association for Clinical Genomic Science (ACGS) recommendations [54]. Variant classification goes as follows: benign variants, possibly benign variants, variants of uncertain significance, likely pathogenic variants, and pathogenic variants. Classification of variants is supported by the evidence categories [53, 54]. It is recommended that gene abbreviations and names are listed according to the HGNC [55] and that reported variant description is in agreement with the standard international nomenclature guidelines of the Human Genome Variation Society (HGVS) [56].

4 Notes

1. Due to a limited number of target regions, some disease-causing variants may be missed with targeted NGS. In addition, due to inclusion of only the sequence of pre-defined genes, the detection of large structural variants and copy number variations is less useful and efficient in targeted NGS compared to WES or WGS [6].
2. Common problems with all NGS sequencing approaches are regions with low (<30%) and high (>60%) GC content [7, 57]. High GC content is often observed in the regions of first exons. High and low GC regions can result in uneven and low coverage that could be due to a bias in PCR amplification, cluster amplification, or due to less effective hybridization [57, 58]. Some new experimental designs have been proposed (amplification-free sequencing, optimized PCR protocols) to overcome this issue [58]; however, other methods, like Sanger sequencing, are usually applied to sufficiently sequence those regions and validate NGS results [58, 59].
3. A unique issue with targeted NGS is the dilemma on which genes to include in the custom-gene panel. Many candidate genes could also mean many VUS, which are hard to clinically-interpret. To maximize sensitivity and minimize the clinical burden, a well-designed phenotype-directed diagnostic gene panel will include all known genes associated with the disease and exclude genes of uncertain significance [60]. However, for research purposes, it is worth considering the inclusion of genes of uncertain significance that were shown to be good candidate genes in the pathway analysis.
4. Hybridization time has a significant effect on the efficiency of enrichment [61]. Extended hybridization time in the enrichment step could contribute to better capture of target regions and consequently result in better coverage of these regions [61, 62].
5. The paired-end sequencing was preferred over single end, as paired-end sequencing improves the accuracy of read alignment and therefore is much more effective in resolving structural variants and detection of insertion–deletion variants [6, 63].
6. Regions with high sequence similarity, like homologous genes (i.e., *HBA1* and *HBA2*) pose a problem in sequencing

specificity, alignment of reads and interpretation of variants [64]. For analysis of genes with high homology, PCR-based enrichment is more suitable for capturing these homologous genes, as specific primers can be designed in target divergent regions [57].

References

1. Pei XM, Yeung MHY, Wong ANN, Tsang HF, Yu ACS, Yim AKY, Wong SCC (2023) Targeted sequencing approach and its clinical applications for the molecular diagnosis of human diseases. Cells 12(3). https://doi.org/10.3390/cells12030493
2. Tomc J, Debeljak N (2021) Molecular pathways involved in the development of congenital Erythrocytosis. Genes (Basel) 12(8). https://doi.org/10.3390/genes12081150
3. Kristan A, Gašperšič J, Režen T, Kunej T, Količ R, Vuga A, Fink M, Žula Š, Anžej Doma S, Preložnik Zupan I, Pajič T, Podgornik H, Debeljak N (2021) Genetic analysis of 39 erythrocytosis and hereditary hemochromatosis-associated genes in the Slovenian family with idiopathic erythrocytosis. J Clin Lab Anal 35(4):e23715. https://doi.org/10.1002/jcla.23715
4. Gašperšič J, Kristan A, Kunej T, Zupan IP, Debeljak N (2021) Erythrocytosis: genes and pathways involved in disease development. Blood Transfus 19(6):518–532. https://doi.org/10.2450/2020.0197-20
5. Gulilat M, Lamb T, Teft WA, Wang J, Dron JS, Robinson JF, Tirona RG, Hegele RA, Kim RB, Schwarz UI (2019) Targeted next generation sequencing as a tool for precision medicine. BMC Med Genet 12(1):81. https://doi.org/10.1186/s12920-019-0527-2
6. Bewicke-Copley F, Arjun Kumar E, Palladino G, Korfi K, Wang J (2019) Applications and analysis of targeted genomic sequencing in cancer studies. Comput Struct Biotechnol J 17:1348–1359. https://doi.org/10.1016/j.csbj.2019.10.004
7. Kristan A, Pajič T, Maver A, Režen T, Kunej T, Količ R, Vuga A, Fink M, Žula Š, Podgornik H, Anžej Doma S, Preložnik Zupan I, Rozman D, Debeljak N (2021) Identification of variants associated with rare hematological disorder Erythrocytosis using targeted next-generation sequencing analysis. Front Genet 12:689868. https://doi.org/10.3389/fgene.2021.689868
8. Illumina (2021) Illumina DNA prep with enrichment, reference guide. https://support.illumina.com/content/dam/illumina-support/documents/documentation/chemistry_documentation/illumina_prep/illumina-dna-prep-with-enrichment-reference-guide-1000000048041-07.pdf
9. Technologies A (2013) Agilent high sensitivity DNA kit guide. https://www.agilent.com/cs/library/usermanuals/Public/G2938-90321_SensitivityDNA_KG_EN.pdf
10. IDT (2022) xGen™ Hyb panel design tool. https://sfvideo.blob.core.windows.net/sitefinity/docs/default-source/user-guide-manual/xgen-hyb-panel-design-tool-user-guide.pdf?sfvrsn=701cff07_2
11. Illumina (2020) MiniSeq™ sequencing system. https://www.illumina.com/content/dam/illumina/gcs/assembled-assets/marketing-literature/zz-archived-files/miniseq-system-spec-sheet-770-2015-039/miniseq-system-spec-sheet-770-2015-039.pdf
12. Ignatov KB, Blagodatskikh KA, Shcherbo DS, Kramarova TV, Monakhova YA, Kramarov VM (2019) Fragmentation through polymerization (FTP): a new method to fragment DNA for next-generation sequencing. PLoS One 14(4):e0210374. https://doi.org/10.1371/journal.pone.0210374
13. Kia A, Gloeckner C, Osothprarop T, Gormley N, Bomati E, Stephenson M, Goryshin I, He MM (2017) Improved genome sequencing using an engineered transposase. BMC Biotechnol 17(1):6. https://doi.org/10.1186/s12896-016-0326-1
14. Quail MA, Swerdlow H, Turner DJ (2009) Improved protocols for the illumina genome analyzer sequencing system. Curr Protoc Hum Genet Chapter 18: Unit 18 12. https://doi.org/10.1002/0471142905.hg1802s62
15. Jassal B, Matthews L, Viteri G, Gong C, Lorente P, Fabregat A, Sidiropoulos K, Cook J, Gillespie M, Haw R, Loney F, May B, Milacic M, Rothfels K, Sevilla C, Shamovsky V, Shorser S, Varusai T, Weiser J, Wu G, Stein L, Hermjakob H, D'Eustachio P (2019) The reactome pathway knowledgebase. Nucleic Acids Res 48(D1):D498–D503. https://doi.org/10.1093/nar/gkz1031
16. Szklarczyk D, Franceschini A, Wyder S, Forslund K, Heller D, Huerta-Cepas J,

Simonovic M, Roth A, Santos A, Tsafou KP, Kuhn M, Bork P, Jensen LJ, von Mering C (2014) STRING v10: protein–protein interaction networks, integrated over the tree of life. Nucleic Acids Res 43(D1):D447–D452. https://doi.org/10.1093/nar/gku1003

17. Consortium TU (2020) UniProt: the universal protein knowledgebase in 2021. Nucleic Acids Res 49(D1):D480–D489. https://doi.org/10.1093/nar/gkaa1100
18. Stelzer G, Rosen N, Plaschkes I, Zimmerman S, Twik M, Fishilevich S, Stein TI, Nudel R, Lieder I, Mazor Y, Kaplan S, Dahary D, Warshawsky D, Guan-Golan Y, Kohn A, Rappaport N, Safran M, Lancet D (2016) The GeneCards suite: from gene data mining to disease genome sequence analyses. Curr Protoc Bioinformatics 54: 1.30.31–31.30.33. https://doi.org/10.1002/cpbi.5
19. Uhlen M, Oksvold P, Fagerberg L, Lundberg E, Jonasson K, Forsberg M, Zwahlen M, Kampf C, Wester K, Hober S, Wernerus H, Björling L, Ponten F (2010) Towards a knowledge-based human protein atlas. Nat Biotechnol 28(12):1248–1250. https://doi.org/10.1038/nbt1210-1248
20. Coordinators NR (2015) Database resources of the National Center for biotechnology information. Nucleic Acids Res 44(D1):D7–D19. https://doi.org/10.1093/nar/gkv1290
21. Tweedie S, Braschi B, Gray K, Jones TEM, Seal RL, Yates B, Bruford EA (2021) Genenames.org: the HGNC and VGNC resources in 2021. Nucleic Acids Res 49(D1):D939–D946. https://doi.org/10.1093/nar/gkaa980
22. Bergant G, Maver A, Lovrecic L, Cuturilo G, Hodzic A, Peterlin B (2018) Comprehensive use of extended exome analysis improves diagnostic yield in rare disease: a retrospective survey in 1,059 cases. Genet Med 20(3):303–312. https://doi.org/10.1038/gim.2017.142
23. Schilbert HM, Rempel A, Pucker B (2020) Comparison of read mapping and variant calling tools for the analysis of plant NGS data. Plants (Basel) 9(4). https://doi.org/10.3390/plants9040439
24. Li H, Durbin R (2009) Fast and accurate short read alignment with burrows-wheeler transform. Bioinformatics 25(14):1754–1760. https://doi.org/10.1093/bioinformatics/btp324
25. Van der Auwera GA, Carneiro MO, Hartl C, Poplin R, Del Angel G, Levy-Moonshine A, Jordan T, Shakir K, Roazen D, Thibault J, Banks E, Garimella KV, Altshuler D, Gabriel S, DePristo MA (2013) From FastQ data to high confidence variant calls: the genome analysis toolkit best practices pipeline. Curr Protoc Bioinformatics 43(1110):11 10 11–11 10 33. https://doi.org/10.1002/0471250953.bi1110s43
26. McLaren W, Gil L, Hunt SE, Riat HS, Ritchie GR, Thormann A, Flicek P, Cunningham F (2016) The Ensembl variant effect predictor. Genome Biol 17(1):122. https://doi.org/10.1186/s13059-016-0974-4
27. Wang K, Li M, Hakonarson H (2010) ANNOVAR: functional annotation of genetic variants from high-throughput sequencing data. Nucleic Acids Res 38(16):e164. https://doi.org/10.1093/nar/gkq603
28. Cingolani P, Platts A, Wang le L, Coon M, Nguyen T, Wang L, Land SJ, Lu X, Ruden DM (2012) A program for annotating and predicting the effects of single nucleotide polymorphisms, SnpEff: SNPs in the genome of Drosophila melanogaster strain w1118; iso-2; iso-3. Fly (Austin) 6(2):80–92. https://doi.org/10.4161/fly.19695
29. Fairley S, Lowy-Gallego E, Perry E, Flicek P (2019) The international genome sample resource (IGSR) collection of open human genomic variation resources. Nucleic Acids Res 48(D1):D941–D947. https://doi.org/10.1093/nar/gkz836
30. Fu W, O'Connor TD, Jun G, Kang HM, Abecasis G, Leal SM, Gabriel S, Rieder MJ, Altshuler D, Shendure J, Nickerson DA, Bamshad MJ, Akey JM (2013) Analysis of 6,515 exomes reveals the recent origin of most human protein-coding variants. Nature 493(7431):216–220. https://doi.org/10.1038/nature11690
31. Karczewski KJ, Francioli LC, Tiao G, Cummings BB, Alföldi J, Wang Q, Collins RL, Laricchia KM, Ganna A, Birnbaum DP, Gauthier LD, Brand H, Solomonson M, Watts NA, Rhodes D, Singer-Berk M, England EM, Seaby EG, Kosmicki JA, Walters RK, Tashman K, Farjoun Y, Banks E, Poterba T, Wang A, Seed C, Whiffin N, Chong JX, Samocha KE, Pierce-Hoffman E, Zappala Z, O'Donnell-Luria AH, Minikel EV, Weisburd B, Lek M, Ware JS, Vittal C, Armean IM, Bergelson L, Cibulskis K, Connolly KM, Covarrubias M, Donnelly S, Ferriera S, Gabriel S, Gentry J, Gupta N, Jeandet T, Kaplan D, Llanwarne C, Munshi R, Novod S, Petrillo N, Roazen D, Ruano-Rubio V, Saltzman A, Schleicher M, Soto J, Tibbetts K, Tolonen C, Wade G, Talkowski ME, Aguilar Salinas CA, Ahmad T, Albert CM, Ardissino D, Atzmon G, Barnard J, Beaugerie L, Benjamin EJ, Boehnke M, Bonnycastle LL, Bottinger EP, Bowden DW, Bown MJ, Chambers JC, Chan

JC, Chasman D, Cho J, Chung MK, Cohen B, Correa A, Dabelea D, Daly MJ, Darbar D, Duggirala R, Dupuis J, Ellinor PT, Elosua R, Erdmann J, Esko T, Färkkilä M, Florez J, Franke A, Getz G, Glaser B, Glatt SJ, Goldstein D, Gonzalez C, Groop L, Haiman C, Hanis C, Harms M, Hiltunen M, Holi MM, Hultman CM, Kallela M, Kaprio J, Kathiresan S, Kim B-J, Kim YJ, Kirov G, Kooner J, Koskinen S, Krumholz HM, Kugathasan S, Kwak SH, Laakso M, Lehtimäki T, Loos RJF, Lubitz SA, Ma RCW, MacArthur DG, Marrugat J, Mattila KM, McCarroll S, McCarthy MI, McGovern D, McPherson R, Meigs JB, Melander O, Metspalu A, Neale BM, Nilsson PM, O'Donovan MC, Ongur D, Orozco L, Owen MJ, Palmer CNA, Palotie A, Park KS, Pato C, Pulver AE, Rahman N, Remes AM, Rioux JD, Ripatti S, Roden DM, Saleheen D, Salomaa V, Samani NJ, Scharf J, Schunkert H, Shoemaker MB, Sklar P, Soininen H, Sokol H, Spector T, Sullivan PF, Suvisaari J, Tai ES, Teo YY, Tiinamaija T, Tsuang M, Turner D, Tusie-Luna T, Vartiainen E, Vawter MP, Ware JS, Watkins H, Weersma RK, Wessman M, Wilson JG, Xavier RJ, Neale BM, Daly MJ, MacArthur DG, Genome Aggregation Database C (2020) The mutational constraint spectrum quantified from variation in 141,456 humans. Nature 581(7809):434–443. https://doi.org/10.1038/s41586-020-2308-7

32. Rentzsch P, Witten D, Cooper GM, Shendure J, Kircher M (2019) CADD: predicting the deleteriousness of variants throughout the human genome. Nucleic Acids Res 47 (D1):D886–D894. https://doi.org/10.1093/nar/gky1016

33. Adzhubei IA, Schmidt S, Peshkin L, Ramensky VE, Gerasimova A, Bork P, Kondrashov AS, Sunyaev SR (2010) A method and server for predicting damaging missense mutations. Nat Methods 7(4):248–249. https://doi.org/10.1038/nmeth0410-248

34. Vaser R, Adusumalli S, Leng SN, Sikic M, Ng PC (2016) SIFT missense predictions for genomes. Nat Protoc 11(1):1–9. https://doi.org/10.1038/nprot.2015.123

35. Pejaver V, Urresti J, Lugo-Martinez J, Pagel KA, Lin GN, Nam HJ, Mort M, Cooper DN, Sebat J, Iakoucheva LM, Mooney SD, Radivojac P (2020) Inferring the molecular and phenotypic impact of amino acid variants with MutPred2. Nat Commun 11(1):5918. https://doi.org/10.1038/s41467-020-19669-x

36. Capriotti E, Calabrese R, Fariselli P, Martelli PL, Altman RB, Casadio R (2013) WS-SNPs&GO: a web server for predicting the deleterious effect of human protein variants using functional annotation. BMC Genomics 14(Suppl 3):S6. https://doi.org/10.1186/1471-2164-14-s3-s6

37. Mi H, Muruganujan A, Ebert D, Huang X, Thomas PD (2019) PANTHER version 14: more genomes, a new PANTHER GO-slim and improvements in enrichment analysis tools. Nucleic Acids Res 47(D1):D419–D426. https://doi.org/10.1093/nar/gky1038

38. Capriotti E, Calabrese R, Casadio R (2006) Predicting the insurgence of human genetic diseases associated to single point protein mutations with support vector machines and evolutionary information. Bioinformatics 22(22):2729–2734. https://doi.org/10.1093/bioinformatics/btl423

39. Choi Y, Chan AP (2015) PROVEAN web server: a tool to predict the functional effect of amino acid substitutions and indels. Bioinformatics 31(16):2745–2747. https://doi.org/10.1093/bioinformatics/btv195

40. Schwarz JM, Cooper DN, Schuelke M, Seelow D (2014) MutationTaster2: mutation prediction for the deep-sequencing age. Nat Methods 11(4):361–362. https://doi.org/10.1038/nmeth.2890

41. Kopanos C, Tsiolkas V, Kouris A, Chapple CE, Albarca Aguilera M, Meyer R, Massouras A (2019) VarSome: the human genomic variant search engine. Bioinformatics 35(11):1978–1980. https://doi.org/10.1093/bioinformatics/bty897

42. Lin H, Hargreaves KA, Li R, Reiter JL, Wang Y, Mort M, Cooper DN, Zhou Y, Zhang C, Eadon MT, Dolan ME, Ipe J, Skaar TC, Liu Y (2019) RegSNPs-intron: a computational framework for predicting pathogenic impact of intronic single nucleotide variants. Genome Biol 20(1):254. https://doi.org/10.1186/s13059-019-1847-4

43. Desmet FO, Hamroun D, Lalande M, Collod-Beroud G, Claustres M, Beroud C (2009) Human splicing finder: an online bioinformatics tool to predict splicing signals. Nucleic Acids Res 37(9):e67. https://doi.org/10.1093/nar/gkp215

44. Shibata A, Okuno T, Rahman MA, Azuma Y, Takeda J, Masuda A, Selcen D, Engel AG, Ohno K (2016) IntSplice: prediction of the splicing consequences of intronic single-nucleotide variations in the human genome. J Hum Genet 61(7):633–640. https://doi.org/10.1038/jhg.2016.23

45. Jaganathan K, Kyriazopoulou Panagiotopoulou S, McRae JF, Darbandi SF, Knowles D, Li YI, Kosmicki JA, Arbelaez J, Cui W, Schwartz GB, Chow ED, Kanterakis E, Gao H, Kia A, Batzoglou S,

Sanders SJ, Farh KK (2019) Predicting splicing from primary sequence with deep learning. Cell 176(3):535–548.e524. https://doi.org/10.1016/j.cell.2018.12.015

46. Ashkenazy H, Abadi S, Martz E, Chay O, Mayrose I, Pupko T, Ben-Tal N (2016) ConSurf 2016: an improved methodology to estimate and visualize evolutionary conservation in macromolecules. Nucleic Acids Res 44(W1): W344–W350. https://doi.org/10.1093/nar/gkw408
47. Altschul SF, Madden TL, Schaffer AA, Zhang J, Zhang Z, Miller W, Lipman DJ (1997) Gapped BLAST and PSI-BLAST: a new generation of protein database search programs. Nucleic Acids Res 25(17):3389–3402. https://doi.org/10.1093/nar/25.17.3389
48. Madeira F, Park YM, Lee J, Buso N, Gur T, Madhusoodanan N, Basutkar P, Tivey ARN, Potter SC, Finn RD, Lopez R (2019) The EMBL-EBI search and sequence analysis tools APIs in 2019. Nucleic Acids Res 47(W1): W636–W641. https://doi.org/10.1093/nar/gkz268
49. Landrum MJ, Lee JM, Benson M, Brown GR, Chao C, Chitipiralla S, Gu B, Hart J, Hoffman D, Jang W, Karapetyan K, Katz K, Liu C, Maddipatla Z, Malheiro A, McDaniel K, Ovetsky M, Riley G, Zhou G, Holmes JB, Kattman BL, Maglott DR (2018) ClinVar: improving access to variant interpretations and supporting evidence. Nucleic Acids Res 46(D1):D1062–D1067. https://doi.org/10.1093/nar/gkx1153
50. Fokkema IF, Taschner PE, Schaafsma GC, Celli J, Laros JF, den Dunnen JT (2011) LOVD v.2.0: the next generation in gene variant databases. Hum Mutat 32(5):557–563. https://doi.org/10.1002/humu.21438
51. Tate JG, Bamford S, Jubb HC, Sondka Z, Beare DM, Bindal N, Boutselakis H, Cole CG, Creatore C, Dawson E, Fish P, Harsha B, Hathaway C, Jupe SC, Kok CY, Noble K, Ponting L, Ramshaw CC, Rye CE, Speedy HE, Stefancsik R, Thompson SL, Wang S, Ward S, Campbell PJ, Forbes SA (2019) COSMIC: the catalogue of somatic mutations in cancer. Nucleic Acids Res 47(D1):D941–D947. https://doi.org/10.1093/nar/gky1015
52. Griffith M, Spies NC, Krysiak K, McMichael JF, Coffman AC, Danos AM, Ainscough BJ, Ramirez CA, Rieke DT, Kujan L, Barnell EK, Wagner AH, Skidmore ZL, Wollam A, Liu CJ, Jones MR, Bilski RL, Lesurf R, Feng YY, Shah NM, Bonakdar M, Trani L, Matlock M, Ramu A, Campbell KM, Spies GC, Graubert AP, Gangavarapu K, Eldred JM, Larson DE, Walker JR, Good BM, Wu C, Su AI, Dienstmann R, Margolin AA, Tamborero D, Lopez-Bigas N, Jones SJ, Bose R, Spencer DH, Wartman LD, Wilson RK, Mardis ER, Griffith OL (2017) CIViC is a community knowledgebase for expert crowdsourcing the clinical interpretation of variants in cancer. Nat Genet 49(2):170–174. https://doi.org/10.1038/ng.3774
53. Richards S, Aziz N, Bale S, Bick D, Das S, Gastier-Foster J, Grody WW, Hegde M, Lyon E, Spector E, Voelkerding K, Rehm HL, on behalf of the ALQAC (2015) Standards and guidelines for the interpretation of sequence variants: a joint consensus recommendation of the American college of medical genetics and genomics and the association for molecular pathology. Genet Med 17(5):405–423. https://doi.org/10.1038/gim.2015.30
54. Ellard S, Baple EL, Berry I, Forrester N, Turnbull C, Owens MM, Eccles DM, Abbs SJ, Scott R, Deans ZC, Lester T, Jo, Campbell, Newman WG (2019) McMullan D ACGS best practice guidelines for variant classification 2019
55. Seal RL, Braschi B, Gray K, Jones TEM, Tweedie S, Haim-Vilmovsky L, Bruford EA (2023) Genenames.org: the HGNC resources in 2023. Nucleic Acids Res 51(D1):D1003–D1009. https://doi.org/10.1093/nar/gkac888
56. den Dunnen JT (2016) Sequence variant descriptions: HGVS nomenclature and Mutalyzer. Curr Protoc Hum Genet 90: 7.13.11–17.13.19. https://doi.org/10.1002/cphg.2
57. Garcia-Garcia G, Baux D, Faugere V, Moclyn M, Koenig M, Claustres M, Roux AF (2016) Assessment of the latest NGS enrichment capture methods in clinical context. Sci Rep 6:20948. https://doi.org/10.1038/srep20948
58. Chen YC, Liu T, Yu CH, Chiang TY, Hwang CC (2013) Effects of GC bias in next-generation-sequencing data on de novo genome assembly. PLoS One 8(4):e62856. https://doi.org/10.1371/journal.pone.0062856
59. De Cario R, Kura A, Suraci S, Magi A, Volta A, Marcucci R, Gori AM, Pepe G, Giusti B, Sticchi E (2020) Sanger validation of high-throughput sequencing in genetic diagnosis: still the best practice? Front Genet 11: 592588. https://doi.org/10.3389/fgene.2020.592588
60. Bean LJH, Funke B, Carlston CM, Gannon JL, Kantarci S, Krock BL, Zhang S, Bayrak-Toydemir P, Committee ALQA (2020)

Diagnostic gene sequencing panels: from design to report-a technical standard of the American College of Medical Genetics and Genomics (ACMG). Genet Med 22(3): 453–461. https://doi.org/10.1038/s41436-019-0666-z

61. Singh RR (2022) Target enrichment approaches for next-generation sequencing applications in oncology. Diagnostics (Basel) 12(7). https://doi.org/10.3390/diagnostics12071539
62. Young JM, Higgins D, Austin JJ (2019) Hybridization enrichment to improve forensic mitochondrial DNA analysis of highly degraded human remains. Front Ecol Evol 7. https://doi.org/10.3389/fevo.2019.00450
63. Nakazato T, Ohta T, Bono H (2013) Experimental design-based functional mining and characterization of high-throughput sequencing data in the sequence read archive. PLoS One 8(10):e77910. https://doi.org/10.1371/journal.pone.0077910
64. Girodon F, Airaud F, Garrec C, Bezieau S, Gardie B (2017) Gene panel sequencing in idiopathic erythrocytosis. Haematologica 102(1):e30. https://doi.org/10.3324/haematol.2016.158337

Chapter 4

Collection and Processing of Samples for Next-Generation Sequencing to Study the Gut Microbiome

Maxwell Bokor, Emily Chiacchiaro, and Sangita Phadtare

Abstract

16S rRNA gene sequencing is commonly used for identification and quantitation of microorganisms in complex biological mixtures, such as the human gut microbiome. The 16S rRNA gene is an excellent target gene for sequencing DNA in a heterogenous sample, as it is a highly conserved part of the transcriptional machinery. Universal PCR primers are used to amplify the conserved regions of 16S. Thus, it is possible to amplify the gene in a wide range of different microorganisms from a single sample. As the 16S rRNA gene consists of both conserved and variable regions, sequencing of the variable regions can be used to differentiate between different bacterial species.

Key words Gut microbiome, 16S rRNA sequencing, Nucleic acid stabilizer, Bar coded primers

1 Introduction

Of note, numerous studies have successfully characterized the microbes in the human gut and uncovered various associations between microbial species, or lack thereof, and their connection to metabolic health, atopic disease, mental health, neurological deficits, cancer, and more. Recent research demonstrates that the protective nature of certain microbes may begin as early as infancy, as demonstrated by the potential link between breastfeeding and allergy prevention [1–3]. Furthermore, some studies suggest that alterations in the gut microbiome may affect weight homeostasis, gene expression, and increase inflammatory biomarkers commonly associated with obesity and metabolic disease [4–7]. In conjunction with the recent rise in mental health disorders, gut microbiota continuously demonstrate a reciprocal relationship with the brain via the gut brain axis. Accordingly, researchers hypothesize that those lacking microbial diversity in the gut may be at higher risk

Authors "Maxwell Bokor" and "Emily Chiacchiaro" have equally contributed to this chapter.

Valerie J. Carabetta and Olaitan Akintunde (eds.), *High Throughput Gene Screening: Methods and Protocols*,
Methods in Molecular Biology, vol. 2866, https://doi.org/10.1007/978-1-0716-4192-7_4,

for psychiatric and neurologic ailments [8]. Studies suggest that the neurologic impact of the gut microbiome ranges from behavioral alterations during short illnesses to more chronic effects, as seen in animal models of depression [9, 10]. Lastly, data also supports that certain species of gut microbes may contribute to the pathogenesis of polyps and subsequently colorectal cancer [11–13]. While the field of gut microbiome research continues to explode, advancing our understanding of its connection to human diseases is imperative.

Our knowledge regarding the gut microbiome has grown exponentially in recent years and accordingly, new technologies to collect, process, and sequence the samples have emerged in parallel [14]. These techniques primarily rely on sequencing of the bacterial 16S rRNA gene and in conjunction with metagenomics, have increased our understanding of the vast variability of the microbiome between individuals [15, 16]. Two key measurements used in the evaluation of these include alpha and beta diversity. Alpha diversity refers to the number and abundance of different organisms within a single microbiome community. Broadly, similarities and differences in alpha diversity contribute to an individual's unique microbiome. On the other hand, beta diversity describes the similarities and differences of organisms between different individual's microbiomes [17, 18]. Both alpha and beta diversity can be examined through the differences in 16S rRNA sequences. Amplifying these sequences allows researchers to uncover the abundance of these variations. Sequencing information can then be synthesized and visualized utilizing bioinformatics open-sourced platforms such as QIIME. Ultimately, knowledge of microbial diversity provides valuable perspectives on both individual and community health, including susceptibility and resistance to disease.

Sequencing the microbiome and visualizing the results remains complex, but a standardized protocol for those participating in this area of research is necessary to ensure the integrity of the results. The smallest of errors in the collection, storage, processing, or analysis of fecal samples can compromise results. In order to translate this basic science research to clinical application, investigators must ensure that their work reflects the true composition of the gut microbial community and is not an artifact of human error. The prospect of microbial manipulation to treat and prevent disease is thrilling, but precision and accuracy must remain a priority. This chapter details one method of processing and analyzing the gut microbiome from stool samples. Each segment is preceded by a brief overview as well as the overarching goal of the section and how it contributes to the entire procedure. By utilizing this technology and methodology, we are able to process high volumes of samples to expand our datasets and better understand the expansive nature of the gut microbiome [19, 20].

2 Materials

2.1 Sample Collection and Storage

1. Stool nucleic acid collection and preservation tubes (*see* **Note 1**).
2. Commode specimen collection system, 800 cc, designed to place atop a toilet to collect stool sample.
3. Culture swabs, preferably with pronged ending attached to cap for easy transport.
4. Ice pack (for example: Polar Tech Industries Ice-Brix Leak-proof gel refrigerant, Premium Poly Pouch).
5. Biohazard collection bag.
6. Nitrile gloves.

2.2 DNA Extraction, Quantification, and Storage

1. DNeasy PowerSoil HTP 96 Kit (Qiagen), or similar (*see* **Note 2**).
2. Nanodrop, Qubit, or equivalent to measure nucleic acid concentrations.
3. Biosafety cabinet.

2.3 PCR Amplification, Gel Electrophoresis and DNA Pooling

1. RNase away, or equivalent for cleaning surfaces.
2. 70% Ethanol.
3. Hot start Taq polymerase master mix (*see* **Note 3**).
4. Primers, with one uniquely barcoded.
5. dH_2O.
6. Thermocycler.
7. 96-well plates.
8. Gel electrophoresis apparatus.
9. DNA loading dye.
10. 100 bp Ladder (e.g., Thermo Fisher Scientific GeneRuler 100 bp DNA Ladder).
11. Agarose.
12. DNA stain, such as ethidium bromide or Syber Safe stain.
13. Primers (*see* **Note 4**).

2.4 DNA Pooling and Preparation for Sequencing

1. SequalPrep™ Normalization Plate (96) Kit.
2. Nuclease-free, filter pipette tips.

3 Methods

3.1 Sample Collection and Storage

The following section details the necessary materials and steps in order to collect a stool sample. Each step must be followed very closely to avoid sample contamination and improper storage. Patients are first consented as per the regulations set by the specific institutional IRB (*Institutional Review Board*) for human research. Then, patients are given instructions and appropriate materials for stool collection and transport based on the method of collection (with or without nucleic acid stabilizer). It is crucial to preserve the integrity of the sample by avoiding contamination during collection and growth of bacteria post-collection. The nucleic acid stabilizer stops bacterial growth at the time of collection. These samples can be stored at room temperature for 2–3 days before storage in a freezer (Fig. 1). For the swab method, samples are collected on the same day and brought to the laboratory with a pre-frozen ice pack. These are stored in a −80 °C freezer immediately upon arrival (Fig. 2). For a comparison of these two methods, see Fig. 3. Appropriate handling of the samples ensures that the ratios of each bacterial species present at the time of collection are preserved. The following describes the processes to be completed by the patients during the stool collection phase. All the necessary materials should be included in the stool collection kit provided to the patient. All procedures are carried out at room temperature unless otherwise specified.

Toilet Hat and Nucleic Acid Stabilizer Method (Fig. 1)

1. Wash hands thoroughly and put on gloves. It is imperative that the sample remains sterile from the external bacterial community.
2. Remove the sample collection tube from the bag. Do not dispose of the bag.

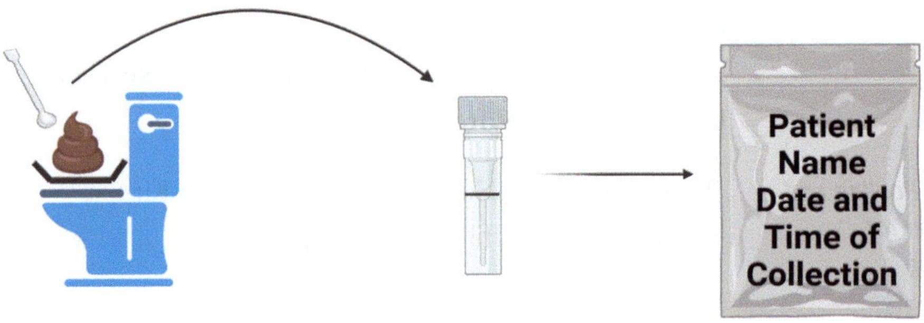

Fig. 1 Toilet hat and stabilizer sample collection method. Created with BioRender.com

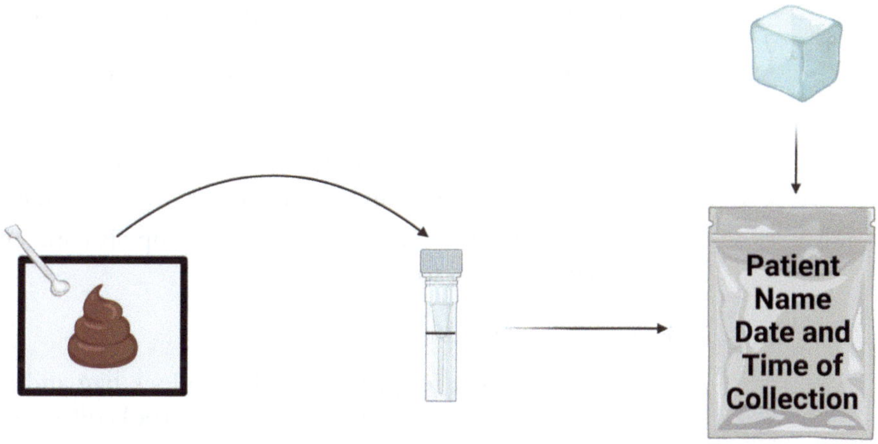

Fig. 2 Swab and ice pack sample collection method. Created with BioRender.com

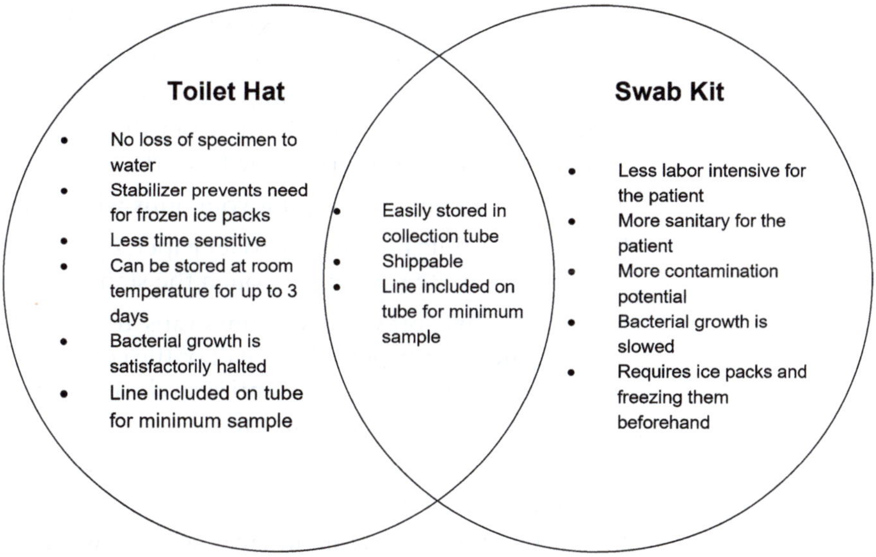

Fig. 3 Venn diagram comparing the toilet-hat-stabilizer and swab-ice-pack sample collection and transport methods

3. Place the toilet hat on the toilet. Produce a sample by moving bowels onto the toilet hat. Do not urinate in the toilet hat.
4. Open the sample collection tube and remove the spatula. Do not allow the spatula to touch anything other than the stool sample. Take extra care to ensure that neither the spatula nor the collection tube is dropped. Do not place stool directly into the collection tube.

5. Fill the collection tube to at least the halfway point and no more than two-thirds of the tube. A line is present on the tube to ensure an adequate amount of sample.
6. Close the tube tightly and shake to disperse the stabilizer.
7. Close the sample tightly in a zip lock bag and write your name, date, and time of collection on the sample.
8. Bring the sample to the next follow-up visit and ensure it is collected by study personnel.

Swab with Ice Pack Method (Fig. 2)

1. Remove the ice pack from the zip lock bag and place it in the freezer at least one day prior to stool collection. Do not collect the sample unless you are sure you can deliver it to a study personnel on the same day.
2. Upon stool collection, place the frozen ice pack within the zip lock bag. Take the zip lock bag, which includes the collection tube and swabs, to the restroom.
3. Wash hands adequately and put on gloves. It is imperative that the sample remains sterile from the external bacterial community. Remove the tube from the zip lock bag.
4. Produce a sample by wiping your anus with toilet paper.
5. While carefully holding the toilet paper, open the tube with other hand. Be sure to not touch the stem or cotton swab.
6. Use the two provided cotton swabs to collect a sample from the toilet paper. About half of the cotton swab should be brown; do not use the entire swab. Do not place stool directly into the collection tube.
7. Place the swabs back into the tube, close tightly, and put it back in the zip lock bag.
8. Put the zip lock bag in an envelope and write your name, date, and time of collection on the sample. One swab will be used for DNA extraction, while the other will be stored in the freezer at either –20 °C or –80 °C.

3.2 DNA Extraction, Quantification, and Storage

This section describes the process of extracting the DNA from the original stool samples, which can then be used for further testing. Please note that all steps should be carried out at room temperature unless otherwise stated. The DNA extraction steps are the same for samples collected by either method described above.

1. Before starting, if using the DNeasy PowerSoil HTP 96 Kit, verify that solution C1 (lysis buffer containing SDS) is completely dissolved; heat to 60 °C if not, and wash solution C5-D has 100% ethanol added.
2. Wear a clean pair of gloves and change them periodically.

3. Add 200 mg of the collected stool sample to the bead plate wells (provided with kit). For preserved stool (stabilizer), use a spatula or cut pipette tip and for not preserved stool (swab kit), use the swab to transfer.
4. Follow the step-by-step detailed instructions provided by the manufacturer of the selected kit.
5. Quantitate the isolated DNA samples using a fluorimeter, such as the Nanodrop or Qubit devices. Record the concentration and the A260/A280 ratio. Ideally the ratio should be between 1.8–2.0 (*see* **Note 5**).
6. Store the DNA at –20 °C.

3.3 PCR Amplification, Gel Electrophoresis and Pooling

PCR amplification is typically carried out using a 96-well plate, with each well containing DNA from a different stool sample, conventional PCR primers, and other components required for a PCR reaction. A unique bar-coded primer is also added to each well, which is critical for identification of each participant's sample. The unique code of the primer will be used to identify the sample following sequencing. Ensure that all materials in the biosafety cabinet are sterile, as any contamination will interfere with results.

1. Prepare the biosafety cabinet. 15–20 min before you begin, UV sterilize the cabinet. Remove all items that will not be used for this procedure and wipe down the hood with 70% ethanol and RNase away. Place supplies for the procedure next to the cabinet and change gloves. As you transfer items into the cabinet, disinfect with ethanol, if possible. Check to make sure all appropriate items are in the cabinet and change gloves again. All of the following steps are to be completed inside the safety cabinet.
2. In the reservoir, mix the appropriate volume of the master mix and nuclease-free water. For example, for a plate of 110 wells, mix 1375 μL of master mix and 1155 μL of nuclease-free water.
3. Add common 1 μL forward primer.
4. Dispense 23 μL of the mixture into each well.
5. Dispense 1 μL of bar-coded primer and 1 μL of gDNA (the sample) into each well.
6. Place the plate in the thermal cycler using the 16 s rRNA program (Fig. 4).

(a) 94 °C, 3 min for 1 cycle

(b) 94 °C, 45 s

(c) 50 °C, 60 s

(d) 72 °C, 90 s, 35 cycles to step b

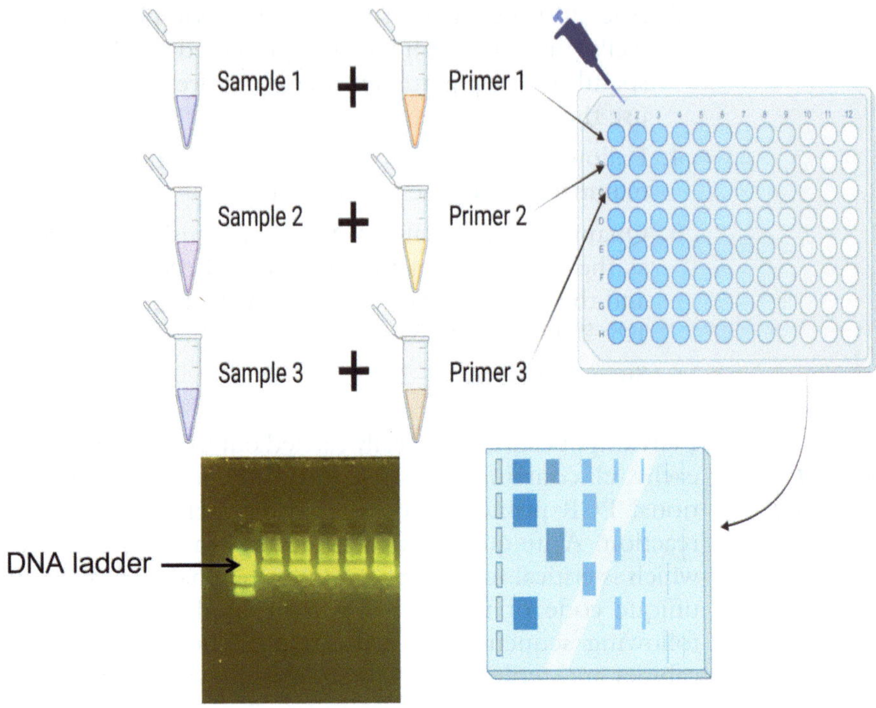

Fig. 4 Example of PCR flowchart. Created with BioRender.com

(e) 72 °C, 10 min, final extension

(f) 4 °C hold.

7. Repeat steps 2–6 to create a second 96-well PCR plate that is identical to the first.
8. Pour a 3% agarose gel. Measure out the agarose into 0.5X TBE or Syber Safe stain, if using, in a flask and gently swirl. Fold paper towel into the top of the flask to act as a stopper and block evaporation.
9. Microwave the flask for 2 min. After 2 min, remove the flask and swirl again. Be sure to use a hot pad when removing the flask. Microwave the flask for 2–4 more minutes, depending on the gel size (*see* **Note 6**). Stop when the agarose has completely dissolved.
10. Pour the contents of the flask into the gel-casting tray. Allow the agarose to cool for 1 min and then add the combs in the appropriate places. Shield the gel from any light and allow it to set for 20–45 min, depending on the size.
11. Submerge the set gel in 0.5X TBE.
12. Add 3 μL of dye into each well of a clean 96-well plate.
13. Pipette 5 μL of the PCR product into each corresponding well and mix well.

14. Load 5 µL of DNA molecular weight markers ladder in the first well. Load 5 µL of the samples to the remaining wells. Run the gels at 130 V (constant) for 30–45 min depending on the size of the gel. Let the dye run about two thirds of way through. Cover the apparatus with cardboard to protect from light. The expected amplicon size is ~390 bp. A sample gel with the DNA ladder and five samples is shown in Fig. 4.

3.4 DNA Pooling and Preparation for Sequencing

This section describes the process for combining all of the derived PCR products from various stool samples to be sequenced together. The unique barcoded primer utilized in the PCR amplification step allows for continuous identification of unique samples, even after combination. This step also standardizes the concentrations of each of these samples. The pooled sample can then be used for sequencing.

1. Transfer the 25 µL of PCR product from the duplicate plate into the corresponding well of the first plate. This should result in a total of 45 µL in each well.
2. Transfer 20 µL of the pooled PCR products to the corresponding wells of the normalization plate.
3. Add 20 µL of SequalPrep™ Normalization Binding buffer to each well. Mix the solution sufficiently with 3–4 aspirations using a pipette in each well.
4. Incubate for one hour at room temperature. During this period, the DNA binds to the plate.
5. Using a micropipette, aspirate the total liquid (40 µL) from the wells. Ensure that you do not scrape the well during this step.
6. Add 50 µL of the SequalPrep™ normalization wash buffer to each well.
7. Aspirate the wash buffer from each well and discard.
8. Add 20 µL of SequalPrep™ normalization elution buffer to each well. Mix the solution sufficiently with 3–4 aspirations in each well.
9. Incubate for 5 min at room temperature.
10. Pool all purified samples into one sterile reservoir or collection tube.

3.5 Sequencing, Analysis, and Coding

Samples are mailed on dry iced and sequenced at a sequencing facility equipped for 16S rRNA sequencing (*see* **Note 4**). After sequencing, the user is now ready to code and analyze the samples using the QIIME2 analysis program. Please refer to the free, open-source QIIME2 website at qiime2.org for more information regarding the standard protocols. You will need to install QIIME2 on your computer. You will need to create a master

A

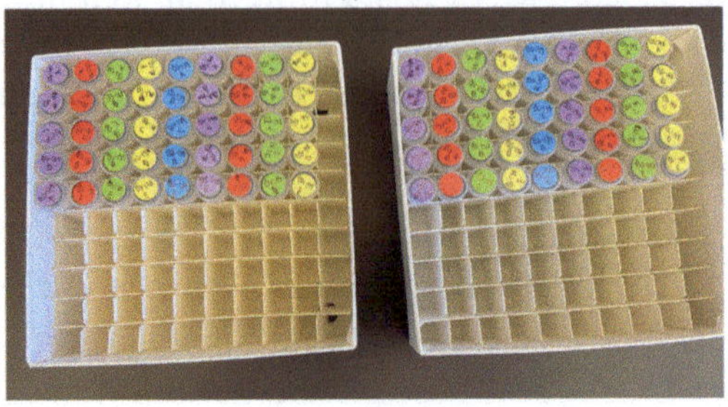

B

SUBJECT CODE	Age	Gender	Ethnicity/Race	Height (cm)	Weight (kg)	Smoking	Alcohol	*H. pylori*	Lactose Intolerant	IBS	HTN	Diabetes
xx1	30	M	white	175.26	131.5	Nonsmoker	N	N	N	N	Y	N
xx2	49	F	africanamerican	157.48	109.8	Nonsmoker	Y	Y	N	N	Y	N
xx3	61	F	white	175.26	152.9	Nonsmoker	N	N	N	N	Y	Y
xx4	54	F	white	165.09	143.8	Nonsmoker	Y	N	N	N	Y	N
xx5	62	M	white	179.07	147.4	Nonsmoker	Y	N	N	N	Y	N
xx6	31	F	other	165.09	101.2	Nonsmoker	Y	N	N	N	N	N
xx7	66	M	white	182.88	117.9	Nonsmoker	N	N	N	N	Y	N
xx8	42	M	other	182.88	140.6	Quit	Y	Y	Y	Y	N	N
xx9	41	F	other	160.02	172.4	Quit	N	Y	N	N	Y	Y
xx10	30	M	white	175.26	131.5	Nonsmoker	N	N	N	N	Y	N

Fig. 5 (**a**) Example of samples arranged in a box for sequencing; (**b**) example of a coding excel prepared

excel that has all the data points and the respective unique primer sequences (barcodes) for each sample. Studies typically include alpha and beta diversity analyses. Example of samples arranged in a box for sequencing (Fig. 5a) and coding excel prepared (Fig. 5b) are shown. An example of an analysis can be followed at https://earthmicrobiome.org/data-and-code/.

4 Notes

1. Collection and preservation tubes should be similar to those available from Norgen Biotek (Cat. 45,630, 45,660, https://norgenbiotek.com/product/stool-nucleic-acid-collection-and-preservation-tubes). A 2 g stool collection tube with the ability to store DNA for up to 2 years and RNA up to 7 days is required. Tubes should be kept tightly sealed at room temperature. The preservative halts growth of Gram-negative bacteria, fungi, and inactivates viruses. RNA stability is enhanced with storage between –20 °C to –80 °C.
2. This chapter is written for use of the Qiagen DNeasy PowerSoil HTP 96 Kit. Any method or kit is suitable that has a 96-well high-throughput extractor for samples of 250 mg or less and can be used to purify DNA from difficult samples types, like stool.
3. We use GoTaq HS Colorless Master Mix from Promega. The DNA polymerase is only activated above 70 °C, which limits the production of unwanted products.

Table 1
Primers used in this study

Primer name	Sequence 5′→3′
Read 1 sequencing primer (forward)	TATGGTAATTGTGTGYCAGCMGCCGCGGTAA
Read 2 sequencing primer (reverse)	AGTCAGCCAGCCGGACTACNVGGGTWTCTAAT
Index sequencing primer	AATGATACGGCGACCACCGAGATCTACACGCT

4. Unique barcoded primers can be ordered from any company that provides sequencing primers for metagenomic analysis. We used Integrative DNA Technologies (IDT), xGen™ Metagenomics Amplicon Panels. Sequencing primers also required are listed in Table 1.
5. We use the protocols provided by the manufacturers for DNA concentrations as low as 1 ng/μ l and as high as 200 ng/μl.
6. For small gels, dissolve 0.6 g agarose into 20 ml of 0.5X TBE or Syber Safe stain in 0.5X TBE (ThermoFisher Scientific), if using. For large gels, dissolve 2.5 g agarose into 135 ml of buffer.

References

1. van den Elsen LWJ, Garssen J, Burcelin R, Verhasselt V (2019) Shaping the gut microbiota by breastfeeding: the gateway to allergy prevention? Front Pediatr 7:47. https://doi.org/10.3389/fped.2019.00047
2. Wegienka G, Ownby DR, Havstad S, Williams LK, Johnson CC (2006) Breastfeeding history and childhood allergic status in a prospective birth cohort. Ann Allergy Asthma Immunol 97(1):78–83. https://doi.org/10.1016/S1081-1206(10)61374-9
3. Sears MR, Greene JM, Willan AR, Taylor DR, Flannery EM, Cowan JO, Herbison GP, Poulton R (2002) Long-term relation between breastfeeding and development of atopy and asthma in children and young adults: a longitudinal study. Lancet 360(9337):901–907. https://doi.org/10.1016/S0140-6736(02)11025-7
4. Sanz Y, Santacruz A, Gauffin P (2010) Gut microbiota in obesity and metabolic disorders. Proc Nutr Soc 69(3):434–441. https://doi.org/10.1017/S0029665110001813
5. Nadal I, Santacruz A, Marcos A, Warnberg J, Garagorri JM, Moreno LA, Martin-Matillas M, Campoy C, Marti A, Moleres A, Delgado M, Veiga OL, Garcia-Fuentes M, Redondo CG, Sanz Y (2009) Shifts in clostridia, bacteroides and immunoglobulin-coating fecal bacteria associated with weight loss in obese adolescents. Int J Obes 33(7):758–767. https://doi.org/10.1038/ijo.2008.260
6. Kalliomaki M, Collado MC, Salminen S, Isolauri E (2008) Early differences in fecal microbiota composition in children may predict overweight. Am J Clin Nutr 87(3):534–538. https://doi.org/10.1093/ajcn/87.3.534
7. Cani PD, Amar J, Iglesias MA, Poggi M, Knauf C, Bastelica D, Neyrinck AM, Fava F, Tuohy KM, Chabo C, Waget A, Delmee E, Cousin B, Sulpice T, Chamontin B, Ferrieres J, Tanti JF, Gibson GR, Casteilla L, Delzenne NM, Alessi MC, Burcelin R (2007) Metabolic endotoxemia initiates obesity and insulin resistance. Diabetes 56(7):1761–1772. https://doi.org/10.2337/db06-1491
8. Yarandi SS, Peterson DA, Treisman GJ, Moran TH, Pasricha PJ (2016) Modulatory effects of gut microbiota on the central nervous system: how gut could play a role in neuropsychiatric health and diseases. J Neurogastroenterol Motil 22(2):201–212. https://doi.org/10.5056/jnm15146
9. Maes M, Twisk FN, Kubera M, Ringel K, Leunis JC, Geffard M (2012) Increased IgA responses to the LPS of commensal bacteria is

associated with inflammation and activation of cell-mediated immunity in chronic fatigue syndrome. J Affect Disord 136(3):909–917. https://doi.org/10.1016/j.jad.2011.09.010

10. Park AJ, Collins J, Blennerhassett PA, Ghia JE, Verdu EF, Bercik P, Collins SM (2013) Altered colonic function and microbiota profile in a mouse model of chronic depression. Neurogastroenterol Motil 25(9):733–e575. https://doi.org/10.1111/nmo.12153
11. Rezasoltani S, Asadzadeh Aghdaei H, Dabiri H, Akhavan Sepahi A, Modarressi MH, Nazemalhosseini Mojarad E (2018) The association between fecal microbiota and different types of colorectal polyp as precursors of colorectal cancer. Microb Pathog 124:244–249. https://doi.org/10.1016/j.micpath.2018.08.035
12. Wang T, Cai G, Qiu Y, Fei N, Zhang M, Pang X, Jia W, Cai S, Zhao L (2012) Structural segregation of gut microbiota between colorectal cancer patients and healthy volunteers. ISME J 6(2):320–329. https://doi.org/10.1038/ismej.2011.109
13. Bashir A, Miskeen AY, Bhat A, Fazili KM, Ganai BA (2015) Fusobacterium nucleatum: an emerging bug in colorectal tumorigenesis. Eur J Cancer Prev 24(5):373–385. https://doi.org/10.1097/CEJ.0000000000000116
14. Allaband C, McDonald D, Vazquez-Baeza Y, Minich JJ, Tripathi A, Brenner DA, Loomba R, Smarr L, Sandborn WJ, Schnabl B, Dorrestein P, Zarrinpar A, Knight R (2019) Microbiome 101: studying, analyzing, and interpreting gut microbiome data for clinicians. Clin Gastroenterol Hepatol 17(2):218–230. https://doi.org/10.1016/j.cgh.2018.09.017
15. Weersma RK, Zhernakova A, Fu J (2020) Interaction between drugs and the gut microbiome. Gut 69(8):1510–1519. https://doi.org/10.1136/gutjnl-2019-320204
16. Clarridge JE 3rd (2004) Impact of 16S rRNA gene sequence analysis for identification of bacteria on clinical microbiology and infectious diseases. Clin Microbiol Rev 17(4):840–862., Table of contents. https://doi.org/10.1128/CMR.17.4.840-862.2004
17. Walters KE, Martiny JBH (2020) Alpha-, beta-, and gamma-diversity of bacteria varies across habitats. PLoS One 15(9):e0233872. https://doi.org/10.1371/journal.pone.0233872
18. Liu YX, Qin Y, Chen T, Lu M, Qian X, Guo X, Bai Y (2021) A practical guide to amplicon and metagenomic analysis of microbiome data. Protein Cell 12(5):315–330. https://doi.org/10.1007/s13238-020-00724-8
19. White B, Sterrett JD, Grigoryan Z, Lally L, Heinze JD, Alikhan H, Lowry CA, Perez LJ, DeSipio J, Phadtare S (2021) Characterization of gut microbiome and metabolome in helicobacter pylori patients in an underprivileged community in the United States. World J Gastroenterol 27(33):5575–5594. https://doi.org/10.3748/wjg.v27.i33.5575
20. O'Neill L, Pandya V, Grigoryan Z, Patel R, DeSipio J, Judge T, Phadtare S (2022) Effects of milkfat on the gut microbiome of patients after bariatric surgery, a pilot study. Obes Surg 32(2):480–488. https://doi.org/10.1007/s11695-021-05805-z

Part II

Epigenome-Based Sequencing Technologies

Chapter 5

Whole Genome Methylation Sequencing via Enzymatic Conversion (EM-seq): Protocol, Data Processing, and Analysis

Nelly N. Olova and Simon Andrews

Abstract

Whole genome bisulfite sequencing (WGBS) has been the gold standard technique for base resolution analysis of DNA methylation for the last 15 years. It has been, however, associated with technical biases, which lead to overall overestimation of global and regional methylation values, and significant artifacts in extreme cytosine-rich DNA sequence contexts. Enzymatic conversion of cytosine is the newest approach, set to replace entirely the use of the damaging bisulfite conversion of DNA. The EM-seq technique utilizes TET2, T4-BGT, and APOBEC in a two-step conversion process, where the modified cytosines are first protected by oxidation and glucosylation, followed by deamination of all unmodified cytosines to uracil. As a result, EM-seq is degradation-free and bias-free, requires low DNA input, and produces high library yields with longer reads, little batch variation, less duplication, uniform genomic coverage, accurate methylation over a larger number of captured CpGs, and no sequence-specific artifacts.

Key words Enzymatic methylome sequencing, WGBS, EM-seq, BS-seq, DNA methylation, TET2, T4-BGT, APOBEC, Bismark, SeqMonk, Nextflow

1 Introduction

DNA methylation is an epigenetic modification with established importance in the study of both healthy development and disease [1]. It is present in various levels and patterns across prokaryotes and eukaryotes including all vertebrates [2], plants [3], some but not all invertebrates including many insects [4], and some fungi [5]. In the last 15 years, the genomic distribution of 5-methylcytosine (5mC) has been successfully characterized at single base resolution through the combination of the bisulfite conversion method [6] with high-throughput, next-generation sequencing [7, 8], establishing whole genome bisulfite sequencing (WGBS) as the gold standard for studying DNA methylation. WGBS has since enabled a comprehensive characterization of the distribution

Valerie J. Carabetta and Olaitan Akintunde (eds.), *High Throughput Gene Screening: Methods and Protocols*, Methods in Molecular Biology, vol. 2866, https://doi.org/10.1007/978-1-0716-4192-7_5,

and dynamics of 5mC in numerous cellular models, tissues, and organisms in fundamental molecular processes and in pathology, which has contributed greatly to our current understanding of its roles.

The original pre-bisulfite library preparation protocol, known as BS-seq or MethylC-seq [9], where DNA is fragmented, end repaired, and ligated to a methylated adapter before bisulfite conversion and amplification, required large amounts of DNA (typically 0.5–4 μg) due to bisulfite-induced DNA degradation [10]. This shortcoming was mitigated with the development of post-bisulfite WGBS protocols [11, 12], where the bisulfite conversion-associated degradation was harnessed to provide the initial fragmentation, and ligation of the adapters after conversion greatly reduced degradation losses. This allowed sequencing of much lower amounts of input DNA, down to single cells [13, 14]. With the ever-increasing use of WGBS, however, it became evident that all approaches relying on bisulfite conversion suffer from various shortcomings, such as uneven and biased genomic coverage, skewed base composition, and high inter-sample and inter-protocol variability, which resulted in errors in methylation signal output and interpretation [12, 15–19]. More recently, a novel enzymatic methyl-sequencing (EM-seq) approach for the detection of 5mC has been developed [20] to replace the use of the highly destructive chemical conversion through sodium bisulfite, which was recognized as the root cause for the biases and artifacts in WGBS libraries [16]. EM-seq utilizes a combination of enzymatic reactions to convert cytosine to uracil (Fig. 1) with similar efficiency to bisulfite conversion, while preserving the DNA backbone and remaining degradation and bias-free [18, 20,

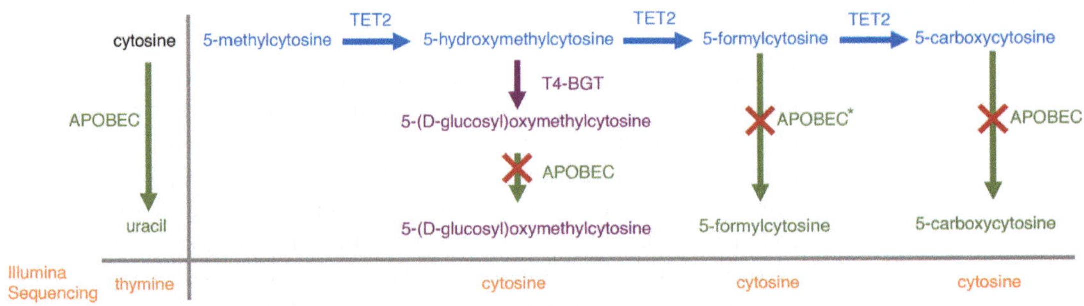

Fig. 1 A schematic of the enzymatic conversion of cytosine and its derivatives, reprinted and modified from [20] under CC BY-NC 4.0 license. First, 5mC is oxidized by TET2 (ten-eleven translocation methylcytosine dioxygenase 2) to 5hmC. 5hmC can be further modified using two possible routes. T4-BGT (T4-phage beta-glucosyltransferase) can glucosylate 5hmC to 5gmC or TET2 can further oxidize 5hmC into 5fC and then 5caC. TET2 and T4-BGT work together so that both enzymes complete the modification of native 5mC and 5hmC to nearly 100%. In a subsequent reaction, APOBEC (apolipoprotein B mRNA editing enzyme catalytic subunit 3A) deaminates unmodified cytosines and converts them into uracils, while 5mC, which has been modified into 5gmC (the major derivative product), 5fC and 5caC, remains resistant to deamination (*APOBEC deaminates trace amounts of 5fC, and 0% of 5caC and 5gmC [20])

21]. As a result, EM-seq delivers higher library yields with longer reads and highly uniform coverage over all sequence contexts, including the extremely C-rich asymmetric unmethylated regions, which constitute the main source of false positive signals in classical BS-seq libraries due to a combination of high degradation coupled with conversion resistance [16]. Thus, EM-seq achieves the highest accuracy in methylation estimation, together with very high reproducibility between samples, little batch variation, less duplication, and a larger number of captured CpGs [18–22]. These features make it superior to WGBS and suitable for all applications, including low input and fragmented material, such as FFPE, cell-free DNA, and single cells [23]. In particular, EM-seq can resolve cases of elusive and questionably low methylation levels, strongly affected by false positive artifacts from the bisulfite, such as in mitochondrial DNA and invertebrate DNA, which have been the subject of disputes for several decades [24–28].

The current protocol features the New England Biolabs (NEB) whole genome library preparation kit, originally used successfully for the classical BS-seq protocol, and more recently, optimized for enzymatic conversion. The principle and steps are identical to the original BS-seq pre-bisulfite protocol, but enzymatic conversion replaces the bisulfite conversion step (Fig. 2, [29]). Therefore, if it is strictly necessary to use bisulfite conversion, such as adding samples to existing public or unpublished BS-seq data, the current protocol is applicable; however, both the bisulfite conversion kit and associated methylated adapters must be purchased separately (*see* **Note 1**). It is important to note that neither WGBS nor EM-seq can distinguish between 5mC and 5-hydroxy-methylcytosine (5hmC) and both modifications appear as positive signals. In EM-seq, unlike WGBS, native 5-formyl-cytosine (5fC) and 5-carboxy-cytosine (5caC) are not converted to uracil and are also read as positive signals (Fig. 1, [20]); however, their contribution to the total positive calls pool is negligible.

An advantage of EM-seq is that the output is identical to WGBS in that unmethylated cytosines appear as thymines, making the analysis pipelines developed for WGBS entirely suitable and directly transferrable to EM-seq. The Bioinformatics Group at the Babraham Institute, Cambridge, were among the first to develop

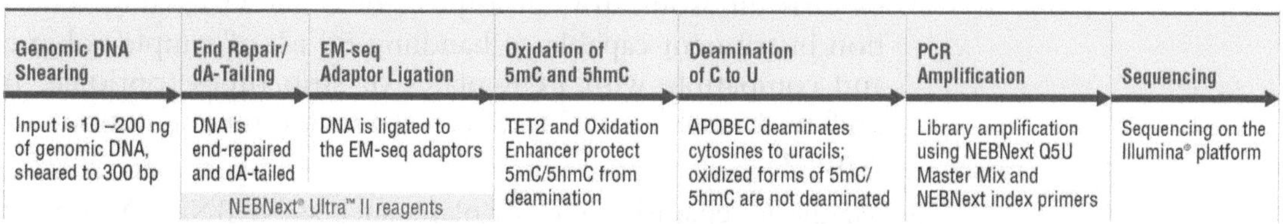

Fig. 2 A schematic of the EM-seq workflow. Library preparation steps are performed with the NEBNext Ultra II kit reagents, followed by a two-step enzymatic conversion and library amplification with a uracil-tolerant DNA polymerase (Q5U). Reprinted from www.neb.com (2023) with permission from New England Biolabs, Inc

tools for the analysis of WGBS data and in the last 10 years, those tools have become some of the best established and highly used worldwide. Each tool has a detailed User Guide and a wealth of associated information on the respective website or GitHub pages. Here, we describe the typical DNA methylation processing and analysis workflow and its capabilities for an end-to-end analysis of classical WGBS data, now fully transferrable for data generated by EM-seq.

2 Materials

Reagents in this list, where indicated, must be purchased from the recommended suppliers to guarantee the quality of outcome. For equipment and plasticware, the manufacturers are suggestions, and any preferred supplier of good-quality consumables or equipment can be used. Please note, it is important to use low retention (i.e., non-stick) nuclease-free plasticware throughout, especially pipette tips.

2.1 Reagents and Kits

1. Genomic DNA isolated in a manual or automated fashion, excluding chloroform-containing protocols [30].
2. Qubit DNA Quantification Assay Kit (High Sensitivity) or Quant-iT PicoGreen dsDNA Assay Kit with Lambda DNA standard (Invitrogen/Thermo Fisher Scientific).
3. NEBNext UltraShear for 24 or 96 reactions (recommended instead of ultra-sonication).
4. NEBNext Enzymatic Methyl-seq Kit for 24 or 96 samples.
5. Formamide—recommended, or optional 0.1 N NaOH.
6. 80% Ethanol, freshly prepared from 100% ethanol and ultra-pure nuclease-free water.
7. Low TE buffer (10 mM Tris pH 8.0, 0.1 mM EDTA), or 10 mM Tris pH 8.0 buffer.
8. High-purity nuclease-free water.

2.2 Specialist Laboratory Equipment

1. Qubit fluorometer or a fluorescent microplate reader, such as a PHERAstar FS instrument.
2. Covaris ultrasonicator, Bioruptor, or other DNA fragmentation instrument capable of handling 50 μL of sample volume and compatible with PCR plates or strip tubes (optional, if preferred instead of UltraShear enzymatic fragmentation, *see* **Note 2**).
3. Magnetic separation rack or plate, such as NEBNext Magnetic Separation Rack or Invitrogen Thermo Fisher DynaMag 96 Side Magnet Well Plate.

4. PCR machine.
5. Bioanalyzer or TapeStation or other fragment analyzer, together with associated reagents and consumables (always use the "high-sensitivity"' DNA assays).
6. Benchtop microcentrifuge for 0.2 mL tubes and plates.
7. Vortexer.
8. Lab fume hood.

2.3 Pipettes and Plasticware

1. A set of standard pipettes with low retention nuclease-free tips of various sizes.
2. Multichannel pipettes: manual 2–20 μL and 20–200 μL 8-channel pipettes (*see* **Note 3**) and an electronic dispensing 100–1200 μL 8-channel pipette with a buffer reservoir.
3. Electronic multi-dispenser pipette, such as Eppendorf Multipette E3 with charging adapter and associated combitips of different sizes (*see* **Note 4**).
4. Pipet-aid/pipettor with stripettes.
5. 0.2 mL PCR strip tubes (clear) or clear 96-well PCR plates.
6. Adhesive seals for PCR plates and seal applicators.
7. Plastic tubes of different sizes—0.5 mL to 5 mL tubes, and 15- or 50-mL conical centrifuge tubes.
8. Ice bucket with ice.

2.4 Analysis Tools and Equipment

All listed tools should be installed, configured, and ready to use.

1. One linux/unix machine (multi-core recommended) with at least 32GB of RAM.
2. FastQC for data quality check: http://www.bioinformatics.babraham.ac.uk/projects/fastqc/
3. FastQ Screen for data purity (contamination) check: https://www.bioinformatics.babraham.ac.uk/projects/fastq_screen/
4. TrimGalore! and CutAdapt [31] for read trimming and Illumina adapter removal: http://www.bioinformatics.babraham.ac.uk/projects/trim_galore/; https://github.com/marcelm/cutadapt
5. Bismark for bisulfite-converted read mapping and methylation extraction [32]: http://www.bioinformatics.babraham.ac.uk/projects/bismark/
6. Bowtie2 is necessary for bismark read mapping [33]: http://bowtie-bio.sourceforge.net/bowtie2/index.shtml
7. Samtools is necessary for bismark to create bam files [34]: http://samtools.sourceforge.net/
8. MultiQC—for a final processing summary report [35]: https://multiqc.info/

9. Nextflow configured with an appropriate executor—for running pipelines [36] (*see* **Note 5**): https://www.nextflow.io/
10. SeqMonk for processed data analysis and visualization: https://www.bioinformatics.babraham.ac.uk/projects/seqmonk/
11. The above tools need Perl, Python, and Java installed, and R for some SeqMonk functionalities.

3 Methods

The NEBNext EM-seq Kit Instruction Manual is very detailed [37] and will be updated through the years. The current protocol includes all steps, highlighting additional important points from the practice and sharing further insights in the **Notes** section. A notable difference is that the current protocol focuses on high-throughput library preparation, for projects with 24 to 96 samples, which are performed manually and not with a liquid handling automated system (i.e., a robot). If less experienced, it is advisable to do 24 to 48 samples at a time, to not risk over-drying the beads during clean-up steps.

3.1 DNA Preparation

1. DNA concentrations must be measured by Qubit or Picogreen dsDNA assay before library preparation, according to respective kit's manufacturer's instructions (*see* **Note 6**). Higher concentrations of DNA will need to be diluted first to less than 200 ng/μL.
2. Dilute the controls according to EM-seq user manual. Unmethylated Lambda and methylated pUC19 control DNAs must be added to each sample (*see* **Note 7**), standardly starting at a 1:1000 mass ratio and going much lower, depending on control genome size and required sequencing depth.
3. Calculate all volumes of DNA and low TE/Tris buffer, including the spike-in unmethylated and methylated controls as per the user manual (*see* **Note 8**). The assay is validated for as little as 10 ng DNA (*see* **Note 9**) and up to 200 ng DNA, made in a 26 μL final volume if using UltraShear, or 50 μL if using ultrasonication (*see* **Note 2**).
4. Prepare each sample by adding the required volumes of low TE/Tris buffer, methylation controls, and sample DNA (*see* **Note 10**). This step should not be performed in haste and should be allowed plenty of time with a large number of samples.

3.2 DNA Shearing

1. On ice, prepare a mixture of the UltraShear reaction buffer and UltraShear enzyme mix, according to the number of samples. Add some extra volume to aid with dispensing as shown below. Mix well and distribute equal volumes to an 8-well strip for the multichannel pipette (*see* **Note 3**).

Reagent	1X RXN volume	24X RXN volume (27×)	96X RXN volume (100×)
UltraShear reaction buffer	14.0 μL	378 μL	1400 μL
UltraShear	4.0 μL	108 μL	400 μL

2. With a multichannel pipette, add 18 μL of the UltraShear Master Mix to each sample, mixing thoroughly by pipetting up and down.
3. Place in a preheated Thermocycler with a heated lid (75 °C) and choose a program according to the desired fragment size as shown in the UltraShear manual. For a 2 × 150 bp paired end (PE) sequencing, which is becoming the standard, use the following: 15 min at 37 °C, 15 min at 65 °C, and hold at 4 °C (*see* **Note 11**). Continue directly with Subheading 3.3 End prep.
4. For Covaris/Bioruptor: follow instrument handling instructions for shearing the ready samples—choose 300–400 bp if aiming for a 2 × 150 bp PE sequencing. Carefully transfer the sheared DNA to new nuclease-free PCR tube strips or a plate for the next step (*see* **Notes 12** and **13**).

3.3 End Prep (Includes End Repair and A-Tailing)

1. On ice, prepare a master mix of the end prep reaction according to the number of samples, adding some extra volume to aid with dispensing as shown below. Note that the master mix composition will differ depending on the fragmentation method used in Subheading 3.2. Mix well and distribute equal volumes to an 8-well strip for the multichannel pipette (*see* **Note 3**).

Reagent	UltraShear / sonication 1X RXN Volume	UltraShear / sonication 24X RXN Volume (27x)	UltraShear / sonication 96X RXN Volume (100x)
Ultra II end prep reaction buffer	None / 7.0 μL	None / 189 μL	None / 700 μL
500 mM DTT (from UltraShear kit)	2.0 μL / none	54 μL / none	200 μL / none
Ultra II end prep enzyme mix	3.0 μL / 3.0 μL	81 μL / 81 μL	300 μL / 300 μL

2. With a multichannel pipette, add 5 μl of the End Prep Master Mix to each sample if UltraShear was used in 3.2, or 10 μl of the End Prep Master Mix if non-enzymatic shearing was used, mixing thoroughly by pipetting up and down.
3. Place in a Thermocycler with a heated lid and run the following program: 30 min at 20 °C, 30 min at 65 °C, and hold at 4 °C. Continue with next step.

3.4 EM-Seq Adapter Ligation

1. On ice, prepare a mixture of the Ligation master mix and Enhancer according to the number of samples. Add some extra volume to aid with dispensing as shown below.

Reagent	1X RXN volume	24X RXN volume (27×)	96X RXN Volume (100×)
Ligation enhancer	1.0 μL	27 μL	100 μL
Ultra II ligation master mix	30.0 μL	810 μL	3000 μL (3.0 mL)

2. Add 2.5 μL of EM-seq Adaptor to each sample.
3. Add 31 μL of the prepared Ligation Master Mix + Enhancer to each sample and mix by pipetting very well to ensure efficient ligation (the mixture is viscous).
4. Incubate at 20 °C for 15 min in a PCR thermal cycler with the heated lid off, then hold at 4 °C. At this point, the samples can be stored at −20 °C overnight or continue with clean-up.

3.5 Clean-Up of Adapter Ligated DNA

1. Remove Sample Purification Beads from storage and vortex until homogenized.
2. Dispense 110 μL beads to each sample (bead ratio: ~1.18X sonication, ~1.3X UltraShear), mix by pipetting very well with care, making sure no beads are lost (the wells are full).
3. Incubate the samples with the beads for at least 5 min at RT.
4. Prepare a fresh stock of 80% ethanol, using nuclease-free water. Mix by inversion and place at RT. Suggested: 40 mL Et-OH + 10 ml H_2O (for 96 samples).
5. Transfer sample strip tubes/plate to a magnetic separation rack/plate and incubate at RT until the solution is completely clear of beads.
6. Keeping the tubes/plate on the magnet, carefully remove the supernatant and discard without disturbing the beads, using a manual multichannel pipette set at 190 μL.

7. With the tubes/plate still on the magnet, add 200 μL of 80% ethanol to the wells without disturbing the bead pellet. This step is performed with a dispensing multichannel pipette going up to 1200 μL (i.e., for 48 samples or half a plate at a time).
8. Quickly remove and discard the ethanol while keeping the samples on the magnet, avoiding aspiration of the bead pellet (with a manual multichannel pipette set at 200 μL). This step should be performed within 2 min (i.e., 10 s per PCR strip/ plate row).
9. Repeat **Steps** 7 and **8**, to perform two 80% ethanol washes in total. Remove as much of the final wash as possible, once with 200 μL tips and a second time with 10 μL tips to remove all drops around the well walls and bottom, without disturbing the bead pellet.
10. Air dry the bead pellets for 2 min at RT. The wells completed first in **Step 9** will dry first and may be ready for **Step 11** by the time the last wells in **Step 9** are completed. The beads should be matte or still glossy, with no ethanol drops visible in the tube.
11. Remove the tubes/plate from the magnet and quickly add 29 μL of Elution Buffer to the bead pellet. A multi-dispenser is most suitable to use, to avoid wells with visible ethanol while not letting others over-dry (*see* **Note 4**). Mix thoroughly with a manual multichannel pipette ensuring all beads are off the walls and well-resuspended. It is critical to *not* substitute the provided Elution Buffer with TE or another EDTA-containing buffer (*see* **Note 14**).
12. Incubate at RT for at least 1 min to let DNA elute from the beads. Centrifuge briefly to collect solution at the bottom of the tubes.
13. Transfer the tubes/plate back to the magnet and incubate at RT for at least 5 min or until the solution of beads is completely clear.
14. Carefully remove 28 μL of the eluate (with a manual multichannel), ensuring absolutely no beads are carried over and transfer to a new microcentrifuge plate or PCR tubes (*see* **Note 15**). Samples can now be stored at –20 °C, if needed.

3.6 Oxidation of 5mC and 5hmC

1. Prepare TET2 Buffer, depending on the kit purchased. For 24 samples, add 100 μL of TET2 Reaction Buffer to each of the three provided tubes of TET2 Reaction Buffer Supplement (each one is enough for 10 samples), and 400 μl Buffer to each provided Supplement tube for 96 samples (each one is enough for 40 samples). Vortex and spin down, store at –20 °C, and use within 4 months.

2. Prepare the oxidation reaction master mix by combining all components as shown below and vortex quickly.

Reagent	1X RXN volume	24X RXN volume (27x)	96X RXN volume (100x)
TET2 reaction buffer + TET2 reaction buffer supplement	10 μL	270 μL	1000 μL (1.0 mL)
Oxidation supplement	1 μL	27 μL	100 μL
DTT	1 μL	27 μL	100 μL
Oxidation enhancer	1 μL	27 μL	100 μL
TET2	4 μL	108 μL	400 μL

3. Add 17 μL of the Oxidation Master Mix to each sample and mix well by pipetting.
4. Freshly prepare the Fe(II) solution by adding 1 μL of 500 mM Fe(II) stock solution to 1249 μL of water.
5. Add 5 μL of the diluted 0.4 mM Fe(II) solution to each sample and mix by pipetting very well (*see* **Note 16**). Discard the remaining 0.4 mM Fe(II) solution after use.
6. Incubate at 37 °C for 1 h in a thermal cycler with a heated lid on (≥ 45 °C), then hold at 4 °C.
7. Transfer the samples on ice and add 1 μL of Stop Reagent to each reaction.
8. Incubate again at 37 °C for 30 min, in a thermal cycler with a heated lid on (≥ 45 °C), then hold at 4 °C. The samples can remain overnight at 4 °C or be transferred to −20 °C.

3.7 Clean-Up of TET2-Oxidized DNA

1. Vortex the Sample Purification Beads until homogenized and dispense 90 μL (bead ratio ~1.8X) to each sample. Mix well by pipetting, making sure no beads are lost.
2. Repeat the rest of the steps as described in Subheading 3.5, resuspending the beads in 17 μl of Elution Buffer and eluting 16 μL of sample to transfer to new PCR tubes/plate. Again, it is critical to not carry over any beads with the elution buffer (*see* **Note 15**).

3.8 Denaturation of DNA with Formamide (See Note 17)

1. Pre-heat a PCR thermal cycler to 85 °C with the heated lid on (105 °C).
2. In a fume hood, add 4 μL formamide to each eluted 16 μL of TET2 oxidized DNA, mix well by pipetting, and centrifuge briefly.
3. Incubate the samples at 85 °C for 10 min in the pre-heated thermal cycler.

4. Immediately place on ice for 2–3 min to ensure DNA has cooled down completely (i.e., is single stranded) before proceeding to the APOBEC conversion step.

3.9 APOBEC Deamination (Conversion of Unmodified C to U)

1. On ice, mix the components as shown below to prepare the APOBEC master mix.

Reagent	1X RXN volume	24X RXN volume (27×)	96X RXN volume (100×)
Nuclease-free water	68 μL	1836 μL	6.8 mL (6800 μL)
APOBEC reaction buffer	10 μL	270 μL	1 mL (1000 μL)
BSA	1 μL	27 μL	100 μL
APOBEC	1 μL	27 μL	100 μL

2. Add 80 μL of deamination master mix to each of the 20 μL denatured DNA samples, mix well by pipetting and centrifuge to collect all liquid.
3. Incubate at 37 °C for 3 h and 4 °C afterwards, in a thermal cycler with a heated lid on (≥45 °C). The samples can remain overnight at 4 °C or be transferred to −20 °C for storage

3.10 Clean-Up of Deaminated DNA

The beads become stickier and more difficult to resuspend in this clean-up. Make sure they are not over-dried at the end; air-drying times may be shorter than in previous clean-ups.

1. Vortex the Sample Purification Beads until homogenized and dispense 65 μL (bead ratio 0.65X) to each sample. Mix very well by pipetting, making sure no beads are lost in the tips.
2. Repeat the rest of the steps as described in Subheading 3.5, resuspending the beads in 21 μL of Elution Buffer and eluting 20 μL of sample to new PCR tubes/plate. It is critical not to over-dry the beads before adding the Elution Buffer

3.11 PCR Amplification

For index pooling refer to Section 3 from NEB's EM-seq Instruction Manual [37].

1. On ice, add 5 μL of EM-seq Index Primer to each of the purified deaminated DNA samples.
2. Dispense 25 μL of Q5U Master Mix to each sample, mix very well by pipetting, and centrifuge briefly.
3. Place the samples in a PCR thermal cycler and run the following program: 98 °C for 30 sec, then cycling at 98 °C for 10 s, 62 °C for 30 s, and 65 °C for 60 s (for 3 to 8 cycles depending

on starting amount, *see* **Note 18**), and a final extension at 65 °C for 5 min, followed by a 4 °C hold. Samples can remain overnight at 4 °C or be stored at −20 °C.

3.12 Clean-Up of Amplified DNA Libraries

1. Add 50 μL of nuclease-free water to each 50 μL sample and mix well by pipetting.
2. Vortex the Sample Purification Beads until homogenized and dispense 65 μL (bead ratio 0.65X) to each sample. Mix well by pipetting, making sure no beads are lost in the tips.
3. Repeat the rest of the steps as described in Subheading 3.5, resuspending the beads in 21 μL of Elution Buffer or, for long-term storage, in 1X TE or low TE. Transfer 20 μL of eluate to new PCR tubes/plate.
4. Proceed to quality control (Subheading 3.13) or store the samples at −20 °C.

3.13 Library Quality Control (QC)

1. Depending on the QC instrument used, aliquot 1 to 3 μL of nuclease-free water in a new strip tube/96-well plate according to the number of samples eluted in Subheading 3.12 (Bioanalyzer is more sensitive and requires the higher dilution factor).
2. Take 1 μL of library with a multichannel pipette from the elution plate, transfer to the QC plate with water and mix well.
3. Follow the Bioanalyzer or TapeStation instructions to measure. The Bioanalyzer is more accurate but is not high throughput, while TapeStation can handle 96-well plates.
4. The resulting trace (library size histogram, see EM-seq Instruction Manual [37]) must be gated from 200 bp to 1200 bp in order to determine library quantity and molarity. Remember to multiply to the dilution factor for obtaining the final library estimates for informing sequencing facilities.

3.14 Raw Data Processing

The main raw data processing steps of whole methylome sequencing are generally the same, irrespective of the protocol used. For the tools discussed here, the classical library preparation approach (including EM-seq) is the "default" mode, meaning no extra options need to be specified, with the only exception being a possible further trimming of reads (*see* **Note 19**). We describe here running raw data processing on Nextflow, a platform for managing and running pipelines [36], which makes it especially easy for wet lab scientists to perform their own data processing, even with limited coding experience (*see* **Note 20**). Babraham Bioinformatics maintain several Nextflow pipelines on the GitHub repository, "nf_bisulfite_WGBS" being designed for classical BS-seq/EM-seq. The pipeline contains the following components (modules), some of which can also be run separately as single program pipelines:

```
FastQC
Trim Galore
FastQ Screen [--bisulfite]
Trimmed FastQC
Bismark
Deduplicate Bismark
Methylation extract (coverage file) [--ignore_r2 2 for PE files]
bismark2report
bismark2summary
MultiQC
```

This section includes a breakdown of the analysis steps: the first three are performed by the user, and the rest are performed automatically by the `nf_bisulfite_WGBS` pipeline, unless some changes in steps are required by the user and cannot be added as module arguments as shown in Subheading 3.14.4, in which case the separate modules will need to be run independently (not covered here).

3.14.1 Accessing the "`nf_bisulfite_WGBS`" Pipeline

1. Create a fork of Babraham's Nextflow pipelines repository (https://github.com/s-andrews/nextflow_pipelines) to the user's own repository.
2. Edit the nextflow.config file to add configurations of the user's computer system or high-performance cluster (help from local bioinformatics team might be needed). This might also include making sure the individual packages are accessible from the local system.
3. Change the paths to the genomes in the nextflow_pipelines/genomes.d folder with their locations on the user's system.
4. From a unix command line, run the following on the user's computer or local cluster (there might be individual systems' variations):

```
module load nextflow
git clone https://github.com/User_profile/nextflow_pipelines.git
cd nextflow_pipelines
git pull
```

```
nextflow run nf_bisulfite_WGBS --genome <name-of-genome> *fastq.gz -bg
```

where `-bg` stands for running the pipeline non-interactively in the background (see Nextflow @ Babraham User Guide for all details).

3.14.2 Generation of a Reference Genome

1. The annotated genome of interest must be downloaded in fasta format, where each sequence is annotated with the actual chromosome name and each chromosome sequence is a separate file, such as the searchable Ensembl FTP sequence database: https://www.ensembl.org/info/data/ftp/index.html
2. The most accurate approach is to map simultaneously to the genome of interest and the control DNA sequences via creating a combined genome, rather than aligning separately. To create a combined genome, add pUC19 and Lambda phage sequences as if they were additional chromosomes to the samples' genome.
3. This step ensures the genomic reference sequence is converted and looks like bisulfite-converted DNA, to make further alignment possible. It is separate from the pipeline because it is performed once per annotated genome. The script takes fasta format sequence files, and after completion, the converted reference genome path must be added to the nextflow_pipelines/genomes.d folder.

```
module load bismark
bismark_genome_preparation --path_to_aligner /path/to/bowtie2/ --verbose /path/to/genome/
```

3.14.3 Raw Data Quality Control with FastQC

FastQC provides an overview with summary graphs and tables to help assess data quality, exported in an HTML report and accessible offline. It is best to do this module separately before running the full pipeline, to assess the need for further read trimming (*see* **Note 19**).

```
nextflow run nf_fastqc *fastq.gz
```

3.14.4 Trimming of Adapters and Low-Quality Bases with TrimGalore!

If FastQC plots show a significant base composition bias at the start and end of your EM-seq library reads (*see* **Note 19**), the pipeline command line should be modified to remove (hard clip) the problematic sequences (the example is for paired end data and 10 bases of biased sequence):

```
nextflow run nf_bisulfite_WGBS --genome <name-of-genome> *fastq.gz -bg --trim_galore_args="--clip_R1 10 --clip_R2 10 --three_prime_clip_R1 10 --three_prime_clip_R2 10"
```

3.14.5 FastQ Screen and FastQC of Trimmed Data

1. FastQ Screen scans the data for possible sequence contamination with DNA from another source or species. In `nf_bisulfite_WGBS` it works by default with the argument `--bisulfite` for converted sequences.
2. FastQC performs data quality assessment as above, but post-trimming.

3.14.6 Read Alignment, Deduplication, and Methylation Extraction with Bismark

1. The alignment to a reference genome finds the genomic locations of the read sequences to yield base pair resolution. A very detailed Bismark User Guide (http://felixkrueger.github.io/Bismark/) of how it works is available online.
2. Deduplication is compulsory for WGBS/EM-seq analysis to remove reads with identical coordinates on the same strand, which are considered the result of PCR over-amplification of the same original read.
3. Methylation extraction extracts and compiles all cytosine calls and genomic coordinates from the aligned reads in CG, CHG, and CHH contexts (H = A, G, or T), and generates lighter files for further analysis. It also removes redundant information from overlapping read 1 (R1) and read 2 (R2) sequences in PE libraries.

3.14.7 Bismark Summary Reports and MultiQC

1. This step generates per sample reports for the completion of the previous bismark steps, such as aligned and duplicated reads, totals of methylated and unmethylated calls and percentages, and M-bias reports.
2. MultiQC provides summaries for all previously generated reports and logs for all samples and generates an HTML output with plots to visualize overall processing parameters and quality across samples.

3.15 Analysis of Methylation with SeqMonk

SeqMonk is a program for the analysis of high-throughput sequencing data. It is genome browser-based, with a graphical user interface, making it particularly suitable for wet lab users. There are comprehensive online guides on how SeqMonk works (https://www.bioinformatics.babraham.ac.uk/training.html), and several YouTube tutorials. This chapter focuses on the basics and principles of DNA methylation analysis, broken down in several steps, and further details can be found in the online tutorials and in the SeqMonk "Help" section, which is also comprehensive.

3.15.1 Create a SeqMonk Project

You must initially create a project using the same genome assembly version used in Subheading 3.14.2, after which the coverage (.cov) files from the bismark methylation extraction step can be directly imported into the SeqMonk project. Note that in methylation analysis "reads" are the individual CpG calls from the sequencing (read length = 1 bp), and "strand" signifies methylated ("+" and red) or unmethylated ("-"and blue).

3.15.2 Probe Generation

In SeqMonk terminology, a "probe" is a region in the genome where we wish to make a quantitative measurement. Probes can be arbitrary tiled regions, or designed around a genomic feature (e.g., genes or CpG islands). Ideally the probes should be designed as windows that cover the entire genome, allowing us to evaluate the data on a global level and facilitating robust comparison between samples. While these windows can be of a fixed base pair length (e.g., 1–5 kb via the "Running window probe generator"), because of the varying density of CpGs in the genome, it is better to use windows with a fixed CpG count (i.e., 50–200 CpGs), so that the number of methylation calls in each window is roughly equal. The latter approach is performed with the "Read position probe generator" by selecting all datasets and choosing a preferred CpG count, and results in even coverage with more uniform "weights" of the probes for smoother plots and more robust downstream analyses. The generated probes are stable and will not change until we generate new probes.

3.15.3 Read Count Quantitation and Evaluation

1. First, we count the number of CpG calls identified from our sequencing, on the set of probes we have generated, to evaluate any need for filtering.
2. Check the distribution of reads per probe (read count histogram, Fig. 3a), and zoom in if some probes have reads into the 100,000 s, as such numbers are clear outliers and most likely mapping artifacts. Since these will affect methylation quantitation, those probes need to be excluded from further analysis.
3. If the probes were generated by fixed number of CpG positions, exploring read length and other metrics compared between the samples will also give a good idea of the data and where (or for which samples) to expect weaknesses.

3.15.4 Filtering

There are many options for filtering and "filter on value" will allow filtering based on the quantitated value of each probe. Thus, selecting probes with less than 500 read counts as in the given example will create a new probe list where all the high-coverage outlier probes are removed (Fig. 3b, c).

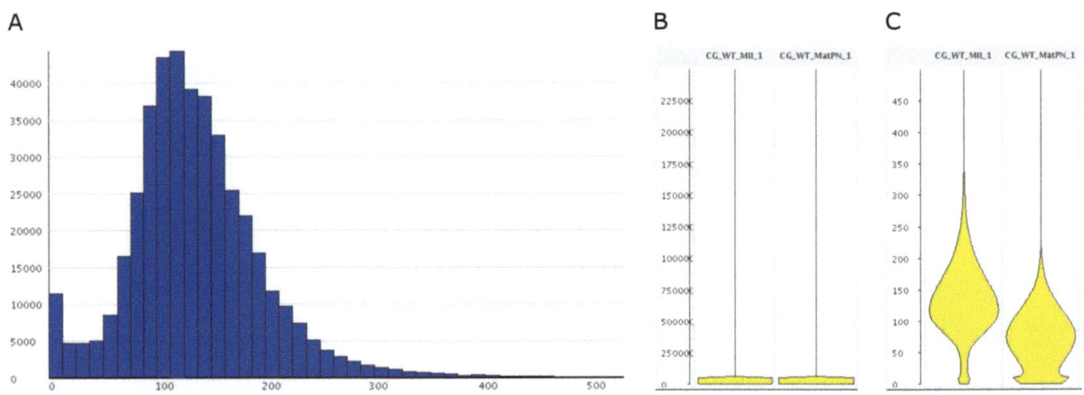

Fig. 3 Data filtering prior to methylation analysis based on read count per probe. (**a**) Read count histogram, zoomed in so that high coverage reads are not visible. (**b**) Bean plot of read probe count between samples before filtering and (**c**) after filtering out outliers with a read count above 500

3.15.5 Methylation Quantitation

The "Bisulfite methylation over features" pipeline (BS-pipeline) is the recommended approach (*see* **Note 21**). It provides data filtering options, such as setting a threshold for minimum CpG observations, i.e., CpG positions covered by at least one call, in order to include a probe (a "horizontal" filtering) and select a minimum depth of coverage for the CpG positions within the probe (a "vertical" filtering). The vertical filtering, although a common practice, has shown a high risk for increasing WGBS coverage biases [16] and should only be used where coverage depth is expected to be extremely high, which is normally associated with enrichment methods such as Reduced Representation sequencing (RRBS), and not with whole genome protocols. The horizontal filtering option, however, is highly recommended since probes with fewer than 5 or even 10 CpG positions within the probe are likely to generate unstable methylation percentages and therefore skew results. This data should not be included in further analyses.

3.15.6 Plotting and Visualizing Global Methylation Levels

Once methylation is quantified, there are multiple ways to explore and visualize the data.

1. Global 5mC levels are the first thing to assess via a number of ways, such as the bean plot (or violin plot, because it also shows density) or box-whisker and star wars plots, which can visualize global methylation levels across all samples.
2. Histograms show methylation density and distribution, while scatter plots provide a one-to-one comparison between the distributions of two samples, such as a control against a condition or treatment (Fig. 4a).
3. Other useful options are heat plots, which are a valuable way to look at many samples at once, or the line plots, where methylation levels are shown as a line connecting the different samples. These are useful if looking at time points, stages, or another type of connected process.
4. The plots can be saved in two formats—a PNG for easy viewing and presentations, and a scalable vector SVG format for generating editable publication quality figures.

3.15.7 Analyzing Genomic Features (e.g., Genes, Promoters, etc.)

1. Feature analysis can be done with the same probes used in the global analysis, by selecting the probes overlapping with a genomic feature ("filter by feature").
2. The same plots as in Subheading 3.15.6 are used for visualization, but they will only represent methylation of the selected genomic feature, such as promoters, and not the whole genome.

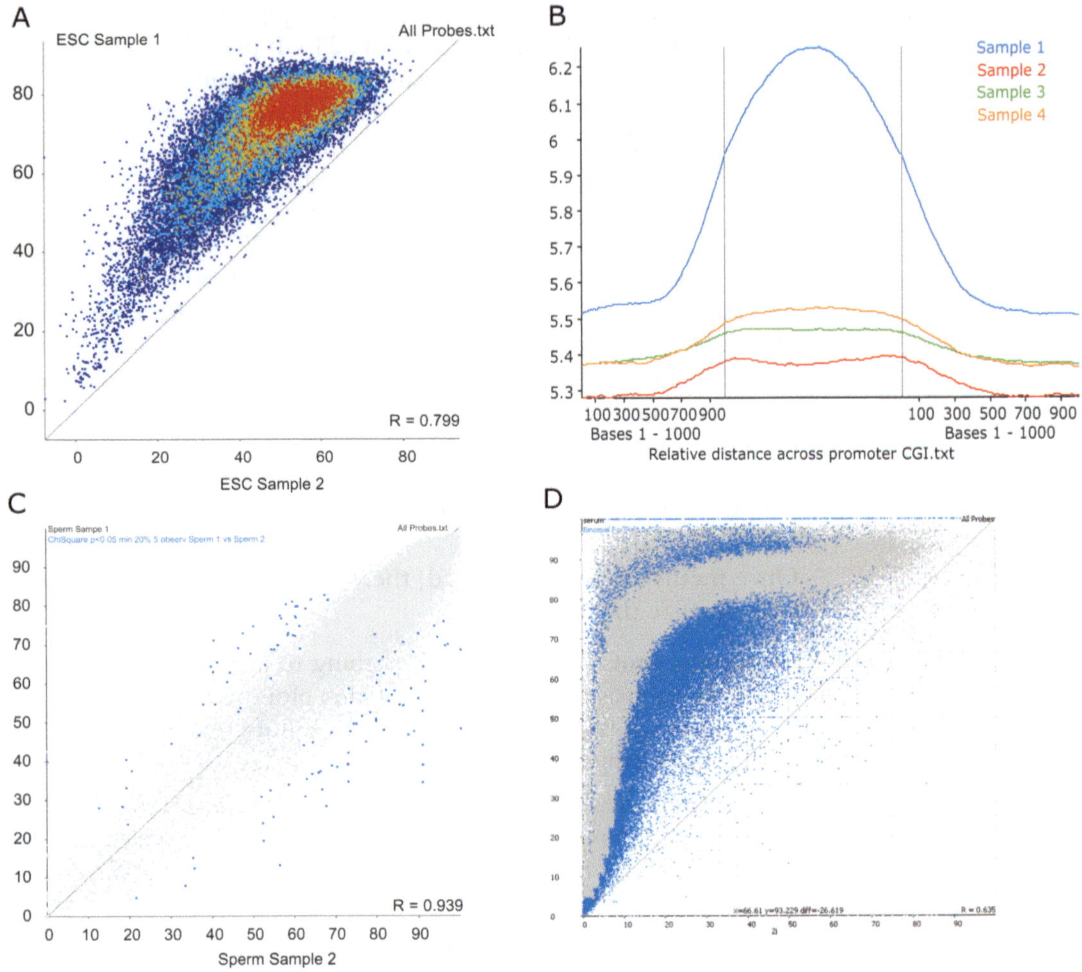

Fig. 4 Examples of plots and analyses from SeqMonk. (**a**) Scatter plot, (**b**) A relative trend plot showing methylation levels across a feature (promoter or CGI), (**c**) A chi-square DMR analysis when global methylation levels between samples do not differ, and (**d**) A binomial differential methylation analysis when global methylation levels are significantly different

3. A more sophisticated analysis will require the generation of new "feature probes," which will delete the previously specified probes. Perform quantitation on the new probes as in Subheading 3.15.5 and visualize as described above.
4. In addition, the averaged methylation trend is commonly investigated over certain genomic features, such as transcriptional start sites (TSS), promoters, CpG islands (CGIs), or gene bodies. Probes are generated to span over the feature, starting before and finishing after for several (kilo)bases. After quantitation, cumulative or relative trend plots are generated (Fig. 4b).

3.15.8 Assessing Differential Methylation

This analysis aims to find regions, which are differentially methylated between two samples (i.e., DMRs). It is the most complex, and often most important methylation analysis, involving several statistical considerations. SeqMonk has multiple statistical tests and approaches that can be employed in accordance with the type of data and research question.

1. For data with no global methylation changes between a sample and control, contingency statistics such as chi-square, Fisher's exact test, logistic regression, or the bisulphite analysis mode of EdgeR [38] are available options for defining thresholds and significance for differential methylation (Fig. 4c).
2. For datasets where total methylation changes globally from the control, such as during reprogramming stages or other conditions vastly affecting global levels, binomial statistics are more useful in defining regions that change more or less than the global average change. For example, here we look for regions that escape the global changes or such that stand out above the average change (Fig. 4d).
3. Always check random hits, especially the top hits, for underlying artifacts by zooming in the gene or region location in the SeqMonk browser and visually inspecting if the actual data (methylation calls) make sense.

3.15.9 Generating and Saving Reports

Reports can be saved at multiple steps of the described analysis workflow.

1. Datastore reports generate summarized information between the different samples, such as total number of methylated and unmethylated cytosines, coverage, and calculated median and mean of global methylation levels.
2. Annotated probe reports contain detailed information about each probe together with the quantitated methylation value and can be imported in R packages for further analysis and visualization.
3. Feature reports are like the annotated reports but summarize data for a specific feature instead.
4. Data can also be exported and saved for the filtered probe lists and from the plots, to be transferred to other packages for visualization or analysis.

4 Notes

1. The EM-seq protocol is based on the NEBNext Ultra II DNA Library Prep Kit for Illumina, which is offered separately without the Enzymatic Conversion Module and is appropriate for

BS-seq. Replace the enzymatic conversion steps described here with bisulfite conversion. It is important to note that the methylated adapters included in the EM-seq kit are different from the methylated adapters used for BS-seq and the latter must be purchased separately from Illumina (see NEBNext Ultra II's instruction manual). A comprehensive analysis of BS-seq performance with different bisulfite conversion protocols can be found here [16].

2. Ultra-sonication is a widely used method for DNA shearing; however, there is evidence it preferentially targets methylated DNA and can be used as a hypomethylating agent [39, 40]. NEBNext UltraShear is a novel enzymatic shearing module developed to circumvent the use of ultra-sonication and aid with high-throughput sample fragmentation, being performed in a PCR strip tube or plate format and followed directly by library preparation steps. Enzymatic shearing different from UltraShear is not recommended with the current protocol, as it often leads to loss of DNA methylation (yields short, methylated fragments below retention size).

3. Multichannel pipettes are the natural choice for PCR strips and 96-well plates; however, it only works if the tips fit properly on all channels and guarantee pipetting accuracy. It is also important to choose a pipette gentle to the thumbs, as the current high-throughput protocol requires heavy manual pipetting and should not impose a health risk; tip ejection should not require extreme force and cause extra pressure to the thumb (ideally tip ejection should not be performed with the thumb). CAPP and Rainin manual multichannel pipettes are of good quality, CAPP being the multichannel gold standard, also gentle to the hand: its tips fit uniformly and allow very accurate dispensing throughout all channels; tip ejection requires the whole palm. Importantly, the reagent for use with a multichannel pipette needs to be dispensed in an 8-tube strip or a special reservoir in advance, which can be fiddly (200 μL 8-strip tubes may need refills) and usually associated with some master-mix waste. *See* **Note 4** for an alternative.

4. Multi-dispensing pipettes (single channel dispensers) can be a very useful alternative to multichannel pipettes and are much more sparing to the hands, but there is a risk of cross-contamination between samples. This risk can be avoided by aiming the tip at the inside of the well wall, without touching it, so that the drop of liquid sticks to the wall, especially with small volumes. Larger volumes can be ejected directly aiming the bottom, again without touching anywhere inside the well. Reagent void volume (which is lost inside the dispensing tip) can be very low.

5. The individual workflow components do not need Nextflow and can be run separately. This chapter, however, describes the use of ready-made Nextflow pipelines for WGBS/EM-seq analysis, which is the simplest and most reproducible approach to perform data processing and quality check, and is recommended for publication purposes.
6. A Nanodrop spectrophotometer is good for initial quantification but is not appropriate for determining the final DNA concentration for library preparation due to its low accuracy.
7. Bisulfite-based analysis traditionally utilized unconverted DNA controls, added to samples in small amounts as a "spike-in" to calculate bisulfite conversion efficiency, while positive controls were not necessary. With the two-step enzymatic conversion, however, an unmethylated control would only provide evidence for the efficacy of the second step, where C is deaminated to U with APOBEC (*see* Fig. 1). A fully methylated control, on the contrary, evidences the efficacy of the TET2 + T4-BGT first reaction, and should appear as fully methylated (i.e., 5mC successfully converted to 5gmC and 5caC) in the post-processing analysis.
8. Calculate in MS Excel and print the sample IDs, buffer volume, and DNA sample volume, both as a list and as a 96-well plate/ PCR strip schematic, which will help reduce pipetting errors. Order the samples by increasing or decreasing pipetting volume rather than by sample ID (e.g., buffer decreasing and DNA increasing) to make pipetting easy and quick without substantial changes to pipet volumes between each sample, and to reduce the risk of errors (two neighboring samples will have the same or similar volumes).
9. When using 10 ng DNA with the current protocol, take note that the enzymatic UltraShear fragmentation outperforms ultra-sonication most notably with lower DNA input, achieving significantly higher usable read counts and CpG capture at 10 ng [41]. For even lower starting amounts, EM-seq has been tested and works well with as little as 100 pg, however, some performance parameters will decrease (note this has only been tested with ultra-sonication) [19, 20]. A specialized low DNA input EM-seq protocol has been optimized by NEB to be released separately.
10. Making a pipetting error is easy with many samples, and it is advisable to use a system that will reduce the likelihood of errors. Using the sample list & plate order from **Note 8**, add the low TE/Tris buffer first, then the unmethylated and methylated control DNAs and sample DNA last. Advisably, preparing the conversion controls together with the low TE/Tris buffer as a master mix and dispensing them at once

altogether, will greatly reduce pipetting effort and possible mistakes (the slightly different sample to control ratios due to different buffer volume per sample will be negligible). Double and triple check correct Sample ID and allocated position in plate before pipetting. Cover completed PCR tubes or plate columns to avoid pipetting in the same place.

11. EM-seq can handle larger insert sizes and some Illumina instruments can now sequence 600 bp of insert length. The current protocol describes the generation of libraries suitable for 150 bases long PE sequencing (insert size >300 bp). If smaller size fragments are required, the described protocol applies, with the exception of the final library clean-up where bead to sample ratios would differ: refer to the NEBNext EM-seq or UltraShear Instruction Manuals [37].
12. Instructions on ultra-sonication parameters should be provided with the instrument. An example of Covaris S2 sonication to mean fragment size between 350 to 450 bp is: 200 cycles per burst, 10% duty-cycle, intensity of 4 or 5, treatment time of 50 s (once) or 40 s (twice).
13. Low TE/10 mM Tris buffer must be used for sonication and not water, as this helps stabilize and solubilize the DNA, mitigating the side effects of biased degradation and loss of methylation from ultra-sonication (*see* **Note 2**).
14. EDTA inhibits the TET2 oxidation reaction; therefore, the elution at this purification stage is with the provided Elution Buffer, or alternatively with EDTA-free Tris-HCl buffer or nuclease-free water.
15. Magnetic beads inhibit the enzymatic reactions that follow and there should be no carryover of beads into the eluate.
16. Fe(II) catalyzes and starts the TET2 oxidation reaction once all other components are present. Adding Fe(II) earlier with the rest of components may compromise the efficiency of the enzymatic reaction.
17. Two options are described in the NEBNext EM-seq Instruction Manual [37]; however, the developers strongly recommend the use of formamide instead of NaOH, therefore only the formamide denaturation is described here. NaOH is more likely to have problems with precise concentration if not freshly prepared and yield incompletely denatured DNA, which results in incomplete enzymatic conversion. A detailed protocol for NaOH denaturation can be found in the kit's instruction manual.
18. EM-seq does not need substantial amplification as does classical BS-seq [16, 18], but requires a minimum of 3 PCR cycles to prepare the adapter-ligated DNA fragments with the correct

ends for Illumina sequencing [20]. PCR cycle number guide: 3–4 cycles for 100–200 ng DNA input, 5–6 cycles for 50 ng DNA input, and 7–8 cycles for 10 ng DNA input.

19. There have been observations of base composition bias in EM-seq reads within the first and last 10 bases, both on read 1 (R1) and read 2 (R2), which do not affect cytosine and likely will not affect the methylation call output. However, any remnants of artificially introduced sequence in the reads, which do not belong to the reference genome sequence, could affect downstream mapping. Default alignment parameters are quite stringent, allowing 2 bases mismatch for a 75 base long read, which means such reads can be excluded at the alignment step, leading to a loss of mappable data. While NEB advises removing an extra 5 bases off both R1 and R2, performing quality control with FastQC as a first step will help assess the need for any extra trimming. A conservative approach, especially when working with a large sample size with variable end-bias, would be trimming (i.e., hard clipping) the first and last 10 bases of each read, in addition to the default trimming of low-quality bases and Illumina adapters.

20. Nextflow needs to be installed on the user's local system, such as a high-performance computer cluster, which is an easily executable self-installing procedure (see Nextflow documenta tion). In addition to the pipeline presented here, Nextflow pipelines are available through the nf-core community framework [42], including "methylseq" (https://nf-co.re/met hylseq/2.4.0), which, in its "Bismark workflow" mode, overlaps to a large extent with the workflow described here, and has a parameter `--em_seq` to clip 8 bases of both read ends (R1 & R2) in case of sequence biases. Importantly, nf-core pipelines do not require any prior software installation, they come with pre-built reference genome indices (to download the required genome), and are built with user-friendly web graphical interfaces.

21. The Bisulfite methylation pipeline (BS-pipeline) estimates methylation values of each CpG position first by averaging the values from all calls associated with that position, and subsequently averaging the methylation values of all CpG positions over the whole probe. The alternative and simplest approach to measure methylation is through "difference quantitation," where the overall percentage of positive (methylated) calls out of total calls is calculated per probe, irrespective of the CpG position they came from. However, the latter approach was shown to highly bias the methylation estimation in WGBS, since unmethylated CpGs generate overall fewer calls and have lower coverage (i.e., depth per position) than the positive,

methylated CpGs [16]. The latter approach will work better with EM-seq; however, this has not been evaluated yet. The BS-pipeline has the added benefit of options for data filtering.

Acknowledgements

The authors thank Louise Williams, Max Fritsch, and Michael Skelly from New England Biolabs Inc. for valuable discussions, critical reading, and provision of materials, as well as Phil Ewels from Seqera Labs for valuable discussions and critical reading of the manuscript.

References

1. Luo C, Hajkova P, Ecker JR (2018) Dynamic DNA methylation: in the right place at the right time. Science 80(361):1336–1340. https://doi.org/10.1126/science.aat6806
2. Schübeler D (2015) Function and information content of DNA methylation. Nature 517: 321–326. https://doi.org/10.1038/nature14192
3. Zhang H, Lang Z, Zhu J (2018) Dynamics and function of DNA methylation in plants. Nat Rev Mol Cell Biol 19:489–506. https://doi.org/10.1038/s41580-018-0016-z
4. Regev A, Lamb MJ, Jablonka E (1998) The role of DNA methylation in invertebrates : developmental regulation or genome defense? Mol Biol Evol 15:880–891
5. Selker EU, Tountas NA, Cross SH et al (2003) The methylated component of the Neurospora crassa genome. Nature 422:893–897. https://doi.org/10.1038/nature01564
6. Frommer M, Mcdonald LE, Millar DS et al (1992) A genomic sequencing protocol that yields a positive display of 5-methylcytosine residues in individual DNA strands. PNAS 89: 1827–1831
7. Cokus SJ, Feng S, Zhang X et al (2008) Shotgun bisulphite sequencing of the Arabidopsis genome reveals DNA methylation patterning. Nature 452:215–219. https://doi.org/10.1038/nature06745
8. Lister R, O'Malley RC, Tonti-Filippini J et al (2008) Highly integrated single-base resolution maps of the epigenome in Arabidopsis. Cell 133:523–536. https://doi.org/10.1016/j.cell.2008.03.029
9. Urich MA, Nery JR, Lister R et al (2015) MethylC-seq library preparation for base-resolution whole-genome bisulfite sequencing. Nat Protoc 10:475–483. https://doi.org/10.1038/nprot.2014.114
10. Tanaka K, Okamoto A (2007) Degradation of DNA by bisulfite treatment. Bioorg Med Chem Lett 17:1912–1915. https://doi.org/10.1016/j.bmcl.2007.01.040
11. Miura F, Enomoto Y, Dairiki R, Ito T (2012) Amplification-free whole-genome bisulfite sequencing by post-bisulfite adaptor tagging. Nucleic Acids Res 40:e136. https://doi.org/10.1093/nar/gks454
12. Raine A, Manlig E, Wahlberg P et al (2017) SPlinted ligation adapter tagging (SPLAT), a novel library preparation method for whole genome bisulphite sequencing. Nucleic Acids Res 45:1–15. https://doi.org/10.1093/nar/gkw1110
13. Smallwood SA, Lee HJ, Angermueller C et al (2014) Single-cell genome-wide bisulfite sequencing for assessing epigenetic heterogeneity. Nat Methods 11:817–820. https://doi.org/10.1038/nmeth.3035
14. Farlik M, Sheffield NC, Nuzzo A et al (2015) Single-cell DNA methylome sequencing and bioinformatic inference of epigenomic cell-state dynamics. Cell Rep 10:1386–1397. https://doi.org/10.1016/j.celrep.2015.02.001
15. Ji L, Sasaki T, Sun X et al (2014) Methylated DNA is over-represented in whole-genome bisulfite sequencing data. Front Genet 5:1–10. https://doi.org/10.3389/fgene.2014.00341
16. Olova N, Krueger F, Andrews S et al (2018) Comparison of whole-genome bisulfite sequencing library preparation strategies identifies sources of biases affecting DNA methylation data. Genome Biol 19:1–19. https://doi.org/10.1186/s13059-018-1408-2

17. Zhou L, Ng HK, Drautz-moses DI et al (2019) Systematic evaluation of library preparation methods and sequencing platforms for high-throughput whole genome bisulfite sequencing. Sci Rep 9:1–16. https://doi.org/10.1038/s41598-019-46875-5
18. Feng S, Zhong Z, Wang M, Jacobsen SE (2020) Efficient and accurate determination of genome - wide DNA methylation patterns in Arabidopsis thaliana with enzymatic methyl sequencing. Epigenetics Chromatin 1–17. doi: https://doi.org/10.1186/s13072-020-00361-9
19. Foox J, Nordlund J, Lalancette C et al (2021) The SEQC2 epigenomics quality control (EpiQC) study. Genome Biol 22:1–30. https://doi.org/10.1186/s13059-021-02529-2
20. Vaisvila R, Ponnaluri VKC, Sun Z et al (2021) Enzymatic methyl sequencing detects DNA methylation at single-base resolution from picograms of DNA. Genome Res 31:1280–1289. https://doi.org/10.1101/gr.266551.120
21. Han Y, Zheleznyakova GY, Marincevic-zuniga Y et al (2022) Comparison of EM-seq and PBAT methylome library methods for low-input DNA. Epigenetics 17:1195–1204. https://doi.org/10.1080/15592294.2021.1997406
22. Morrison J, Koeman JM, Johnson BK et al (2021) Evaluation of whole - genome DNA methylation sequencing library preparation protocols. Epigenetics Chromatin 14:1–15. https://doi.org/10.1186/s13072-021-00401-y
23. Chatterton Z, Lamichhane P, Ahmadi Rastegar D et al (2023) Single-cell DNA methylation sequencing by combinatorial indexing and enzymatic DNA methylation conversion. Cell Biosci 13:1–11. https://doi.org/10.1186/s13578-022-00938-9
24. Nass MMK (1973) Differential methylation of mitochondrial and nuclear DNA in cultured mouse, hamster and virus-transformed hamster cells in vivo and in vitro methylation. J Mol Biol 155–175
25. van der Wijst MGP, Rots MG (2015) Mitochondrial epigenetics: an overlooked layer of regulation? Trends Genet 31:353–356. https://doi.org/10.1016/j.tig.2015.03.009
26. Shao Z, Han Y, Zhou D (2023) Optimized bisulfite sequencing analysis reveals the lack of 5-methylcytosine in mammalian mitochondrial DNA. BMC Genomics 24:439. https://doi.org/10.1186/s12864-023-09541-9
27. Raddatz G, Guzzardo PM, Olova N et al (2013) Dnmt2-dependent methylomes lack defined DNA methylation patterns. Proc Natl Acad Sci USA 110:8627–8631. https://doi.org/10.1073/pnas.1306723110
28. Krauss V, Reuter G (2011) DNA methylation in drosophila — a critical evaluation, 1st edn. Elsevier Inc.
29. Williams L, Bei Y, Church HE, et al (2019) Enzymatic methyl-seq: the next generation of methylome analysis
30. Soriano-Tarraga C, Jimenez-Conde J, Giralt-Steinhauer E et al (2013) DNA isolation method is a source of global DNA methylation variability measured with LUMA. Experimental analysis and a systematic review. PLoS One 8:1–8. https://doi.org/10.1371/journal.pone.0060750
31. Martin M (2013) Cutadapt removes adapter sequences from high-throughput sequencing reads. EMBnet J 7:2803–2809. doi:https://doi.org/10.14806/ej.17.1.200
32. Krueger F, Andrews SR (2011) Bismark: a flexible aligner and methylation caller for bisulfite-Seq applications. Bioinformatics 27:1571–1572. https://doi.org/10.1093/bioinformatics/btr167
33. Langmead B, Salzberg SL (2012) Fast gapped-read alignment with bowtie 2. Nat Methods 9:357–359. https://doi.org/10.1038/nmeth.1923
34. Li H, Handsaker B, Wysoker A et al (2009) The sequence alignment/map format and SAMtools. Bioinformatics 25:2078–2079. https://doi.org/10.1093/bioinformatics/btp352
35. Ewels P, Magnusson M, Lundin S, Käller M (2016) MultiQC: summarize analysis results for multiple tools and samples in a single report. Bioinformatics 32:3047–3048. https://doi.org/10.1093/bioinformatics/btw354
36. Tommaso PDI, Chatzou M, Floden EW et al (2017) Nextflow enables reproducible computational workflows. Nat Biotechnol 35:316–319. https://doi.org/10.1038/nbt.3820
37. New England Biolabs I (2020) NEBNext enzymatic methyl-seq Kit, pp 1–32
38. Chen Y, Pal B, Visvader JE, Smyth GK (2017) Differential methylation analysis of reduced representation bisulfite sequencing experiments using edgeR. F1000Research 6:1–42. https://doi.org/10.12688/f1000research.13196.1
39. Garafutdinov RR, Galimova AA, Sakhabutdinova AR (2019) The influence of CpG (5′-d

(CpG)-3′ dinucleotides) methylation on ultrasonic DNA fragmentation. J Biomol Struct Dyn 37:3877–3886. https://doi.org/10.1080/07391102.2018.1533888

40. Hidvégi N, Gulyás A, Dobránszki J (2022) Ultrasound, as a hypomethylating agent, remodels DNA methylation and alters mRNA transcription in winter wheat (Triticum aestivum L.) seedlings. Physiol Plant:1–16. https://doi.org/10.1111/ppl.13777

41. NEB (2023) Performance data NEBNext UltraShear™

42. Ewels PA, Peltzer A, Fillinger S et al (2020) The nf-core framework for community-curated bioinformatics pipelines. Nat Biotechnol 38:276–278. https://doi.org/10.1038/s41587-020-0435-1

Chapter 6

MNase-seq to Identify Genome-Wide DNA-Protein Interactions

Benjamin R Carone

Abstract

Identification of the occupancy of transcription factors (TFs) and nucleosomes across the genome yields insights into the regulation of gene expression patterns. While several independent techniques can be performed and then analyzed in composite to reveal this chromatin landscape, the use of micrococcal nuclease (MNase) digestion can resolve the footprints of nearly all chromatin proteins simultaneously. The protocol below describes the use of MNase to identify chromatin footprints of both TFs and nucleosomes in two vastly different cell types, Mouse embryonic stem cells (mESCs) and sperm, with differing levels of chromatin compaction.

Key words Chromatin, Micrococcal nuclease, Nucleosome occupancy, MNase digestion, MNase-seq, Next-generation sequencing

1 Introduction

Deoxyribonucleic acid (DNA) is the primary heritable genetic material for all organisms. Encoded within the sequence of DNA are instructions for both the transcription of genes and the coding of proteins which are required for essential functions within the cell. However, the sequence of nucleic acids alone is not sufficient to direct the location, magnitude, and timing of gene expression patterns in the genome. To help organize the transcriptional patterns of genes, DNA associates with regulatory proteins and protein complexes, chromatin. These DNA-protein interactions are required for the establishment and maintenance of gene expression patterns within both prokaryotic and eukaryotic cells.

In eukaryotes, chromatin patterning at the primary level is facilitated by the wrapping of ~147 bp of DNA around an octamer of histone proteins, termed nucleosomes [1]. This packaging occurs regularly throughout the genome with relatively consistent spacing between nucleosomes and only a small region of unbound

Valerie J. Carabetta and Olaitan Akintunde (eds.), *High Throughput Gene Screening: Methods and Protocols*, Methods in Molecular Biology, vol. 2866, https://doi.org/10.1007/978-1-0716-4192-7_6,

DNA, ~10–18 bp in between referred to as linker DNA [2]. The most significant exception to this pattern of regular spacing is the transcriptional start site (TSS), in which the DNA in the promoter region upstream of the TSS remains unbound by nucleosomes, the nucleosome free region (NFR). The presence and size of this unbound region has been demonstrated to correlate with specific DNA sequences, the binding of transcription factors, and the overall level of gene expression [3]. Moreover, a careful look at the interplay of these protein interactions at the promoter region yields insights into the mechanisms behind transcriptional regulation of these genes [4]. Therefore, the identification of the global chromatin pattern within cells often yields insights into the transcriptional regulation of genes.

Several methodologies have been developed to measure DNA-protein interactions within the genome, including affinity-based approaches such as chromatin immunoprecipitation (ChIP) [5], DamID [6], and DNA protection-based approaches. For the latter approach, the addition of exogenously added enzymes or chemicals can detect the accessibility of the DNA by creating breaks in the DNA backbone in regions which are not protected through DNA-protein interactions. For enzymes that induce dsDNA breaks, several different molecules are commonly used, including restriction enzymes [7], DNase I [8], micrococcal nuclease (MNase) [9], and DNA transposases [10]. In all these cases, chromatin is exposed to the DNA cleavage enzyme(s) which results in digestion of dsDNA that is not associated with proteins, which can be subsequently analyzed by gel electrophoresis, PCR amplification, or next-generation sequencing (NGS).

Of all the DNA cleavage enzymes, MNase is the most widely utilized to identify nucleosome positions at high resolution within the genome. MNase preferentially binds to and creates dsDNA breaks within the linker regions of DNA (Fig. 1). Despite primarily making cuts at sites of unoccupied DNA, MNase also demonstrates a preference for AT-rich DNA sequences [11]. When performed on formaldehyde crosslinked or native chromatin, this digestion yields a combination of mono, di, and tri-nucleosomal fragments appearing as a ladder in gel electrophoresis (Fig. 2). Variations in the amount of MNase used and the length of the exposure to MNase results in differences in fragment length and the proportion of mono-nucleosome to multi-nucleosome fragments. Furthermore, differential solubility of chromatin fragments to salt solutions following MNase digestion can yield different populations of nucleosomal DNA fragments [12]. Moreover, the isolation of and mapping of mono-nucleosomes from various levels of digestion can generate significant variation in positions of mapped nucleosomes. Taken together, early experiments in genome-wide mapping of nucleosomes varied significantly in methodology and reproducibility. In contrast, recent approaches use multiple levels of

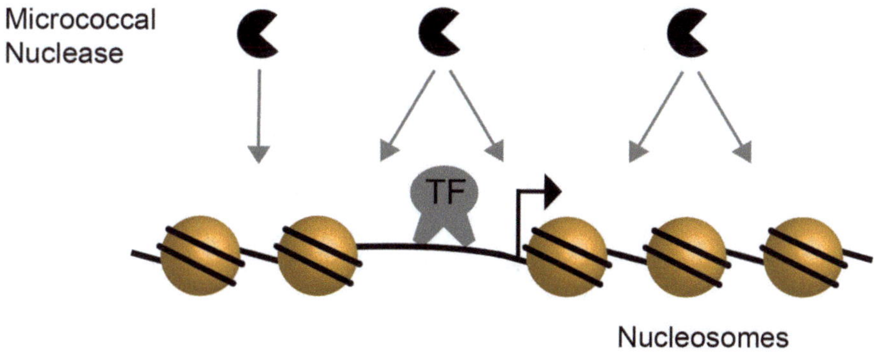

Fig. 1 Principle behind chromatin accessibility by MNase. Micrococcal nuclease can bind to regions of DNA that are not protected by nucleosomes or transcription factors (TF). These regions include areas such as linker DNA, in between adjacent nucleosomes, as well as area surrounding TFs within the NFR present in the promoter of most eutherian genes. NFR: nucleosome free regions

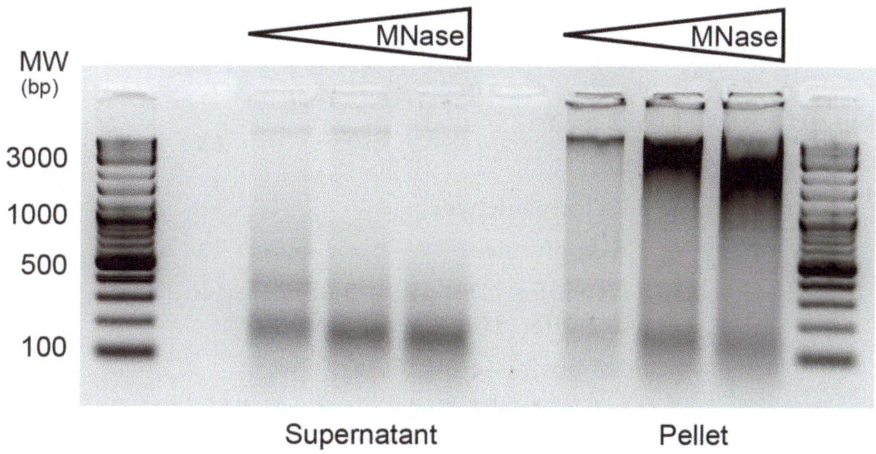

Fig. 2 Typical digestion patterns of eutherian chromatin with MNase. Lanes 1 and 10 contain NEB 1Kb Plus DNA ladder. Mouse embryonic stem cells (mESCs) were digested with various amounts of MNase, 2, 10, 50 units from left to right for lanes 3, 4, and 5, and the same for 7, 8, and 9. Following digestion, and before DNA extraction, insoluble chromatin was pelleted by centrifugation at 4000× g for 5 min. DNA was extracted from the supernatant (lanes 3, 4, and 5), the remaining pellets were resuspended in DNA extraction buffer, and DNA was isolated (lanes 7, 8, and 9)

MNase titration, with collection of all digestion fragments regardless of size to generate the most consistent and complete nucleosome mapping data [13].

2 Materials

2.1 MNase Digestion

1. 10 cm dish.
2. 15 mL conical tubes.
3. 1.5 mL microcentrifuge tubes.

4. 37% formaldehyde.
5. 2.5 M glycine.
6. 37 °C water bath.
7. Phosphate buffered saline (PBS).
8. 0.05% Trypsin/Ethylenediaminetetraacetic acid (EDTA).
9. Cell Lysis Buffer: (10 mM Tris–HCl pH 7.5, 3 mM $MgCl_2$, 0.5% NP-40, 10 mM NaCl, Protease Inhibitor (EDTA-free complete mini—1 tablet (Roche)).
10. 1 M $CaCl_2$.
11. 0.5 M EGTA.
12. Micrococcal Nuclease (MNase).
13. Proteinase K (20 mg/mL).
14. 10% SDS solution in water.
15. 25:24:1 phenol:chloroform:isoamyl alcohol.
16. Ethanol (EtOH).

2.2 Isolation and MNase Digestion of Sperm DNA

1. M2 media.
2. PBS.
3. 37% formaldehyde.
4. 2.5 M glycine.
5. TLS buffer (450 μL of 0.05% Trypsin EDTA, 450 μL of cell lysis buffer, 100 μL of 0.1 M DTT).
6. Fetal bovine serum (FBS).
7. 1 mL syringe with 26-guage needle.
8. 0.5 mM EGTA.
9. 5 M NaCl.
10. Micrococcal Nuclease (MNase).
11. Proteinase K (20 mg/mL).
12. 25:24:1 phenol:chloroform:isoamyl alcohol.
13. Ethanol (EtOH).

2.3 Ilumina Paired-End Sequencing

1. Calf intestinal alkaline phosphatase (CIP).
2. End it Repair Kit (recommended, EPICENTER Biotechnologies).
3. Thermocycler.
4. 25:24:1 phenol:chloroform:isoamyl alcohol.
5. EtOH.
6. exo Klenow (recommended, EPICENTRE Biotechnologies).

7. Fast-Link ligation kit (recommended, EPICENTRE Biotechnologies).
8. Ampure Beads (Agencourt).
9. Platinum Pfx (Invitrogen).

3 Methods

The following protocol will take the user from the preparation of a biological sample to the preparation of NGS libraries for mapping of TF and nucleosome positions. The type of chromatin preparation from various cell types as well as the amount of MNase digestion on chromatin can have a dramatic effect on final mapping results. Before proceeding with protocol, it is recommended to *see* **Note 4** on levels of digestion.

3.1 MNase Digestion of Mouse Embryonic Stem Cells (mESCs), HeLa, or Fibroblasts

1. Starting from one confluent 10 cm dish about 3×10^6 cells, detach adherent cells using trypsin with a time and concentration dependent on cell type to completely lift all cells from plate. A concentration of 0.05% Trypsin/EDTA in PBS for 3 min at 37 °C will detach most cells.
2. Transfer cells to a 15 mL conical tube with 10 mL of cell-specific growth media.
3. Add 270 μL of 37% formaldehyde (1% final concentration) and incubate for 15 min at 37 °C (*see* **Note 1**).
4. Quench the reaction with 1 mL of 2.5 M glycine for 5 min at 37 °C. Cells can be used immediately after crosslinking or saved for later (*see* **Note 2**).
5. Pellet cells 1000 × g for 5 min, wash 1× with PBS, resuspend cell pellet in 1.5 mL cell lysis buffer.
6. Transfer 1.5 mL of cell suspension into a new microcentrifuge tube and add 5 μL of 1 M $CaCl_2$ and rotate at 4 °C for 15 min (*see* **Note 3**).
7. Split into 3 tubes, 500 μL each and incubate the sample for 4 min at 37 °C in a water bath.
8. Add 2, 10, or 50 Units of MNase to tubes, mix by pipetting, and incubate the samples for 4 min at 37 °C in a water bath (*see* **Note 4**).
9. Stop the MNase reaction with addition of 10 μL of 0.5 M EGTA and put on ice (*see* **Note 5**).
10. Add 10 μL of 20 mg/mL Proteinase K and 20 μL of 10% SDS overnight at 65 °C, followed by standard 25:24:1 phenol:chloroform:isoamyl alcohol (PCI) and EtOH extraction of DNA.

11. Check the DNA fragment sizes using gel electrophoresis (low-resolution) or a capillary approach (high-resolution) to identify levels of DNA digestion before moving forward. Results should be similar to the digestion patterns for mono, di, tri-nucleosomes shown for the supernatant fractions in Fig. 2 (*see* **Note 6**).

3.2 MNase Digestion of Mammalian Sperm

Mammalian sperm is highly compacted by the replacement of histones with protamines during the process of spermatogenesis. Furthermore, in many eutherians, disulfide bonds formed between cysteines within protamines will further compact chromatin and prevent accessibility of MNase to open DNA regions. The following protocol is adapted from Carone et al. [14] as a method to perform MNase digestion on mouse sperm. More or less decondensation of sperm chromatin may be required for sperm from different organisms with lower and higher levels of chromatin compaction, such as humans or rats, respectively.

3.2.1 Isolating Sperm from Mus Musculus

1. Dissect out the cauda epididymis and vas deferens and place into 1 mL of 37 °C pre-gassed standard M2 media in microcentrifuge tube. Cut or pull apart this tissue to allow the sperm to swim out.
2. Incubate the sperm for 30 min at 37 °C.
3. Centrifuge the sample for 1 min at 110 × *g*.
4. Take supernatant and place into new tube avoiding any contaminating pieces of epididymal tissue.
5. Centrifuge the sperm sample at 4000 × *g* for 3 min and discard the supernatant.
6. Resuspend the sperm sample in 1 mL of water for 5 min at 37 °C. This step lyses any residual epithelial cells but leaves the sperm intact.
7. Centrifuge the sperm sample at 4000 × *g* for 3 min and discard supernatant.
8. Resuspend sample in 1 mL of PBS.
9. Add 27 μL of formaldehyde (or to a final concentration of 1%) and incubate for 15 min at 37 °C.
10. Add 100 μL of 2.5 M glycine and incubate for 5 min at 37 °C. Cells can be used immediately after crosslinking or saved for later (*see* **Note 2**).

3.2.2 MNase Digestion of Mus Sperm

Note 6 is critical, so read before proceeding.

1. Centrifuge the sperm sample at 4000 × *g* for 3 min and discard the supernatant.

2. Resuspend the pellet in 1 mL of PBS, record OD_{600} value of sperm sample for later (*see* **Note 7**).
3. Centrifuge the sperm sample at 4000 × *g* for 3 min and discard the supernatant.
4. Resuspend in 1 mL TLS buffer for 5 min at 37 °C.
5. Add 10 μL of FBS to inactivate the trypsin and incubate for 5 min at 37 °C.
6. Pipette up and down using a 26-gauge needle in a 1 mL syringe 20 times.
7. Centrifuge the sperm sample at 4000 × *g* for 3 min and discard the supernatant.
8. Resuspend in 1 mL of PBS supplemented with 2 mM $CaCl_2$ (final concentration).
9. Incubate the sample for 4 min at 37 °C in a water bath.
10. For the MNase digestion, add 8 units of enzyme to a sperm sample that is at an OD_{600} of 0.80, so about 10 U/OD on average. This produces 80% mono-nucleosomes and 20% di-nucleosomes. The digestion should be 4 min at 37 °C in a water bath.
11. The reaction is immediately stopped with the addition of 20 μL of 0.5 M Ethylene glycol-bis(2-aminoethylether)-N,N,N′,N′--tetraacetic acid (EGTA) and put on ice.
12. Add NaCl to a final concentration of 600 mM and incubate sample for 2 h at 4 °C, preferably rotating (*see* **Note 8**).
13. Add 10 μL of 20 mg/mL Pro K and 20 μL of 10% SDS overnight at 65 °C followed by standard 25:24:1 phenol:chloroform:isoamyl alcohol and EtOH extraction of DNA.
14. Check the DNA fragment sizes using gel electrophoresis (low-resolution) or a capillary approach (high-resolution) to identify levels of DNA digestion before moving forward. Sperm samples will likely **NOT** be like the digestion patterns for mono, di, tri-nucleosomes shown for the supernatant fraction in Fig. 2, but rather will most likely show primarily mono-nucleosomal fragments.

3.3 Library Preparation Protocol for Illumina Sequencing (Retaining Small Fragments)

The following protocol is used to prepare NGS libraries for Illumina sequencing. Kits from different manufacturers could be used for this protocol, with the most important considerations being: (1) At no point in the process should DNA fragment sizes <25 bp (original digestion fragment size) be excluded from the protocol by DNA purification and (2) To resolve fragment sizes post-sequencing, paired-end sequencing must be performed.

3.3.1 Creating Blunt-End DNA Fragments (End IT Repair Kit)

1. Combine the following in a 0.5 mL PCR tube:
   5 μL 10× buffer
   5 μL dNTP mix (2.5 mM)
   5 μL 10 mM ATP
   1 μL enzyme mix
   1000 ng CIP-treated and PCl purified nucleosomal DNA fragments.
2. Fill tube to 50 μL with ddH_20.
3. Incubate 1 h at 23 °C in a thermocycler.
4. Purify DNA with clean up with PCI Extraction and EtOH extraction.

3.3.2 Adding Single 3′ Adenine Tail

1. Combine the following in a 0.5 mL PCR tube:
   (a) 30 μL DNA solution
   (b) 5 μL 10× Klenow buffer
   (c) 1 μL 10 mM dATP
   (d) 14 μL ddH_20
   (e) 1 μL Klenow –exo
2. Incubate at 23 °C for 45 min in a thermocycler.
3. Purify DNA with clean up with PCI Extraction and EtOH extraction.

3.3.3 Addition of Illumina Adaptors by Ligation (Fast-Link Ligation Kit, See *Note 9)*

1. Combine the following in a 0.5 mL PCR tube:
   10 μL DNA solution
   1.5 μL 10× ligation buffer
   0.75 μL 10 mM ATP
   2 μL Illumina adaptors
   1 μL ligase
2. Incubate at 23 °C for 60 min in a thermocycler.
3. Add to the same tube:
   7.5 μL ddH20
   1 μL 10× buffer
   0.5 μL 10 mM ATP
   1 μL ligase.
4. Incubate at 16 °C for 12–16 h.
5. Clean up with Ampure beads, as per manufacturer instructions and at a 2:1 ratio of beads to sample.

3.3.4 PCR Amplification of DNA Fragments with Adapters

1. Combine in the following in a 0.5 mL PCR tube:
   30 μL DNA template
   1 μL each Illumina primer
   10 μl 10× Pfx buffer (Invitrogen)
   3 μL 10 mM dNTPs
   2 μL 50 mM $MgSO_4$
   1 μL Platinum Pfx
   53 μL ddH20.
2. Using a thermocycler, run the following program:
   1. 94oC for 2 min
   2. 94oC for 15 s
   3. 65oC for 1 min
   4. 68oC for 30 s
   5. Repeat from step 2, 18 times.
   6. 68oC for 5 min.
3. Clean up with Ampure beads as per manufacturer instructions and at a ratio of 2:1 beads to sample.

3.4 Sequencing and Analysis of Mapping Data

Check the DNA fragment sizes using gel electrophoresis (low-resolution) or a capillary approach (high-resolution) to identify sizes of prepared library before sequencing. Fragment sizes should range from around 100 bp to 500 bp, with a strong peak at the mono-nucleosome plus adapter fragment around 250 bp. Even at lower levels of digestion, the PCR amplification will enrich for <300 bp fragments. Most NGS library preparation protocols indicate a tight size restriction for sequencing of fragments, but here it is important to specify that there should be no size selection of DNA library fragments.

Various software packages, including Bowtie2 [15] or HiSat2 [16], are capable of mapping DNA fragments back to the genome and there is no specific preference for which package to utilize. The only consideration is that after mapping the paired-end reads to the genome, before subsequent analysis, read-length distributions should be identified, and TF and nucleosomes should be partitioned into separate classes.

4 Notes

1. Can also be performed without crosslinking with minimal final differences, except for solubility of chromatin. It is not a concern if the DNA is extracted from the entire digest. Do not separate the soluble and pellet fractions of digest.

2. To save for later, cells can be pelleted at 1000× *g* for 5 min, resuspended in 1 mL PBS, transferred to microcentrifuge tube, pelleted again (discard supernatant) and flash frozen with liquid nitrogen.
3. MNase enzyme activity is dependent on the addition of $CaCl_2$. Adding this during the cell lysis step and during mixing allows ideal MNase conditions in place prior to addition of enzyme. Mixing can be accomplished using either a lateral shaker or rotator (tubes are tumbling).
4. The final degree of chromatin digestion is very important to the type and quality of DNA fragments that will be used to identify nucleosome mapping. As described in detail by Henikoff et al. [13], a lower level of digestion (when the tri-nucleosomal fragment is clearly visible as in lane 1 of Fig. 2) will more clearly enrich for bound TFs, while a higher level of digestion (when the mono-nucleosomal fragment (~150 bp) represents >85% of DNA as in lane 3 of Fig. 2) will more clearly delineate NFR regions and first downstream and upstream nucleosome positions. It is recommended to collect digests from both a low and high level of digestion, perform paired-end NGS on both samples individually, and discriminate fragment sizes using bioinformatics approaches post-sequencing.
5. MNase can be purchased from various suppliers as either a liquid or solid form. If the manufacturer has recommended concentrations for performing digestion of chromatin, use those values as a starting point. However, typically we purchase or prepare an MNase stock solution at 10 U/μL, then aliquot this into multiple tubes and store at -80 °C long term. Using the suggested units in the protocol above as a starting point, researchers should titrate their MNase to achieve digestion patterns similar to lanes 1, 2, and 3 in Fig. 2 and then add this amount of MNase for future experiments. Typically, a specific lot of stock solution can be relied on to perform with similar efficiencies after being stored at -80 °C for several years.
6. RNA will obscure the DNA banding patterns on gel. It is advisable and highly recommended to RNase treat following DNA extraction. Banding patterns are clearly defined when run for 1 h at 90 V on a 10 cm 2% Agarose in TAE buffer gel.
7. Here you need to take an OD_{600} reading of your sample so that you can add the correct amount of MNase to your sample later in the protocol.
8. Sperms are refractory to lysis and may not completely break apart allowing soluble chromatin to reverse crosslinking and leave the cell. Incubation in high salt after MNase digestion improves yield of digested DNA fragments.

9. Depending on which platform is used for sequencing, different adapters will be required for different samples and should be identified based on the NGS instrument to be used. For non-sized selected MNase-seq, 100 million paired-end reads are the minimum number of reads for a mammalian genome with ~200 million being the ideal median number of DNA fragments sequenced.

References

1. Noll M, Kornberg RD (1977) Action of micrococcal nuclease on chromatin and the location of histone H1. J Mol Biol 109(3):393–404
2. Luger K et al (1997) Crystal structure of the nucleosome core particle at 2.8 a resolution. Nature 389(6648):251–260
3. Chen X et al (2008) Integration of external signaling pathways with the core transcriptional network in embryonic stem cells. Cell 133(6):1106–1117
4. Radman-Livaja M, Rando OJ (2010) Nucleosome positioning: how is it established, and why does it matter? Dev Biol
5. O'Neill LP, Turner BM (1996) Immunoprecipitation of chromatin. Methods Enzymol 274:189–197
6. Greil F, Moorman C, van Steensel B (2006) DamID: mapping of in vivo protein-genome interactions using tethered DNA adenine methyltransferase. Methods Enzymol 410: 342–359
7. Lipchitz L, Axel R (1976) Restriction endonuclease cleavage of satellite DNA in intact bovine nuclei. Cell 9(2):355–364
8. Noll M (1974) Internal structure of the chromatin subunit. Nucleic Acids Res 1(11): 1573–1578
9. Hewish DR, Burgoyne LA (1973) Chromatin sub-structure. The digestion of chromatin DNA at regularly spaced sites by a nuclear deoxyribonuclease. Biochem Biophys Res Commun 52(2):504–510
10. Buenrostro JD et al (2013) Transposition of native chromatin for fast and sensitive epigenomic profiling of open chromatin, DNA-binding proteins and nucleosome position. Nat Methods 10(12):1213–1218
11. Allan J et al (2012) Micrococcal nuclease does not substantially bias nucleosome mapping. J Mol Biol 417(3):152–164
12. Bhargava BL, Yasaka Y, Klein ML (2011) Computational studies of room temperature ionic liquid–water mixtures. Chem Commun 47(22):6228–6241
13. Henikoff JG et al (2011) Epigenome characterization at single base-pair resolution. Proc Natl Acad Sci USA 108(45):18318–18323
14. Carone BR et al (2014) High-resolution mapping of chromatin packaging in mouse embryonic stem cells and sperm. Dev Cell 30(1):11–22
15. Langmead B, Salzberg SL (2012) Fast gapped-read alignment with Bowtie 2. Nat Methods 9(4):357–359
16. Kim D et al (2019) Graph-based genome alignment and genotyping with HISAT2 and HISAT-genotype. Nat Biotechnol 37(8): 907–915

Chapter 7

ATAC-seq for Characterizing Host and Pathogen Genome Accessibility During Virus Infection

Dawei Liu, Timothy R. Howard, and Ileana M. Cristea

Abstract

Chromatin regulation provides a mechanism through which cells dynamically and rapidly regulate their gene expression profiles, playing a pivotal role in diverse biological processes and disease states. The Assay for Transposase-Accessible Chromatin with high-throughput sequencing (ATAC-seq) is a method that enables genome-wide detection of accessible chromatin regions, providing information on nucleosome positioning and the epigenetic regulation of the chromatin structure. ATAC-seq has been used in various biological contexts, and several reports have demonstrated its application to studying infections with viral or bacterial pathogens. The ability to characterize changes in viral or bacterial genome accessibility during infections provides insights into both pathogen replication and host defense mechanisms. Viral genomes undergo dynamic changes in their structural landscape to facilitate replication and evade host immune responses. Additionally, host cells encode DNA sensors, which are specialized proteins that bind to viral genomes to initiate innate immune responses and sometimes, to suppress viral gene expression. ATAC-seq enables the systematic detection of key structural changes on the viral genome mediated by either viral or host proteins, offering mechanistic insights into virus-host interactions. Here, we describe an ATAC-seq method optimized for studying changes in chromatin accessibility in both host and viral genomes. We have previously applied this method to demonstrate a systematic decrease in the genome accessibility of herpes simplex virus type I (HSV-1) enabled by a host antiviral factor, the interferon-gamma inducible protein 16 (IFI16) during infection of human fibroblasts. This protocol can be adapted to various biological contexts involving the introduction of foreign DNA, making it a valuable tool for a broad range of research endeavors.

Key words ATAC-seq, qPCR, Infection, Chromatin accessibility, Normalization, Cell lysis, Transposition, Library generation, Virus infection

1 Introduction

Epigenetic regulation is a major mechanism through which cells dynamically regulate their gene expression profiles in response to various conditions or stimuli, such as during development or changing environmental cues [1, 2]. This regulation is achieved through the compaction or relaxation of the chromatin structure by mechanisms including histone modification, nucleosome

Valerie J. Carabetta and Olaitan Akintunde (eds.), *High Throughput Gene Screening: Methods and Protocols*,
Methods in Molecular Biology, vol. 2866, https://doi.org/10.1007/978-1-0716-4192-7_7,

positioning, DNA methylation, and the action of chromatin remodeling complexes [3–5]. The open, permissive, or closed state of the chromatin can dictate the accessibility of the genome to transcription factors and, thus, the transcriptional activity of the chromatin region. Hence, chromatin accessibility has become a readout of chromatin organization and regulatory capacity, providing a means for understanding dynamic, genome-wide changes in gene regulation [6]. Techniques such as DNase-seq (DNase I hypersensitive sites sequencing) [7], MNase-seq (micrococcal nuclease digestion and sequencing) [8], NOMe-seq (nucleosome occupancy and methylome sequencing) [9], and FAIRE-seq (Formaldehyde-Assisted Isolation of Regulatory Elements and sequencing) [10] have been developed to elucidate the structural landscapes of chromatin. Among the chromatin profiling methods, the assay for transposase-accessible chromatin with high-throughput sequencing (ATAC-seq) has gained substantial recognition due to its ability to interrogate the dynamic landscape of chromatin accessibility with high sensitivity and resolution [11]. ATAC-seq leverages the hyperactive Tn5 transposase enzyme to simultaneously fragment chromatin and integrate sequencing adapters into open chromatin regions. Subsequent high-throughput sequencing and mapping of the insertion ends allows for a systematic characterization of the active regulatory elements, including promoters, enhancers, and other DNA regulatory regions.

Since its inception in 2013 by Buenrostro et al. [11], ATAC-seq has proven to be a versatile and powerful method for various applications (Fig. 1). A common application is nucleosome mapping, allowing for the precise determination of nucleosome positions on chromatin [12]. Another application is transcription factor binding analysis, enabling the characterization of transcription factors and their respective binding locations [13]. The reads obtained from the accessible regions of the chromatin can also facilitate the discovery of novel enhancer regions, allowing for in-depth analysis of regulatory elements driving gene expression patterns [14, 15]. From a medically relevant standpoint, ATAC-seq can help elucidate the alterations in chromatin accessibility in the contexts of cancer [16], neurodegeneration [17], and other diseases [18–20], and the distinct chromatin landscapes in certain disease states hold potential for biomarker discovery [21].

An important human disease research area where ATAC-seq holds great promise is that of viral infections (Fig. 2). During infections, viral genomes can undergo dynamic changes in their structural landscape to facilitate efficient replication [22]. For example, they can recruit viral or host transcription factors for viral gene expression and interact with chromatin remodeling complexes to temporally regulate the distinct stages of the virus replication cycle [23]. In some cases, viral genomes can also undergo chromatinization by seizing control of host histones to regulate

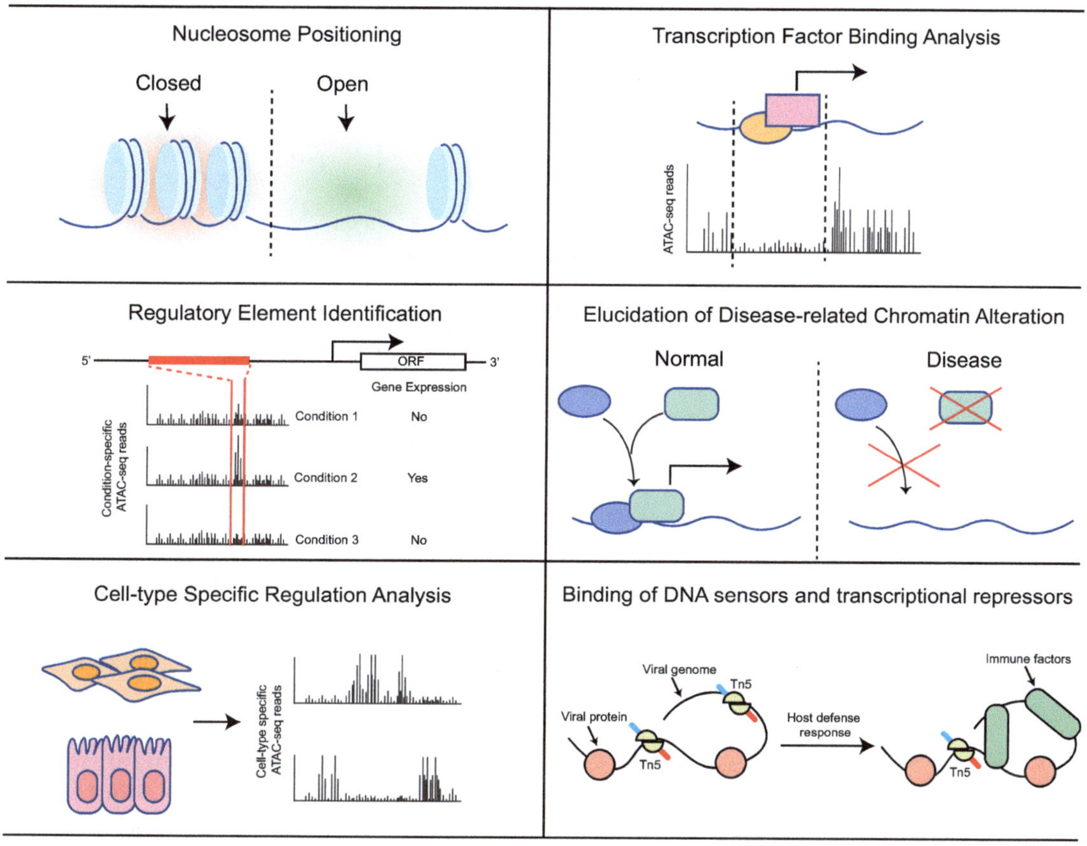

Fig. 1 Examples of applications enabled by ATAC-seq

latency and evade immune detection [24, 25]. In these instances, ATAC-seq can serve as a valuable tool to elucidate key structural changes in the viral genome, thus providing mechanistic understanding of viral replication processes and revealing regulatory elements that promote or inhibit viral infections.

ATAC-seq can also be utilized to investigate host responses to viral infections. Cells encode various transcriptional regulators, such as histone deacetylases and DNA methyltransferases, which can alter both the host and viral genomic landscape during infections [26–28]. On the other hand, the presence of the foreign nucleic acids can lead to the activation of DNA sensors, which can bind to viral DNA and elicit potent expression of antiviral interferons and cytokines [29, 30]. Some DNA sensors also have the capacity to directly suppress viral transcription, likely through direct occupancy of the viral DNA or recruitment of other transcriptional repressors [31–38]. One example of such a DNA sensor is the interferon gamma-inducible protein 16 (IFI16), which we have previously shown to be sufficient to suppress transcription in vitro, as well as capable of systematically reducing viral genome accessibility and a subsequent global decrease in viral gene

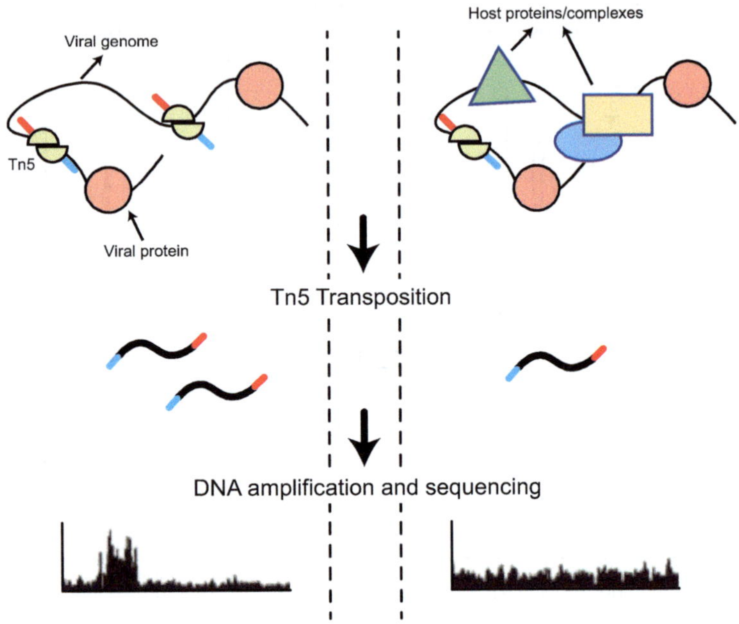

Fig. 2 Workflow schematics illustrating changes in ATAC-seq reads as a result of altered viral genome accessibility mediated by host proteins

expressions [32, 33]. In these cases, changes to chromatin accessibility revealed by ATAC-seq can unravel novel biology about intrinsic and innate immune responses and provide a comprehensive view of the regulatory landscape associated with DNA sensing during infection. Furthermore, by integrating ATAC-seq data with other genomic, transcriptomic, and proteomic datasets, we can obtain multidimensional information on the interplay between cellular defense responses and viral-induced pathogenesis, which can facilitate the identification of targets for therapeutic intervention. The versatile nature of ATAC-seq positions it as a valuable tool for dissecting the complex network of DNA sensing pathways during viral infections and for advancing our understanding of virus-host interactions.

Here, we provide a protocol for conducting ATAC-seq experiments in the context of virus infection, emphasizing its application to deciphering the DNA sensing landscape during viral infection. Adapting the original ATAC-seq method and including modifications from Ackermann et al. [39], we outline the experimental design, sample preparation, library construction, and normalization strategies. This method can be utilized to not only gain insights into the complex interplay between viruses and the host immune system, but also to shed light onto other conditions where foreign DNA is introduced into host cells, informing about possible strategies for therapeutic intervention.

2 Materials

2.1 Cell Preparation

1. Phosphate buffered saline (PBS).
2. Trypsin for cell culture (if adherent cells are used).
3. Cell culture media (e.g., Dulbecco's Modified Eagle Medium).
4. Hemacytometer.
5. Microcentrifuge tubes.

2.2 Cell Lysis, Transposition, and DNA Purification

1. Nuclease-free water.
2. Refrigerated centrifuge.
3. ATAC wash buffer: 10 mM Tris-HCl, pH 7.5, 10 mM NaCl, 3 mM MgCl2, 0.1% (v/v) NP-40, 0.1% (v/v) Tween 20, 0.01% (v/v) digitonin (*see* **Note 1**).
4. Tagment DNA TDE1 Enzyme and Buffer Kits (Illumina).
5. Transposition reaction mix: 1× Tagment DNA buffer, 33% (v/v) PBS, 0.1% (v/v) Tween 20, 0.01% (v/v) digitonin, 5% (v/v) Tagment DNA enzyme 1 (*see* **Note 2**).
6. Thermomixer.
7. DNA purification kit (*see* **Note 3**).

2.3 Assessing Amplification Cycles by quantitative, real-time polymerase chain reaction (qPCR) and Library Generation by PCR

1. PCR tubes.
2. PCR primers (*see* **Note 4**).
3. PCR thermocycler.
4. qPCR master mix [e.g., SYBR Green Master Mix (Thermo Fisher Scientific)].
5. NEBNext High-Fidelity 2× PCR Master Mix (New England Biolabs).
6. qPCR instrument.
7. Qubit fluorometer and respective assay kit.
8. Agilent Bioanalyzer High-Sensitivity DNA Analysis kit.
9. Bioanalyzer.

3 Methods

3.1 Cell Preparation and Infection

1. Typically, 50,000–100,000 cells are used for this protocol. It is advisable to seed three times as many cells (i.e., 150,000–300,000 cells) in the appropriate tissue culture vessel to arrive at an appropriate density for infection or for the type of perturbation that one wishes to apply (*see* **Note 5**).

2. Harvest and count cells as per standard protocols, including but not limited to trypsinization and scraping for adherent cells or centrifugation for cells in suspension. Split 50,000–100,000 cells for the experiment.
3. Pellet the cells by centrifuging at 350× g for 5 min. Wash the cells once with 250 μL of ice-cold 1× PBS. Centrifuge again to remove supernatant. Place the cell pellet on ice.
4. Pellet the remaining cells by centrifuging at 350× g for 5 min. The cell pellet can be frozen and stored at −20 °C. This cell pellet can be used for parallel qPCR quantification, such as viral genome quantification for normalization purposes.

3.2 Cell Lysis, Transposition, and DNA Purification

1. Wash the cells with 1 mL ATAC wash buffer by pipetting gently three times to resuspend and invert three times.
2. Centrifuge at 500× g for 5 min at 4 °C.
3. Remove and discard the supernatant.
4. Repeat steps 1–3 two more times for a total of three washes.
5. Resuspend the cell pellet in 50 μL of transposition reaction mix. Pipet gently six times to ensure homogenous mixing.
6. Incubate at 37°C for 30 min on a thermomixer at 1000 rpm.
7. Purify the transposed DNA with a DNA purification kit.
8. Elute the DNA in 10 μL of elution buffer.

3.3 Assessing Amplification Cycles by qPCR and Library Generation by PCR

1. Combine the following in a PCR tube for each sample for a total of 50 μL:

   10 μL of purified transposed DNA.

   10 μL of nuclease-free H_2O.

   2.5 μL of PCR Primer 1 (25 μM).

   2.5 μL of barcoded PCR primer 2 (25 μM)

   25 μL of NEBNext High-Fidelity 2× PCR Master Mix.
2. Perform PCR amplification using the following program:

1 cycle	72 °C for 5 min 98 °C for 30 s
5 cycles	98 °C for 10 s 63 °C for 30 s 72 °C for 1 min

To minimize PCR bias during PCR library generation, the number of amplification cycles need to be determined using qPCR to avoid saturation.

3. Remove PCR tubes from the thermocycler. Remove 5 μL of the reaction mix and combine the following to make a qPCR reaction mix in a final volume of 15 μL (*see* **Note 6**):

   5 μL of PCR mix.

   1.5 μL of nuclease-free water.

   0.5 μL of PCR Primer 1 (25 μM).

   0.5 μL of barcoded PCR primer 2 (25 μM).

   7.5 μL of SYBR Green qPCR Master Mix

4. Perform qPCR using the following program:

1 cycle	98 °C for 30 s
20 cycles	98 °C for 10 s 63 °C for 30 s 72 °C for 1 min

5. Plot R_n values against cycle number to obtain the number of cycles N needed to achieve one-third of the maximum R_n. Depending on the sample amount, the number of cycles needed can vary greatly, ranging from 8 to more than 20 cycles.
6. Run PCR for the remaining 45 μL of the sample for the appropriate number of cycles as determined by qPCR.

1 cycle	98 °C for 30 s
N cycles	98 °C for 10 s 63 °C for 30 s 72 °C for 1 min

7. Purify the amplified DNA library using DNA purification kit. Elute in 20 μL of elution buffer.
8. Make a 1:4 dilution of the library by adding 1 μL of purified library to 3 μL of nuclease-free water.
9. Use 1 μL of the dilution to measure DNA concentration using Qubit fluorometer.
10. To assess the quality of the purified library, run 1 μL of the dilution on the bioanalyzer using the Bioanalyzer High-Sensitivity DNA Analysis kit (*see* **Note 7**).
11. Pool all libraries at equimolar concentrations and perform 50 bp paired-end sequencing.

3.4 Normalization of Reads by Both Sequencing Depth and Viral Genome Content

One key feature to consider when working with infected human cells is the changing genome content due to viral replication throughout the course of infection. Therefore, it is necessary to normalize sequencing reads by both the sequencing depth, as is the

Table 1
Example calculations for normalization factors that account for both sequencing depth and changes in viral genome content

Sample name	Sample 1	Sample 2	Sample 3
Mapped reads to human genome	A	B	C
Mapped reads to viral genome	D	E	F
Total mapped reads	A + D	B + E	C + F
Sequencing depth normalization factor	$1e^6/(A + D)$	$1e^6/(B + E)$	$1e^6/(C + F)$
Viral genome normalization (from qPCR)	1 (reference)	X	Y
Final normalization factor	$1e^6/(A + D)$	$X*1e^6/(B + E)$	$Y*1e^6/(C + F)$

case for all ATAC-seq experiments, as well as the viral genome content, which may vary drastically across different sample conditions. Here we describe one method of normalization, which integrates both the sequencing depth and the changes in viral genome copies across samples, which we used in [32]. In this method, the normalization factor is derived from the product of the normalization factor for sequencing depth and the normalization factor for viral genome copies. An example of a calculation is shown in Table 1.

1. To obtain the normalization factor for sequencing depth, obtain the total mapped reads, given by the sum of mapped reads to the human genome and mapped reads to the viral genome. The normalization factor is then $1e^6/$ total mapped reads (Table 1, row 5).
2. To obtain the normalization factor for viral genome copies, use qPCR to measure the relative quantity of viral genomes using the ΔΔCt method and specific primers targeting the viral genome (*see* **Note 8**). By first normalizing to a reference sample, a normalization factor is determined that accounts for differences in viral genome content (Table 1, row 6).
3. Finally, multiply both normalization factors to obtain the final normalization factor for ATAC-seq (Table 1, row 7).

3.5 Key Considerations

1. It is crucial to perform qPCR to determine the minimum number of cycles required to amplify the library. This helps to reduce qPCR bias that come during later cycles of qPCR, which may interfere with the biological signal of interest.
2. If ATAC-seq is applied to study viral or bacterial infections, it is important to first perform a pilot experiment where only uninfected samples are used to determine the efficiency of transposition. Specifically, one needs to ensure proper tagmentation is

achieved with the Bioanalyzer. The histogram of DNA fragment sizes (determined from the paired-end sequencing reads) should exhibit a strong nucleosome banding pattern corresponding to the length of DNA wrapped around a single nucleosome. These banding patterns will likely not be present with infected cells, as viral or bacterial genomes could obscure them.

3. Based on our experience and suggestions provided by the Encyclopedia of DNA Elements (ENCODE) Consortium [40], ATAC-seq experiments should be assessed with the following standards in mind:
   (a) *Replicates*: experiments should be performed with three or more biological replicates, and at least two (preferably three) technical replicates are required.
   (b) *Read length:* prior to any trimming should be a minimum of 45 base pairs.
   (c) *Aligned reads*: each replicate should have 50 million non-duplicate, non-mitochondrial aligned reads for paired-ended sequencing.
   (d) *Alignment rate:* also known as percentage of mapped reads, should be greater than 95%, though values >80% may be acceptable.
   (e) *Library Complexity:* measured using the Non-Redundant Fraction (NRF) and PCR Bottlenecking Coefficients 1 and 2, or PBC1 and PBC2. The preferred values are as follows: NRF > 0.9, PBC1 > 0.9, and PBC2 > 3 (*See* **Note 9**).

4 Notes

1. This buffer can be made in advance without the detergents and stored at room temperature for long-term.
2. Make the transposition reaction mix fresh, preferably during the centrifugation processes in Subheading 3.2.
3. Any conventional DNA purification kit will suffice (e.g., DNA Clean & Concentrator Kit from Zymo Research). However, it has been noted that column purification does not remove primer dimers and large DNA fragments >1000 bp. An alternative method is to use magnetic bead-based DNA purification (e.g., Agencourt AMPure XP magnetic beads from Beckman Coulter). Depending on the research question, this may or may not interfere with the interpretation of the data. Therefore, it is advised that the user assess the quality of their purified libraries first before deciding which DNA purification method to use.

4. These primer sequences are provided in the study by Buenrostro et al. [11].
5. Aim for the 150,000–300,000 cells on the day of harvest. This is critical if the perturbation takes more than a day to finish. For example, if you are performing an infection time course and plan to harvest cells at 3, 6, and 18 h post infection (hpi), you should seed cells less than the amount required to account for cell growth during this period. Additionally, the harvesting of the cells should be synchronized. Thus, the infection should be first performed for wells that will be infected for 18 h, followed by 6, and 3 h, such that all the cells can be harvested at the same time. For viruses with longer replication cycles, one should consider that viruses can arrest the cell cycle, and the extent of cell growth should be monitored at the time points of interest.
6. Making a master mix containing all the common reagents is recommended. The only reagent not shared among all samples should be the barcoded PCR primer 2 and each individual, partially amplified library.
7. This is not an essential step. However, when one is working with viral DNA, it is important to perform this step to ensure that proper tagmentation is achieved, as many viral DNAs do not exhibit nucleosomal banding patterns (*see* Subheading 3.5).
8. It is sometimes preferable to design multiple primer pairs targeting different regions of the viral genomes to obtain an average of ratios of viral genome quantities. This can help minimize variation in using just one pair of primers.
9. NRF is defined as the number of distinct, uniquely mapping reads divided by the total number of reads. PBC1 = $M_1/M_{distinct}$ where, as specified in ENCODE, M_1 is the "number of genomic locations where exactly one read maps uniquely," and $M_{distinct}$ is the "number of distinct genomic locations to which some read maps uniquely." PBC2 = M_1/M_2, where M_2 is the "number of genomic locations where two reads map uniquely." For more detailed information on these parameters, one should consult the ENCODE Terms and Definitions (https://www.encodeproject.org/data-standards/terms/#library).

Acknowledgments

Funding was provided by NIH grant NIGMS R01-GM114141 (I.M.C), the Princeton Catalysis Initiative (I.M.C.), and the NIH NIGMS Training Grant T32-GM007388 (T.R.H.).

References

1. Allis CD, Jenuwein T (2016) The molecular hallmarks of epigenetic control. Nat Rev Genet 17:487–500
2. Mitchener MM, Muir TW (2022) Oncohistones: exposing the nuances and vulnerabilities of epigenetic regulation. Mol Cell 82:2925–2938
3. Joshi P, Greco TM, Guise AJ et al (2013) The functional interactome landscape of the human histone deacetylase family. Mol Syst Biol 9:672
4. Sabari BR, Zhang D, Allis CD et al (2017) Metabolic regulation of gene expression through histone acylations. Nat Rev Mol Cell Biol 18:90–101
5. Lu C, Coradin M, Porter EG et al (2021) Accelerating the field of epigenetic histone modification through mass spectrometry–based approaches. Mol Cell Proteomics 20: 100006
6. Klemm SL, Shipony Z, Greenleaf WJ (2019) Chromatin accessibility and the regulatory epigenome. Nat Rev Genet 20:207–220
7. Song L, Crawford GE (2010) DNase-seq: a high-resolution technique for mapping active gene regulatory elements across the genome from mammalian cells. Cold Spring Harb Protoc 2010:pdb.prot5384
8. Zaret K (2005) Micrococcal nuclease analysis of chromatin structure. CP. Mol Biol 69
9. Kelly TK, Liu Y, Lay FD et al (2012) Genome-wide mapping of nucleosome positioning and DNA methylation within individual DNA molecules. Genome Res 22:2497–2506
10. Giresi PG, Kim J, McDaniell RM et al (2007) FAIRE (formaldehyde-assisted isolation of regulatory elements) isolates active regulatory elements from human chromatin. Genome Res 17:877–885
11. Buenrostro JD, Giresi PG, Zaba LC et al (2013) Transposition of native chromatin for fast and sensitive epigenomic profiling of open chromatin, DNA-binding proteins and nucleosome position. Nat Methods 10:1213–1218
12. Schep AN, Buenrostro JD, Denny SK et al (2015) Structured nucleosome fingerprints enable high-resolution mapping of chromatin architecture within regulatory regions. Genome Res 25:1757–1770
13. Baek S, Goldstein I, Hager GL (2017) Bivariate genomic Footprinting detects changes in transcription factor activity. Cell Rep 19: 1710–1722
14. Daugherty AC, Yeo RW, Buenrostro JD et al (2017) Chromatin accessibility dynamics reveal novel functional enhancers in *C. elegans*. Genome Res 27:2096–2107
15. Riegel D, Romero-Fernández E, Simon M et al (2023) Integrated single-cell profiling dissects cell-state-specific enhancer landscapes of human tumor-infiltrating CD8+ T cells. Mol Cell 83:622–636.e10
16. Corces MR, Granja JM, Shams S et al (2018) The chromatin accessibility landscape of primary human cancers. Science 362:eaav1898
17. Morabito S, Miyoshi E, Michael N et al (2021) Single-nucleus chromatin accessibility and transcriptomic characterization of Alzheimer's disease. Nat Genet 53:1143–1155
18. Wang J, Zibetti C, Shang P et al (2018) ATAC-Seq analysis reveals a widespread decrease of chromatin accessibility in age-related macular degeneration. Nat Commun 9:1364
19. Sahinyan K, Blackburn DM, Simon M-M et al (2022) Application of ATAC-Seq for genome-wide analysis of the chromatin state at single myofiber resolution. eLife 11:e72792
20. Wang S, Wang Y, Li X et al (2023) ATAC-seq reveals the roles of chromatin accessibility in the chondrocytes of Kashin–Beck disease compared with primary osteoarthritis. Front Genet 14:1169417
21. Dirks RAM, Stunnenberg HG, Marks H (2016) Genome-wide epigenomic profiling for biomarker discovery. Clin Epigenetics 8: 122
22. Lieberman PM (2008) Chromatin organization and virus gene expression. J Cell Physiol 216:295–302
23. Van Leeuwen H, Okuwaki M, Hong R et al (2003) Herpes simplex virus type 1 tegument protein VP22 interacts with TAF-I proteins and inhibits nucleosome assembly but not regulation of histone acetylation by INHAT. J Gen Virol 84:2501–2510
24. Cotter MA, Robertson ES (1999) The latency-associated nuclear antigen tethers the Kaposi's sarcoma-associated herpesvirus genome to host chromosomes in body cavity-based lymphoma cells. Virology 264:254–264
25. Barbera AJ, Chodaparambil JV, Kelley-Clarke B et al (2006) The nucleosomal surface as a docking station for Kaposi's sarcoma herpesvirus LANA. Science 311:856–861
26. Herbein G, Wendling D (2010) Histone deacetylases in viral infections. Clin Epigenetics 1: 13–24
27. Birdwell CE, Queen KJ, Kilgore PCSR et al (2014) Genome-wide DNA methylation as an

epigenetic consequence of Epstein-Barr virus infection of immortalized keratinocytes. J Virol 88:11442–11458
28. Guise A, Budayeva H, Diner B et al (2013) Histone deacetylases in herpesvirus replication and virus-stimulated host defense. Viruses 5: 1607–1632
29. Ma Z, Ni G, Damania B (2018) Innate sensing of DNA virus genomes. Annu Rev Virol 5: 341–362
30. Diner BA, Lum KK, Cristea IM (2015) The emerging role of nuclear viral DNA sensors. J Biol Chem 290:26412–26421
31. Diner BA, Lum KK, Toettcher JE et al (2016) Viral DNA sensors IFI16 and cyclic GMP-AMP synthase possess distinct functions in regulating viral gene expression, immune defenses, and apoptotic responses during herpesvirus infection. mBio 7:e01553–e01516. https://doi.org/10.1128/mBio.01553-16
32. Howard TR, Lum KK, Kennedy MA et al (2022) The nuclear DNA sensor IFI16 indiscriminately binds to and diminishes accessibility of the HSV-1 genome to suppress infection. mSystems 7(3):e00198–22
33. Liu D, Lum KK, Treen N et al (2023) IFI16 phase separation via multi-phosphorylation drives innate immune signaling. Nucleic Acids Res 51(13):6819–6840. https://doi.org/10.1093/nar/gkad449
34. Orzalli MH, DeLuca NA, Knipe DM (2012) Nuclear IFI16 induction of IRF-3 signaling during herpesviral infection and degradation of IFI16 by the viral ICP0 protein. Proc Natl Acad Sci 109:E3008–E3017
35. Orzalli MH, Conwell SE, Berrios C et al (2013) Nuclear interferon-inducible protein 16 promotes silencing of herpesviral and transfected DNA. Proc Natl Acad Sci 110:E4492–E4501
36. Merkl PE, Knipe DM (2019) Role for a filamentous nuclear assembly of IFI16, DNA, and host factors in restriction of Herpesviral infection. mBio 10(1):e02621–e02618
37. Cabral JM, Oh HS, Knipe DM (2018) ATRX promotes maintenance of herpes simplex virus heterochromatin during chromatin stress. Elife 7:e40228
38. Merkl PE, Orzalli MH, Knipe DM (2018) Mechanisms of host IFI16, PML, and Daxx protein restriction of herpes simplex virus 1 replication. J Virol 92:e00057–e00018
39. Ackermann AM, Wang Z, Schug J et al (2016) Integration of ATAC-seq and RNA-seq identifies human alpha cell and beta cell signature genes. Mol Metab 5:233–244
40. Luo Y, Hitz BC, Gabdank I et al (2020) New developments on the encyclopedia of DNA elements (ENCODE) data portal. Nucleic Acids Res 48:D882–D889

Part III

Transcriptome-Based Sequencing Technologies

Chapter 8

RNA Sequencing Protocols for Short-Read Sequencing

Laura Vasquez-Velez, Veera D'Mello, and Patricia Soteropoulos

Abstract

RNA sequencing (RNA-seq) methodologies allow the discovery of novel variants and transcripts. These comprise three general steps: (1) capture of RNA species of interest, (2) conversion of RNA to complementary DNA (cDNA), and (3) modification of cDNA to fit the sequencing platform. Here we describe four different library preparation protocols for short-read sequencing: cDNA synthesis with poly (A) selection, library preparation with ribosomal depletion, and cDNA synthesis with SMART® (Switching Mechanism at 5′ end of RNA Template) technology for low and Pico inputs.

Key words RNA-seq, Library preparation, cDNA, Ribosomal depletion, Poly(A) selection, Short-read sequencing

1 Introduction

RNA sequencing (RNA-seq) comprises several methodologies that support the study of transcription levels, gene expression, and allows the discovery of novel variants and transcripts [1–3]. Generally, RNA-seq library preparation methods consist of three steps: (1) capture of RNA species of interest from total RNA samples or cells;(2) conversion of RNA to complementary DNA (cDNA); and (3) modification of cDNA to fit the sequencing platform [2]. Ribosomal RNA (rRNA) represents over 80% of total RNA in the cell [1, 3], although it does not account for transcriptomics information in most cases [2]. Thus, library preparation starts with the capture of RNA species important for transcriptomics analysis, separating them from abundant rRNA from the total RNA samples. Two main techniques can be used for this goal: mRNA selection and enrichment, and rRNA depletion.

In eukaryotic organisms, protein-coding messenger RNA (mRNA) and long-non-coding RNA (lncRNA) present segments of repeated adenine nucleotides or polyadenylated tails (poly (A) tails) at the 3′ end of transcripts [1, 2]. This structural feature allows for the capture of mRNA from total RNA samples, using

Valerie J. Carabetta and Olaitan Akintunde (eds.), *High Throughput Gene Screening: Methods and Protocols*, Methods in Molecular Biology, vol. 2866, https://doi.org/10.1007/978-1-0716-4192-7_8,

either magnetic beads bearing oligo-dT molecules [1, 2], or oligo-dT priming followed by reverse transcription [2]. Importantly, RNA samples must be high quality, without fragmentation or degradation, and the RNA input concentrations of 10 ng or more is recommended for poly(A) selection-based RNA library preparation methods.

Nonetheless, this methodology does not apply to non-coding RNA or prokaryotic mRNA samples, since these lack poly(A) tails [4, 5], and partially or highly degraded eukaryotic mRNA transcripts [2]. For partially degraded RNA transcripts or those lacking poly(A) tails, RNA-seq libraries are prepared using rRNA depletion methods. Most often, rRNA depletion is mediated by hybridization single-strand DNA (ssDNA) probes that bind to known sequences of ribosomal RNA. After successful hybridization, the heteroduplex fragments of rRNA-ssDNA are removed enzymatically [1, 2].

The next step after isolation of the RNA of interest is its conversion to the more stable cDNA. In most cases, a fragmentation step is required either before or after cDNA synthesis [2, 5]. The synthesis of cDNA is initiated with reverse transcription, mediated by reverse transcriptase, and using either oligo-dT primers (transcripts with poly(A) tails), or random primers that hybridize to the RNA. Next the first strand of cDNA serves as template for second strand cDNA synthesis. After this, the ends of dsDNA are repaired, and the 3′ ends get an adenine overhang, preparing the libraries for adaptor ligation, making the cDNA fragments available as substrate for PCR amplification [4]. The final steps of library preparation for RNA-seq are library amplification and the addition of molecular barcodes [2]. Depending on the transcriptomic analysis goals, researchers could favor conserving the strand specificity of the transcripts to be sequenced. Several commercial kits address this by replacing deoxythymidine triphosphate (dTTP) with deoxyuridine triphosphate (dUTP) during second strand cDNA synthesis and later in the library preparation process, the labeled second strand gets selectively degraded by uracil-DNA glycosylases (UDGs) [5, 6].

A different approach to cDNA synthesis is offered by the SMART® (Switching Mechanism at 5′ end of RNA Template) technology, where reverse transcriptase has terminal transferase characteristics, adding non-templated nucleotides to the 3′ end of the newly synthetized cDNA [3, 7]. In the same reaction, a template switching oligo hybridizes to the non-templated nucleotides in the cDNA and serves as template for the reverse transcriptase to add known sequences to the cDNA. The known sequences also serve as primer sites for subsequent amplification of the cDNA. This way the cDNA strand is primed at both 3′ and 5′ ends [8].

This chapter comprises four basic protocols for RNA-seq. Each protocol can be used with specific sets of sample conditions, and sample sources as outlined below. Protocol 1 is for library construction from eukaryotic mRNA, with total RNA inputs above 10 ng. This protocol is adapted from New England Biolabs (NEB) and enriches eukaryotic polyadenylated mRNA transcripts as it employs poly-dT beads for selection. Protocol 2 is for library construction from eukaryotic or prokaryotic samples with total RNA input from 1 to 500 ng using the Illumina Stranded Total RNA library prep kit. This method employs rRNA depletion to enrich mRNA and other non-coding RNAs. Protocol 3 is for library preparation from eukaryotic, intact to partially degraded RNA using the Takara kit for ultra-low input samples ranging from 10 pg to 10 ng. Finally, protocol 4 is for library preparation from highly degraded mammalian samples, such as formalin-fixed paraffin-embedded (FFPE) tissue with sample inputs of 500 pg to 1 ng.

2 Materials

2.1 Protocol 1: cDNA Synthesis with Poly(A) Selection and Enrichment

1. NEBNext Poly(A) mRNA Magnetic Isolation Module (New England Biolabs).
2. NEBNext Ultra II RNA Library Prep Kit for Illumina (New England Biolabs).
3. NEBNext Multiplex Oligos for Illumina (New England Biolabs, Dual index. Primers set 1 or Dual index primers set 2).
4. SPRIselect Magnetic beads for size selection (Beckman Coulter Life Sciences).
5. Magnetic rack or plate.
6. 80% freshly prepared Ethanol.
7. Microcentrifuge.
8. Vortex mixer.
9. Thermal Cycler.
10. Pipette tips 10 μL, filtered.
11. Pipette tips 200 μL, filtered.
12. Thin wall 200 μL PCR tubes.
13. Bioanalyzer (Agilent Technologies).
14. TapeStation D1000 assay.

2.2 Protocol 2: Library Prep with Illumina Stranded Total RNA Prep, and Ligation with Ribo-Zero Plus Microbiome

1. Illumina Stranded Total RNA Prep with Ribo-Zero Plus Microbiome (Illumina).
2. RNAClean XP magnetic beads (Beckman Coulter Life Sciences).
3. AMPure XP magnetic beads (Beckman Coulter Life Sciences).
4. Magnetic rack or plate.

5. 80% freshly prepared Ethanol.
6. Microcentrifuge.
7. Vortex mixer.
8. Thermal Cycler.
9. Pipette tips 10 μL, filtered.
10. Pipette tips 200 μL, filtered.
11. Thin wall 200 μL PCR tubes.
12. Bioanalyzer (Agilent Technologies).
13. TapeStation HS RNA assay.
14. TapeStation D1000 assay.

2.3 Protocol 3: cDNA Synthesis with Smarter Technology for Ultra-low Input

1. Takara SMART-Seq v4 Ultra Low Input RNA Kit for Sequencing (Takara Bio).
2. NEBNext Ultra II FS DNA Library Prep Kit for Illumina (New England Biolabs).
3. NEBNext Multiplex Oligos for Illumina (New England Biolabs, Dual index primers set 1 or Dual index primers Set 2).
4. RNAClean XP magnetic beads (Beckman Coulter Life Sciences).
5. AMPure XP magnetic beads (Beckman Coulter Life Sciences).
6. Magnetic rack or plate.
7. 80% freshly prepared Ethanol.
8. Microcentrifuge.
9. Vortex mixer.
10. Thermal Cycler.
11. Pipette tips 10 μL, filtered.
12. Pipette tips 200 μL, filtered.
13. Thin wall 200 μL PCR tubes.
14. Bioanalyzer (Agilent Technologies).

2.4 Protocol 4: Library Prep with SMARTer Stranded Total RNA-seq Kit v3 for Pico Input, Mammalian

1. SMARTer Stranded Total RNA-seq Kit v3—Pico Input Mammalian (Takara Bio).
2. AMPure XP magnetic beads (Beckman Coulter Life Sciences).
3. Magnetic rack or plate.
4. 80% freshly prepared Ethanol.
5. Microcentrifuge.
6. Vortex mixer.
7. Thermal Cycler.
8. Pipette tips 10 μL, filtered.

9. Pipette tips 200 μL, filtered.
10. Thin wall 200 μL PCR tubes.
11. Bioanalyzer (Agilent Technologies).
12. TapeStation HS RNA assay.
13. TapeStation D1000 assay.

3 Methods

3.1 Protocol 1: cDNA Synthesis with Poly(a) Selection and Enrichment

3.1.1 Sample and Time Considerations

1. For eukaryotic total RNA, with 10 ng to 1 μg of input RNA. Must be intact mRNA (RIN > 7; *see* **Note 1**).
2. Schedule:
   - Day 1: mRNA isolation, fragmentation, first strand synthesis, second strand synthesis, cDNA clean-up.
   - Day 2: end repair, adaptor ligation, user enzyme treatment, and size selection.
   - Day3: library enrichment, index ligation, final library clean-up, and quality control.

3.1.2 mRNA Isolation, cDNA Synthesis and Purification

This initial step should be performed in one day. mRNA is selectively captured from the total RNA by the oligo dT beads, that bind to the poly(A) tails, characteristic to mature eukaryotic mRNA.

1. Dilute 10 ng to 1 μg of total RNA in 50 μL of nuclease-free water (*see* **Note 2**). Keep on ice.
2. Prepare the first strand reaction buffer and random primer mix (*see* **Note 3**). This will be used immediately after mRNA isolation to elute RNA from oligo-dT beads, followed by fragmentation and random priming. Combine all the components in a nuclease-free microcentrifuge tube according to the number of reactions as follows:

Component	Volume per one library (μL)
NEBNext first Strand synthesis reaction buffer	8
NEBNext random primers	2
Nuclease-free water	10
Total volume	20

3. Mix all components by pipetting ten times, centrifuge briefly.
4. Place the tube on ice until needed.
5. Isolate the mRNA using an appropriate kit based on sample type, following manufacturer's instructions.

6. Prepare the oligo dT beads for mRNA capture and isolation. Combine the following components (up to ten samples/library) in a 1.5 mL nuclease-free tube (*see* **Note 4**):

Component	Volume per one library (μL)
Oligo dT beads	20
RNA binding buffer	100
Total volume per reaction	120

7. Wash the beads by pipetting up and down at least six times.
8. Place the tube on the magnetic rack and incubate until the solution is clear.
9. Keeping the tube on the magnetic rack, remove and discard the supernatant from the tube. Do not disturb the beads.
10. Remove the tube from the magnetic rack.
11. Add 100 μL (per sample) of RNA binding buffer to the beads and wash by pipetting up and down at least six times.
12. Repeat **steps 8–10**.
13. Add 50 μL (per sample) of RNA binding buffer to the beads and mix by pipetting up and down until beads are homogeneous.
14. Add 50 μL of the beads and RNA binding buffer mix to each RNA sample (i.e., 50 μL aliquots). Mix by pipetting up and down at least six times.
15. Place the samples in a thermal cycler at 65 °C for 5 min and cool to 4 °C with the heated lid set at ≥75 °C.
16. Remove samples from the thermal cycler when temperature reaches 4 °C.
17. Mix by pipetting up and down at least six times. Place the samples on the bench and incubate at room temperature for 10 min to allow mRNA to bind to the beads.
18. Place the samples on the magnetic rack and incubate until the solution is clear (at least 2 min, *see* **Note 5**), then discard supernatant carefully. Remove samples from the magnetic rack.
19. Wash the beads by adding 200 μL of wash buffer into the tubes to remove the unbound RNA. Gently pipette the entire volume up and down at least six times to mix thoroughly. Place samples on the magnetic rack and incubate until the solution is clear, discard supernatant carefully. Remove samples from the magnetic rack.
20. Repeat **step 19**.
21. Add 50 μL of Tris–HCl buffer (included in the kit) to each sample. Gently pipette the entire volume up and down at least six times to mix thoroughly.

22. Place the samples in a thermal cycler at 80 °C for 2 min, then cool to 25 °C with the heated lid set at ≥90 °C to do the first elution of the mRNA from the beads.
23. Remove the samples from the thermal cycler when the temperature reaches 25 °C.
24. Add 50 μL of RNA binding buffer to each sample to allow mRNA to rebind to the beads. Gently pipette the entire volume up and down at least six times to mix thoroughly.
25. Incubate the samples at room temperature for 5 min.
26. Place the samples on the magnetic rack and incubate until de solution is clear, then discard the supernatant carefully. Remove samples from the magnetic rack.
27. Wash the beads by adding 200 μL of wash buffer. Gently pipette the entire volume up and down at least six times to mix thoroughly.
28. Centrifuge the samples briefly to collect the liquid from the wall and lid of each tube.
29. Repeat **step 26**.
30. Briefly spin down the samples and place them again on the magnetic rack. Using a 10 μL tip, remove all the wash buffer. Do not disturb the beads that contain the mRNA. Avoid letting the beads dry out before adding the elution buffer.
31. Remove the samples from the magnetic rack.
32. To elute the mRNA from the beads and fragment, add 11.5 μL of the first strand reaction buffer and random primer mix (prepared in **step 2**) and continue the protocol. Avoid transferring any magnetic beads. Vortex briefly.

3.1.3 Fragmentation of mRNA

1. Incubate the samples in a thermal cycler with the heated lid set at 105 °C as follows:
   - 15 min at 94 °C.
   - Hold at 4 °C.
2. Immediately transfer the tubes to ice for 1 min.
3. Quicky spin down the samples in a microcentrifuge to collect the liquid from the sides of the tube and place on the magnetic rack right away until the solution is clear.
4. Collect the fragmented mRNA by transferring 10 μL of the supernatant to new nuclease-free microcentrifuge tubes. If the supernatant volume is less than 10 μL, bring up the volume to 10 μL by adding the first strand synthesis buffer and random primer mix from **step 2**.
5. Place the samples on ice until ready to use.

3.1.4 First Strand and Second cDNA Synthesis

Here mRNA undergoes reverse transcription guided by the random primers mixed during fragmentation for first strand synthesis. The first strand of cDNA is then used as a template for a second strand, producing dsDNA.

1. Combine fragmented and primed RNA with 8 μL of nuclease-free water and 2 μL of NEBNext first strand synthesis enzyme mix.
2. Mix by pipetting up and down at least 10 times, and spin briefly.
3. Incubate the samples in a thermal cycle with the heated lid set at ≥80 °C as follows:
   - 10 min at 25 °C.
   - 50 min at 42 °C (*see* **Note 6**).
   - 15 min at 70 °C.
   - Hold at 4 °C.
4. Place the samples on ice.
5. Prepare the second strand synthesis master mix by combining:

Component	Volume per one library (μL)
NEBNext ultra second Strand synthesis reaction buffer	8
NEBNext ultra second Strand synthesis enzyme mix	4
Nuclease-free water	48

6. Add 60 μL of the second strand master mix to the first strand synthesis product.
7. Mix by pipetting up and down at least ten times, and centrifuge briefly.
8. Incubate in a thermal cycler for 1 h at 16 °C with the heated lid off (or open).
9. Purify the double stranded cDNA by adding 144 μL of resuspended SPRIselect Beads to each sample. Mix by pipetting up and down at least six times.
10. Incubate samples for 5 min at room temperature.
11. Briefly spin down and place samples on a magnetic rack. After the solution is clear (~5–10 min) remove and discard 210 μL of supernatant. Do not disturb the beads that contain the dsDNA.
12. Keeping the samples on the magnetic rack, wash the beads with 200 μL of freshly prepared 80% ethanol, incubate for 30 s at room temperature, and then discard the supernatant.

13. Repeat **step 12**.
14. Spin samples briefly and return them to the magnetic rack. Using a 10 μL pipette tip, remove all remaining ethanol in the tubes. Do not disturb the beads.
15. Air dry the beads for 2–5 min on the magnetic rack. When beads are still glossy but there is no liquid on the tubes, move to the following step.
16. Remove samples from the magnetic rack and wash beads with 53 μL of nuclease-free water to elute the dsDNA. Mix well on a vortex mixer or by pipetting up and down 10 times.
17. Incubate for 2 min at room temperature.
18. Spin down briefly to collect all liquid from the tube walls and lid.
19. Place samples on the magnetic rack until the solution is clear.
20. Remove 50 μL of supernatant and transfer to new nuclease-free PCR tubes. At this point it is safe to stop. Samples can be stored in −20 °C.

3.1.5 End Prep of cDNA, Adaptor Ligation, and Library Enrichment (see **Note 7**)

The dsDNA fragments are repaired by adding adenosine to the 3′ ends, which serve as anchors for the adaptor attachment.

1. On ice, prepare the End prep master mix by combining the following (*see* **Note 8**):

Component	Volume per one library (μL)
NEBNext ultra II end prep reaction buffer	7
NEBNext ultra II end prep enzyme mix	3

2. Add 10 μL of the end prep master mix to each product of the second strand synthesis reaction (total volume ~ 60 μL).
3. Set a 100 μL or 200 μL pipette to 50 μL and then pipette the entire volume up and down at least ten times to mix. Centrifuge briefly.
4. Incubate the samples in thermal cycler with the heated lid set at ≥75 °C as follows:
   - 30 min at 20 °C.
   - 30 min at 65 °C.
   - Hold at 4 °C.
5. Before beginning adaptor ligation, *see* **Note 9**. The adaptor molecules attach to the dsDNA fragments, forming hairpin loop at each end of the strands.

6. Dilute adaptor 1:5 for total RNA inputs between 250 and 1000 ng. Dilute adaptor 1:25 for total RNA inputs between 100 and 249 ng. Dilute adaptor 1:100 for total RNA inputs between 10 and 99 ng (*see* **Note 10**).
7. On ice, prepare the ligation reaction master mix by combining the following (*see* **Note 11**):

Component	Volume per one library (μL)
NEBNext ligation enhancer	1
NEBNext ultra II ligation reaction master mix	30

8. Add 2.5 μL of diluted adaptor to each sample.
9. Add 31 μL of the ligation reaction master mix to each sample (total volume ~ 93.5 μL).
10. Set a 100 μL or 200 μL pipette to 80 μL and then pipette the entire volume up and down at least ten times to mix. Centrifuge briefly.
11. Incubate the samples in a thermal cycler at 20 °C for 15 min with the heated lid off.
12. Add 3 μL of USER enzyme to the ligation reaction product. The USER enzyme cuts the hairpin loop from the adapter removing the uracil group, leaving each end of the dsDNA fragment prepared for the ligation of the Illumina adaptors and the barcodes. Mix well and briefly centrifuge.
13. Incubate the samples in a thermal cycler at 37 °C for 15 min with the heated lid set at ≥45 °C. Proceed immediately to the next step.
14. Purify the adaptor ligation reaction by adding 87 μL of resuspended SPRIselect beads to each sample. Mix by pipetting up and down at least six times.
15. Incubate samples for 10 min at room temperature.
16. Briefly centrifuge and place samples on a magnetic rack. After the solution is clear (~5–10 min) remove and discard 175 μL of supernatant that contains the unwanted fragments. Do not disturb the beads.
17. Keeping the samples on the magnetic rack, wash the beads with 200 μL of freshly prepared 80% ethanol, incubate for 30 s at room temperature, then discard the supernatant.
18. Repeat **step 17**.
19. Centrifuge samples briefly and return them to the magnetic rack. Using a 10 μL pipette tip, remove all remaining ethanol in the tubes. Do not disturb the beads.

20. Air dry the beads for 2–5 min on the magnetic rack. When beads are still glossy but there is no liquid on the tubes, move to the following step.
21. Remove samples from the magnetic rack and wash beads with 25 μL of nuclease-free water to elute the dsDNA (*see* **Note 12**). Mix well on a vortex mixer or by pipetting up and down ten times.
22. Incubate for 2 min at room temperature.
23. Centrifuge briefly to collect all liquid from the tube walls and lid.
24. Place samples on the magnetic rack until the solution is clear.
25. Remove 22 μL of supernatant and transfer to new nuclease-free PCR tubes. At this point, it is safe to stop. Samples can be stored at −20 °C.

3.1.6 PCR Enrichment of Adaptor Ligated DNA

In this step, the forward (i7) and reverse (i5) primers hybridize to the dsDNA fragments through the adaptor sequences added in the previous reaction. Each strand is labeled with the sequences that allow the attachment to the Illumina flow cell, and the index barcodes for sample multiplexing.

1. Set up the PCR reaction by combining the following components:

Component	Volume per one library (μL)
Adaptor-ligated DNA	22
NEBNext ultra II Q5 master mix	25
Index (X) primer/i7 primer	1.5
Universal PCR primer/i5 primer	1.5
Total volume	50

2. Mix well by pipetting up and down at least ten times. Centrifuge briefly.
3. Place samples on a thermal cycler with the heated lid set to 105 °C and perform PCR amplification using the following PCR cycling conditions:
   - 30 s at 98 °C
   - 7–15 cycles* of:

     10 s at 98 °C

     75 s at 65 °C.
   - 5 min at 65 °C
   - Hold at 4 °C.

*The number of PCR cycles should be adjusted based on RNA input:

- 1000 ng, 7–8 cycles are recommended.
- 100 ng, 11–12 cycles are recommended.
- 10 ng, 14–15 cycles are recommended.

4. Purify libraries by adding 45 μL of resuspended SPRIselect beads. Mix by pipetting up and down at least six times.
5. Incubate samples for 5 min at room temperature.
6. Briefly spin down and place samples on a magnetic rack. After the solution is clear (~5 min) remove and discard 85 μL of supernatant that contains the unwanted fragments. Do not disturb the beads.
7. Keeping the samples on the magnetic rack, wash the beads with 200 μL of freshly prepared 80% ethanol, incubate for 30 s at room temperature, then discard the supernatant.
8. Repeat **step** 7.
9. Spin samples briefly and return them to the magnetic rack. Using a 10 μL pipette tip, remove all remaining ethanol in the tubes. Do not disturb the beads.
10. Air dry the beads for 2–5 min on the magnetic rack. When beads are still glossy but there is no more liquid in the tubes, move to the following step.
11. Remove samples from the magnetic rack and wash beads with 28 μL of nuclease-free water to elute the DNA target. Mix well on a vortex or by pipetting up and down 10 times.
12. Incubate for 2 min at room temperature.
13. Centrifuge briefly to collect all liquid from the tube walls and lid.
14. Place samples on the magnetic rack until the solution is clear.
15. Remove 25 μL of supernatant and transfer to new nuclease-free PCR tubes.
16. Proceed to quality control of the libraries and assess the concentration, and fragment size distribution (*see* **Note 14**). Libraries can present small fragment amplification at 120 base pairs. These fragments can be removed by repeating the last clean-up process (**steps 5–11**). After a second clean up, the library concentration will decrease, and the fragment size will change, so the QC must be reassessed. To maintain a high concentration of the libraries, the elution volume in **step 11** can be reduced to 24 μL and transferring 22 μL to new PCR tubes.

3.2 Protocol 2: Library Prep with Illumina Stranded Total RNA Prep, Ligation with Ribo-Zero Plus Microbiome

This kit can be used with samples from both eukaryotic and prokaryotic cells. If used for eukaryotic samples, the microbiome probe pool is not necessary.

3.2.1 Sample and Time Considerations

1. Eukaryotic or prokaryotic, intact total RNA (1–500 ng input).
2. Schedule:

   Day 1: Probe hybridization, rRNA depletion, probe removal, RNA purification, fragmentation, first strand and second strand cDNA synthesis.

   Day 2: Adenylate 3′ ends, anchor ligation, and anchor-ligated DNA purification.

   Day 3: Library amplification and quality control.

3.2.2 Probe Hybridization, rRNA Depletion and RNA Purification

1. If using the microbiome depletion probe pool, dilute 25–500 ng of total RNA to a volume of 10 μL.
2. On ice, prepare the hybridize probe master mix combining the following:

Component	Volume per one library (μL)
Depletion buffer (DPB)	3.6
Depletion probe Pool (DP1)	1.2
Microbiome depletion probe Pool (DPM)	1.2

3. Add 5 μL of hybridize probe master mix to each sample.
4. Mix by pipetting at least ten times. Centrifuge briefly.
5. Place the samples in a preheated, hot-lid thermal cycler with the heated lid set at 100 °C, and set the following program:
   - 2 min at 95 °C
   - Decrease 0.1 °C/s until the temperature reaches 37 °C.
   - Hold at 37 °C.
6. On ice, prepare the rRNA depletion master mix by combining the following:

Component	Volume per one library (μL)
RNA depletion buffer (RDB)	4.8
RNA depletion enzyme (RDE)	1.2

7. Add 5 μL of rRNA depletion master mix to each sample.
8. Mix by pipetting at least ten times. Centrifuge briefly.
9. Place the samples in a preheated, hot-lid thermal cycler with the heated lid set at 100 °C, set the following program:
   - 15 min at 37 °C
   - Hold at 4 °C.
10. On ice, prepare the probe removal master mix combining:

Component	Volume per one library (μL)
Probe removal buffer (PRB)	7.7
Probe removal enzyme (PRE)	3.3

11. Add 10 μL of probe removal master mix to each sample.
12. Mix by pipetting at least ten times. Centrifuge briefly.
13. Place the samples in a preheated hot-lid thermal cycler with the heated lid set at 100 °C, and follow the program:
    - 15 min at 37 °C
    - 15 min at 70 °C
    - Hold at 4 °C.
14. Place RNAClean XP beads at room temperature at least 30 min before RNA purification. Vortex to resuspend.
15. Purify the ribosomal depleted RNA by adding 60 μL of resuspended RNAClean XP beads to each sample. Mix by pipetting at least ten times.
16. Incubate at room temperature for 5 min.
17. Place samples on a magnetic rack for 5 min.
18. Briefly centrifuge and place samples on a magnetic rack. After the solution is clear (~5–10 min), remove and discard 80 μL of supernatant. Do not disturb the beads.
19. Keeping the samples on the magnetic rack, wash the beads with 175 μL of freshly prepared 80% ethanol, incubate for 30 s at room temperature, then discard the supernatant.
20. Repeat **step 19**.
21. Spin samples briefly and return them to the magnetic rack. Using a 10 μL pipette tip, remove all remaining ethanol in the tubes. Do not disturb the beads.
22. Air dry the beads for 2–5 min on the magnetic rack. When beads are still glossy but there is no more liquid on the tubes, move to the following step.

23. Remove samples from the magnetic rack and wash beads with 10.5 μL of elution buffer (ELB) to rehydrate. Mix well on a vortex mixer or by pipetting up and down ten times.
24. Incubate for 2 min at room temperature.
25. Centrifuge briefly to collect all liquid from the tube walls and lid.
26. Place samples on the magnetic rack until the solution is clear.
27. Remove 8.5 μL of supernatant and transfer to new nuclease-free PCR tubes (*see* **Note 13**).

3.2.3 Fragmentation of RNA, First and Second Strand cDNA Synthesis

1. Add 8.5 μL of Elute, Prime, Fragment 3HC Mix (EPH3) to each sample. In this step, the RNA is fragmented and primed with random hexamers in preparation for the first strand cDNA synthesis.
2. Mix by pipetting at least ten times.
3. Place the samples in a preheated, hot-lid thermal cycler with the heated lid set at 100 °C, following the program:
   - 2 min at 94 °C
   - Hold at 4 °C.
4. On ice, prepare the synthesize first strand cDNA master mix by combining the following:

Component	Volume per one library (μL)
First Strand synthesis mix (FSA)	9
Reverse transcriptase (RVT)	1

5. Add 8 μL of first strand cDNA master mix to each sample.
6. Mix by pipetting at least ten times. Centrifuge briefly.
7. Place the samples in a preheated hot-lid thermal cycler with the heated lid set at 100 °C, and follow the program:
   - 10 min at 25 °C.
   - 15 min at 42 °C.
   - 15 min at 70 °C.
   - Hold at 4 °C.
8. Add 25 μL of Second Strand Marking master mix (SMM). In this step, the RNA template is eliminated, and the replacement is synthetized to obtain blunt-ended double stranded cDNA fragments. Instead of deoxythymidine triphosphate (dTTP), the new strand has deoxyuridine triphosphate (dUTP).
9. Mix by pipetting at least ten times. Centrifuge briefly.
10. Place the samples in a preheated, hot-lid thermal cycler with the heated lid set at 40 °C, and set up the following program:

- 1 h at 16 °C.
- Hold at 4 °C for ≤16 h.

11. Place AMPure XP Beads at room temperature for at least 30 min before cDNA purification. Vortex to resuspend.
12. Purify cDNA by adding 90 μL of resuspended AMPure XP Beads to each sample. Mix by pipetting up and down at least six times.
13. Incubate the samples for 5 min at room temperature.
14. Briefly centrifuge and place samples on a magnetic rack. After the solution is clear (~5–10 min) remove and discard 130 μL of supernatant. Do not disturb the beads.
15. Keeping the samples on the magnetic rack, wash the beads with 175 μL of freshly prepared 80% ethanol, incubate for 30 s at room temperature, then discard the supernatant.
16. Repeat **step 15**.
17. Centrifuge samples briefly and return them to the magnetic rack. Using a 10 μL pipette tip, remove all remaining ethanol in the tubes. Do not disturb the beads.
18. Air dry the beads for 2–5 min on the magnetic rack. When beads are still glossy but there is no more liquid in the tubes, move to the following step.
19. Remove samples from the magnetic rack and wash beads with 19.5 μL of resuspension buffer to rehydrate. Mix well on a vortex mixer or by pipetting up and down ten times.
20. Incubate for 2 min at room temperature.
21. Centrifuge briefly to collect all liquid from the tube walls and lid.
22. Place samples on the magnetic rack until the solution is clear.
23. Remove 17.5 μL of the supernatant and transfer to new nuclease-free PCR tubes. Samples can be stored at −20 °C for up to 7 days.

3.2.4 Adenylate 3′ Ends, Ligate Anchors and Amplify Libraries

The blunt-ended cDNA fragments are adenylated on the 3′ ends. The anchor sequences will anneal to the adenosine overhang to prepare the fragments for adaptor and index attachments.

1. Add 12.5 μL of A-tailing mix (ATL4) to each sample.
2. Using a 200 μL pipette, pipette at least ten times to mix. Centrifuge briefly.
3. Place the samples in a preheated, hot-lid thermal cycler with the heated lid set at 100 °C, and set up the following program:
   - 30 min at 37 °C.
   - 5 min at 70 °C
   - Hold at 4 °C.

4. Place the AMPure XP Beads at room temperature for at least 30 min in preparation for the anchor-ligated DNA purification. Vortex to resuspend.
5. Prepare the anchors by adding the following volumes to each sample in the order listed:

Order of addition	Reagent	Volume for sample input ≤100 ng	Volume for sample input >100 ng
1	Resuspension buffer (RSB)	2.5	0
2	RNA index anchor plate	2.5	5
3	Ligation mix (LIGX)	2.5	2.5

6. Set a 200 μL pipette to 100 μL and pipette at least ten times to mix. Centrifuge briefly.
7. Place the samples in a preheated, hot-lid thermal cycler with the heated lid set at 100 °C, and set the following program:
   - 10 min at 37 °C.
   - Hold at 4 °C.
8. Stop the ligation by adding 5 μL of stop ligation buffer (STL) to each sample.
9. Pipette at least ten times to mix.
10. Purify the anchor-ligated DNA by adding 34 μL of resuspended AMPure XP Beads to each sample. Mix by pipetting up and down at least six times.
11. Incubate the samples for 5 min at room temperature.
12. Briefly centrifuge and place samples on a magnetic rack. After the solution is clear (~5–10 min), remove and discard 67 μL of supernatant. Do not disturb the beads.
13. Keeping the samples on the magnetic rack, wash the beads with 175 μL of freshly prepared 80% ethanol, incubate for 30 s at room temperature, then discard the supernatant.
14. Repeat **step 13**.
15. Spin samples briefly and return them to the magnetic rack. Using a 10 μL pipette tip, remove all remaining ethanol in the tubes. Do not disturb the beads.
16. Air dry the beads for 2–5 min on the magnetic rack. When beads are still glossy but there is no more liquid on the tubes, move to the following step.

17. Remove samples from the magnetic rack and wash beads with 22 μL of resuspension buffer (RSB) to rehydrate. Mix well on a vortex mixer or by pipetting up and down ten times.
18. Incubate for 2 min at room temperature.
19. Spin down briefly to collect all liquid from the tube walls and lid.
20. Place samples on the magnetic rack until the solution is clear.
21. Remove 20 μL of supernatant and transfer to new nuclease-free PCR tubes. Samples can be stored at −20 °C for up to 7 days.

3.2.5 Amplify Library

In this section, the premixed forward (i7) and reverse (i5) primers in the Index Adapter plate hybridize to the anchor-ligated cDNA fragments. Each DNA fragment is labeled with sequences that allow the attachment to the Illumina flow cell, the sequencing primers, and the index barcodes for sample multiplexing.

1. Using a new pipette tip for each sample, pierce the foil covering the index adapter plate wells and transfer 2.5 μL of index adapter to each sample. Use a unique well for each sample.
2. Add 20 μL on enhanced primer mix (EPM), and 7.5 μL of nuclease-free water to each sample.
3. Mix by pipetting at least ten times. Centrifuge briefly.
4. Place the samples in a preheated, hot-lid thermal cycler with the heated lid set at 100 °C, and follow the program:
   - 30 s at 98 °C.
   - 15–17 cycles* of:

     10 s at 98 °C.

     30 s at 60 °C.

     30 s at 72 °C.
   - 5 m at 72 °C.
   - Hold at 4 °C for ≤16 h.

   *The number of PCR cycles should be adjusted based on RNA input:
   - 25 ng: 17 cycles
   - 50 ng: 16 cycles
   - 100 ng: 15 cycles
5. Place the AMPure XP Beads at room temperature for at least 30 min before library purification. Vortex to resuspend.
6. Purify the indexed libraries by adding 50 μL of resuspended AMPure XP Beads to each sample. Mix by pipetting up and down at least six times.
7. Incubate the samples for 5 min at room temperature.

8. Briefly centrifuge and place samples on a magnetic rack. After the solution is clear (~5–10 min) remove and discard 90 μL of supernatant. Do not disturb the beads.
9. Keeping the samples on the magnetic rack, wash the beads with 175 μL of freshly prepared 80% ethanol, incubate for 30 s at room temperature, then discard the supernatant.
10. Repeat **step 9**.
11. Spin samples briefly and return them to the magnetic rack. Using a 10 μL pipette tip, remove all remaining ethanol in the tubes. Do not disturb the beads.
12. Air dry the beads for 2–5 min on the magnetic rack. When beads are still glossy but there is no more liquid in the tubes, move to the following step.
13. Remove samples from the magnetic rack and wash beads with 17 μL of resuspension buffer to rehydrate. Mix well on a vortex mixer or by pipetting up and down ten times.
14. Incubate for 2 min at room temperature.
15. Centrifuge briefly to collect all liquid from the tube walls and lid.
16. Place samples on the magnetic rack until the solution is clear.
17. Remove 15 μL of supernatant and transfer to new nuclease-free PCR tubes. Samples can be stored at −20 °C for up to 7 days.

3.3 Protocol 3: cDNA Synthesis with Smarter® Technology for Ultra-low Input

This methodology employed by Takara uses as principle the template-switching technology of the reverse transcriptase. Every strand of mRNA is copied from both 3′ and 5′ ends, first by the 3′ end of the original mRNA strand, using the poly(A) present in mature mRNAs. After reverse transcription, the first strand of cDNA is primed at the 3′ end (complementary to the 5′ end of the mRNA) with non-templated nucleotides.

3.3.1 Sample and Time Considerations

1. Eukaryotic, intact to partially degraded RNA (10 pg to 10 ng).
2. Schedule:
   - **Day 1:** cDNA synthesis and amplification.
   - **Day 2:** Fragmentation, adaptor ligation, and library amplification.

3.3.2 First-Strand cDNA Synthesis

In this section, the 3′ SMART-Seq CDS primer II A hybridizes to the poly(A) tail in the mRNA, serving as a primer for reverse transcription. Later, the SMARTScribe reverse transcriptase (RT) adds non-templated nucleotides to the 3′ end of the new first strand of cDNA. The SMART-Seq v4 oligonucleotide hybridizes to the non-templated nucleotides, giving the RT a new template to continue reverse transcription. The cDNA carries SMART adapter sequences at both ends.

1. Combine up to 9.5 μL of total RNA samples, and negative and positive controls with 1 μL of 10× reaction buffer (1 part RNase inhibitor +20 parts 10× lysis buffer). The resultant volume should be 10.5 μL.
2. Place samples on ice. Add 1 μL of 3' SMART-seq CDS primer II A for input concentrations of 10 pg. Or, add 2 μL of 3' SMART-seq CDS primer II A for input concentrations of 100 pg to 10 ng of total RNA.
3. Incubate samples at 72 °C in a preheated, hot lid thermal cycler for 3 min.
4. While samples are incubating, prepare RT master mix by combining:

Component	Volume per one library (μL)
Ultra low first-Strand buffer	4
SMART-seq v4 oligonucleotide	1
RNase inhibitor	0.5

5. After incubation, place samples on ice for at least 2 min but less than 10 min.
6. Preheat a thermal cycler to 42 °C.
7. Add 2 μL of SMARTScribe Reverse Transcriptase per sample or control to the RT master mix. Mix with a vortex mixer and centrifuge briefly.
8. Add 7.5 μL of the RT master mix to each sample or control. Vortex and centrifuge briefly.
9. Place the samples and controls in a thermal cycler with heated lid, preheated to 42 °C. Run the following program:
   - 90 min at 42 °C.
   - 10 min at 70 °C.
   - Hold at 4 °C.

At this point, it is safe to stop. Samples can be stored at 4 °C overnight.

3.3.3 cDNA Amplification

In this section, the PCR primer II A binds to the flanking sequences added during the first-strand synthesis by the SMART-seq CDS primer II A and the SMART-Seq v4 oligonucleotide serves as the second primer for the cDNA amplification.

1. Add 30 μL of PCR master mix (25 μL of SeqAmp PCR buffer, 1 μL of PCR primer II A, 3 μL of nuclease-free water, and 1 μL of SeqAmp DNA polymerase). Mix with a vortex mixer and centrifuge briefly.
2. Place the samples in a preheated thermal cycler with heated lid and run the following program:
   - 1 min at 95 °C.
   - 7–18 cycles* of:

     10 s at 98 °C

     30 s at 65 °C

     3 min at 68 °C.
   - 10 min at 72 °C.
   - Hold at 4 °C.

   *The number of PCR cycles should be adjusted based on RNA input:
   - 10 ng: 7–8 cycles
   - 1 ng: 10–11 cycles
   - 100 pg: 14–15 cycles
   - 10 pg: 17–18 cycles.

   At this point, it is safe to stop. Samples can be stored at 4 °C overnight.
3. Place the AMPure XP Beads at room temperature for at least 30 min before amplified cDNA purification. Vortex to resuspend.
4. Add 1 μL of 10× lysis buffer to each PCR product.
5. Purify amplified cDNA by adding 50 μL of resuspended AMPure XP Beads. Mix by pipetting up and down at least six times.
6. Incubate samples for 8 min at room temperature.
7. Briefly centrifuge and place samples on a magnetic rack. After the solution is clear (~5–10 min) remove and discard 90 μL of supernatant. Do not disturb the beads.
8. Keeping the samples on the magnetic rack, wash the beads with 200 μL of freshly prepared 80% ethanol, incubate for 30 s at room temperature, then discard the supernatant.
9. Repeat **step 8**.
10. Centrifuge samples briefly and return them to the magnetic rack. Using a 10 μL pipette tip, remove all remaining ethanol in the tubes. Do not disturb the beads.
11. Air dry the beads for 2–5 min on the magnetic rack. When beads are still glossy but there is no more liquid in the tubes, move to the following step.

12. Remove samples from the magnetic rack and wash beads with 17 μL of nuclease-free water to elute the dsDNA. Mix well on a vortex or by pipetting up and down ten times.
13. Incubate for 2 min at room temperature.
14. Centrifuge briefly to collect all liquid from the tube walls and lid.
15. Place samples on the magnetic rack until the solution is clear.
16. Remove 15 μL of supernatant and transfer to new nuclease-free PCR tubes.

3.3.4 Library Preparation

For cDNA inputs of 100 pg to 100 ng (*see* **Note 15**).

1. Adjust the cDNA sample volume to 26 μL adding nuclease-free water.
2. Prepare the Fragmentation master mix by combining the following:

Component	Volume per one library (μL)
NEBNext ultra II FS reaction buffer	7
NEBNext ultra II FS enzyme mix	2

2. Add 9 μL of Fragmentation master mix to each cDNA sample (final volume ~ 35 μL).
3. Vortex reaction for 5 s. Centrifuge briefly.
5. Place samples in a thermal cycler with heated lid set to 75 °C. Run the following program:
   - 10 min at 37 °C*
   - 30 min at 65 °C.
   - Hold at 4 °C.

*Fragmentation occurs at 37 °C. 10 min incubation period is optimized to get a fragment size of 300–700 bp.

4. Dilute adaptor 1:25 for cDNA inputs <5 ng. Dilute adaptor 1: 10 for cDNA inputs up to 99 ng.
5. On ice, prepare the Ligation Reaction master mix by combining the following:

Component	Volume per one library (μL)
NEBNext ligation enhancer	1
NEBNext ultra II ligation reaction master mix	30

6. Add 2.5 μL of diluted adaptor to each sample.
7. Add 31 μL of the Ligation Reaction master mix to each sample (total volume ~ 68.5 μL).
8. Set a 100 μL or 200 μL pipette to 80 μL and then pipette the entire volume up and down at least ten times to mix. Centrifuge briefly.
9. Incubate the samples in a thermal cycler at 20 °C for 15 min with the heated lid off.
10. Add 3 μL of USER enzyme to the ligation reaction product. Mix well and centrifuge.
11. Incubate the samples in a thermal cycler at 37 °C for 15 min with the heated lid set at ≥45 °C. Proceed immediately to the purification of the ligation reaction.
12. Purify the adaptor ligation reaction by adding 57 μL of resuspended SPRIselect beads to each sample. Mix by pipetting up and down at least six times.
13. Incubate samples for 10 min at room temperature.
14. Briefly centrifuge and place samples on a magnetic rack. After the solution is clear (~5–10 min) remove and discard 120 μL of supernatant that contains the unwanted fragments. Do not disturb the beads.
15. Keeping the samples on the magnetic rack, wash the beads with 200 μL of freshly prepared 80% ethanol, incubate for 30 s at room temperature, then discard the supernatant.
16. Repeat **step 15**.
17. Spin samples briefly and return them to the magnetic rack. Using a 10 μL pipette tip, remove all remaining ethanol in the tubes. Do not disturb the beads.
18. Air dry the beads for 2–5 min on the magnetic rack. When beads are still glossy but there is no more liquid on the tubes, move to the following step.
19. Remove samples from the magnetic rack and wash beads with 25 μL of nuclease-free water to elute the DNA. Mix well on a vortex mixer or by pipetting up and down ten times.
20. Incubate for 2 min at room temperature.
21. Spin down briefly to collect all liquid from the tube walls and lid.
22. Place samples on the magnetic rack until the solution is clear.
23. Remove 22 μL of supernatant and transfer to new nuclease-free PCR tubes. At this point, it is safe to stop. Samples can be stored at −20 °C.

3.3.5 PCR Enrichment of Adaptor Ligated DNA

In this section, the forward (i7) and reverse (i5) primers hybridize to the dsDNA fragments through the adaptor sequences added in the previous reaction. Each strand is labeled with the sequences that allow for the attachment to the Illumina flow cell, the sequencing primers to anneal, and contain the index barcodes for sample multiplexing.

1. Set up the PCR reaction by combining the following components:

Component	Volume per one library (μL)
Adaptor-ligated cDNA	22
NEBNext ultra II Q5 master mix	25
Index (X) primer/i7 primer	1.5
Universal PCR primer/i5 primer	1.5
Total volume	50

2. Mix well by pipetting up and down at least ten times. Centrifuge briefly.
3. Place the samples on a thermal cycler with the heated lid set to 105 °C and perform PCR amplification using the following PCR cycling conditions:
   - 30 s at 98 °C.

     3–13 cycles* of:

     10 s at 98 °C

     75 s at 65 °C.
   - 5 min at 65 °C.
   - Hold at 4 °C.

   *The number of PCR cycles should be adjusted based on cDNA input:
   - 100 ng, 3–4 cycles
   - 50 ng, 4–5 cycles
   - 10 ng, 6–7 cycles
   - 5 ng, 7–8 cycles
   - 1 ng, 8–9 cycles
   - 0.5 ng, 8–10 cycles
   - 0.1 ng, 12–13 cycles.
4. Purify libraries by adding 45 μL of resuspended SPRIselect beads. Mix by pipetting up and down at least six times.
5. Incubate samples for 5 min at room temperature.

6. Briefly spin down and place samples on a magnetic rack. After the solution is clear (~5 min) remove and discard 85 μL of supernatant that contains the unwanted fragments. Do not disturb the beads.
7. Keeping the samples on the magnetic rack, wash the beads with 200 μL of freshly prepared 80% ethanol, incubate for 30 s at room temperature, then discard the supernatant.
8. Repeat **step** 7.
9. Centrifuge the samples briefly and return them to the magnetic rack. Using a 10 μL pipette tip, remove all remaining ethanol in the tubes. Do not disturb the beads.
10. Air dry the beads for 2–5 min on the magnetic rack. When beads are still glossy but there is no more liquid in the tubes, move to the following step.
11. Remove samples from the magnetic rack and wash beads with 28 μL of nuclease-free water to elute the DNA target. Mix well on a vortex mixer or by pipetting up and down ten times.
12. Incubate for 2 min at room temperature.
13. Spin down briefly to collect all liquid from the tube walls and lid.
14. Place samples on the magnetic rack until the solution is clear.
15. Remove 25 μL of supernatant and transfer to new nuclease-free PCR tubes.
16. Proceed to quality control of the libraries and assess the concentration. Libraries can present small fragment amplification at 120 base pairs. These fragments can be removed by repeating the last clean-up process (**steps 4–11**). After a second clean up, the library concentration will decrease, and the fragment size will change, thus the quality must be reassessed. To maintain a high concentration of the libraries, the elution volume in **step 11** can be reduced to 24 μL and transfer 22 μL to new PCR tubes.

3.4 Protocol 4: Library Prep with SMARTer Stranded Total RNA-seq Kit v3 for Pico Input Mammalian

In this method, total RNA is converted to cDNA and later probed with mammalian ribosomal sequences. Then, the cDNA from rRNA is depleted.

3.4.1 Sample and Time Considerations

1. Mammal only, highly degraded RNA (FFPE samples) to intact RNA (250 pg to 10 ng).
2. Schedule (*see* **Note 16**):

- Day 1: Fragmentation if required (severely degraded samples), and first strand synthesis.
- Day 2: cDNA purification, ribosomal cDNA depletion, and final library purification.

3.4.2 First-Strand cDNA Synthesis

Fragments of total RNA are randomly primed using the SMART Pico oligos mix v3, allowing for reverse transcription of all RNA species including rRNA. The SMARTScribe Reverse Transcriptase (RT) adds non-templated nucleotides to the 3′ end of the new first strand of cDNA. The Pico v3 SMART UMI adapter (included in the SMART UMI TSO mix v3) hybridizes to the non-templated nucleotides, giving the RT a new template to continue reverse transcription. Also, eight random nucleotides are added by the Pico v3 SMART UMI adapter, labelling the cDNA with unique molecular identifiers for downstream data analysis.

There are two options before proceeding. Option 1 (with fragmentation) is to be used when RNA samples have a RIN ≥ 4. Option 2 (without fragmentation) is to be used when samples are severely degraded (RIN ≤ 4) or FFPE samples. Proceed directly to the first-strand cDNA synthesis (**step** 7) for option 2.

1. Combine the following components in ice:

Component	Volume for one library (μL)
Total RNA	1–8
SMART Pico oligos mix v3	1
5× first-Strand buffer	4
Nuclease-free water	0–7
Total volume	13

2. Incubate the samples at 94 °C in a preheated, hot-lid thermal cycler for 2–4 min depending on the RIN of the input RNA. Incubate for 2 min for samples with RIN of 4; 3 min for samples with RIN 5–6, and 4 min for samples with RIN ≥ 7.
3. Place samples on ice for 2 min.
4. Add 7 μL of First-Strand Synthesis master mix (4.5 μL of SMART UMI-TSO mix v3 + 0.5 μL of RNase inhibitor +2 μL of SMARTScribe II Reverse Transcriptase) to each RNA sample.
5. Vortex for 2 s. Centrifuge briefly.
6. Incubate samples in a preheated, hot-lid thermal cycler with the following program:

- 180 min at 42 °C.
- 10 min at 70 °C.
- Hold at 4 °C.

Samples can be left in the thermal cycler at 4 °C until the next step.

7. Combine the following components in ice:

Component	Volume per one library (μL)
Total RNA	1–8
SMART Pico Oligos mix v3	1
Nuclease-free water	0–7
Total volume	9

8. Incubate samples at 72 °C in a preheated, hot-lid thermal cycler for 3 min.
9. Place samples on ice for 2 min.
10. Add 11 μL of First-Strand Synthesis master mix (4 μL of 5× First-Strand buffer +4.5 μL SMART UMI-TSO mix v3 + 0.5 μL of RNase inhibitor +2 μL of SMARTScribe II Reverse Transcriptase) to each RNA sample.
11. Vortex for 2 s. Centrifuge briefly.
12. Incubate samples in a preheated, hot-lid thermal cycler with the following program:
    - 180 min at 42 °C.
    - 10 min at 70 °C.
    - Hold at 4 °C.

Samples can be left in the thermal cycler at 4 °C until the next step.

3.4.3 Addition of Illumina Adapters and Indexes

Adapters and barcodes are added to the cDNA. The 5' PCR primer hybridizes with the Pico v3 SMART UMI adapter sequence while the 3' PCR primer binds to the sequences associated with the random primer.

1. Add 28 μL of PCR 1 master mix (2 μL of nuclease-free water +25 μL SeqAmp CB PCR buffer (2X) + 1 μL of SeqAmp DNA polymerase) to each sample.
2. Add 2 μL of the premixed SMARTer RNA Unique Dual Index primers.
3. Mix and centrifuge briefly.

4. Place the samples in a preheated, hot-lid thermal cycler. Perform PCR using the following program:
   - 1 min at 94 °C.
   - 5 or 10 cycles* of:

     15 s at 98 °C

     15 s at 55 °C

     30 s at 68 °C.
   - 2 min at 68 °C.
   - Hold at 4 °C.

   *Use 5 cycles for 0.25–10 ng regular RNA samples. Use 10 cycles for 0.5–10 ng FFPE RNA samples. Samples can be safely stored at −20 °C for up to 2 weeks.
5. Place AMPure XP beads at room temperature for at least 30 min before cDNA amplification. Vortex to resuspend.
6. Purify amplified cDNA by adding 40 μL of resuspended AMPure XP Beads. Mix by pipetting up and down at least six times.
7. Incubate samples for 8 min at room temperature.
8. Briefly centrifuge and place samples on a magnetic rack. After the solution is clear (~5–10 min) remove and discard the supernatant. Do not disturb the beads.
9. Keeping the samples on the magnetic rack, wash the beads with 200 μL of freshly prepared 80% ethanol, incubate for 30 s at room temperature, then discard the supernatant.
10. Repeat **step 9**.
11. Spin samples briefly and return them to the magnetic rack. Using a 10 μL pipette tip, remove all remaining ethanol in the tubes. Do not disturb the beads.
12. Air dry the beads for 2–5 min on the magnetic rack. When beads are still glossy but there is no more liquid in the tubes, move to the following step.
13. Remove samples from the magnetic rack and wash beads with 52 μL of nuclease-free water to rehydrate. Mix well on a vortex mixer or by pipetting up and down ten times.
14. Incubate for 5 min at room temperature.
15. Centrifuge briefly to collect all liquid from the tube walls and lid.
16. Place samples on the magnetic rack until the solution is clear.
17. Remove 50 μL of supernatant and transfer to new nuclease-free PCR tubes.

18. Add 40 μL of AMPure XP beads. Mix by pipetting up and down at least six times.
19. Incubate samples for 8 min at room temperature. During incubation time proceed immediately to the ribosomal depletion.

3.4.4 Depletion of Ribosomal cDNA with ZapR v3 and R-Probes v3

In this section, the library fragments that originated from rRNA (18S and 28S) and mitochondrial rRNA (m12S and m16S) are cut by ZapR v3 in the presence of R-probes v3 (mammalian-specific).

1. Thaw R-probes v3 and ZapR buffer at room temperature.
2. Place R-probes v3 on ice as soon as it is thawed and keep the ZapR buffer at room temperature.
3. Keep ZapR v3 on ice always and return to the freezer immediately after use.
4. Place empty PCR tubes on ice (*see* **Note 17**).
5. Preheat the thermal cycler to 72 °C.
6. Upon completion of the 8-min incubation period in **step 19** of Subheading 3.4.3, briefly centrifuge the sample tubes to collect the liquid at the bottom.
7. Place the tubes on the magnetic rack for 5 min or longer, until the solution is clear.
8. During the 5-min incubation period, pipette into the pre-chilled PCR tube(s) a sufficient volume of R-probes v3 for the number of samples (1.5 μL per sample, plus 10% to account for pipetting errors. Keep the PCR tube(s) containing the R-probes v3 on ice and immediately return the remaining unused R-probes v3 to a − 70 °C freezer.
9. Incubate PCR tubes containing R-probes v3 at 72 °C, following the program:
   - 2 min at 72 °C.
   - Hold at 4 °C.
10. Leave the R-probes in the thermal cycler at 4 °C for at least 2 min, but not more than 10–15 min.
11. Once the 5-min incubation period on the magnetic rack is complete and the samples are clear, pipette the supernatant and discard, while keeping the tubes sitting on the magnetic rack.
12. Wash the beads with 200 μL of freshly prepared 80% ethanol, incubate for 30 s at room temperature, then discard the supernatant.
13. Repeat **step 12**.

14. Spin samples briefly and return them to the magnetic rack. Using a 10 μL pipette tip, remove all remaining ethanol in the tubes. Do not disturb the beads.
15. Leave the open sample tubes at room temperature on the magnetic rack until the pellets appear dry.
16. While the beads are drying, prepare the ZapR master mix, by combining the following reagents at room temperature in the order shown:

Component	Volume per one library (μL)
Nuclease-free water	16.8
10× zap buffer	2.2
ZapR v3	1.5
R-probes v3 (from thermal cycler)	1.5
Total volume	22

17. Add 22 μL of ZapR master mix to each sample of dried beads from **step 15**. Remove the sample tubes from the magnetic rack, mix thoroughly to resuspend the beads.
18. Incubate at room temperature for 5 min.
19. Centrifuge briefly to collect all liquid from the tube walls and lid.
20. Place samples on the magnetic rack until the solution is clear.
21. Pipette 20 μL of supernatant into new PCR tubes. Do not disturb the beads.
22. Incubate the samples in a preheated, hot-lid thermal cycler using the following program:
    - 60 min at 37 °C.
    - 10 min at 72 °C.
    - Hold at 4 °C.

    Proceed immediately to the next step.
23. Set up PCR 2, the final RNA-seq library amplification. Add 80 μL of PCR 2 master mix (26 μL of nuclease-free water +50 μL of SeqAmp CB PCR buffer +2 μL of PCR2 primers v3 + 2 μL of SeqAmp DNA polymerase) to each sample.
24. Mix and centrifuge briefly.
25. Place the samples in a preheated, hot-lid thermal cycler. Perform PCR using the following program:

- 1 min at 94 °C.
- 9–16 cycles* of:

  15 s at 98 °C

  15 s at 55 °C

  30 s at 68 °C.
- Hold at 4 °C.

*The number of PCR cycles should be adjusted based on RNA input:

- 5–10 ng, 12 cycles for regular RNA; 10 cycles for FFPE RNA
- 0.5–1 ng, 14–15 cycles for regular RNA; 13 cycles for FFPE RNA
- 0.25 ng, 16 cycles for regular RNA.

Samples can be stored at −20 °C for up to 2 weeks.

3.4.5 Purification of Final RNA-seq Library with AMPure XP Beads

1. Place AMPure XP beads at room temperature for at least 30 min before purification of final RNA-seq library.
2. Add 100 μL of AMPure XP beads. Mix by pipetting up and down at least six times.
3. Incubate samples for 8 min at room temperature.
4. Briefly spin down and place samples on a magnetic rack. After the solution is clear (~5–10 min) remove and discard 190 μL of supernatant. Do not disturb the beads.
5. Keeping the samples on the magnetic rack, wash the beads with 200 μL of freshly prepared 80% ethanol, incubate for 30 s at room temperature, then discard the supernatant.
6. Repeat **step 5**.
7. Spin samples briefly and return them to the magnetic rack. Using a 10 μL pipette tip, remove all remaining ethanol in the tubes. Do not disturb the beads.
8. Air dry the beads for 2–5 min on the magnetic rack. When beads are still glossy but there is no more liquid in the tubes, move to the following step.
9. Remove samples from the magnetic rack and wash beads with 22 μL of nuclease-free water to rehydrate. Mix well on a vortex mixer or by pipetting up and down ten times.
10. Incubate for 5 min at room temperature.
11. Centrifuge briefly to collect all liquid from the tube walls and lid.

12. Place samples on the magnetic rack until the solution is clear.
13. Remove 20 μL of supernatant and transfer to new nuclease-free PCR tubes.
14. Add 20 μL of AMPure XP beads. Repeat **steps 3–8**.
15. Once the beads are dry, add 15 μL of nuclease-free water to rehydrate. Mix well on a vortex mixer or by pipetting up and down ten times.
16. Incubate for 5 min at room temperature.
17. Centrifuge briefly to collect all liquid from the tube walls and lid.
18. Place samples on the magnetic rack until the solution is clear.
19. Remove 13 μL of supernatant and transfer to new nuclease-free PCR tubes.

4 Notes

1. The RIN or RNA integrity number indicates the degree of degradation of a total RNA sample. It ranges from 1 (highly degraded or fragmented total RNA) to 10 (total RNA intact). The RIN number is the result of an algorithm that considers the ratio, area, and intensities in the entire electrophoretic trace of an RNA sample [9].
2. Total RNA aliquots can be stored at −80 °C while preparing the first strand synthesis reaction buffer and random primers mix. Samples can be thawed on ice before the first washing of the oligo dT beads (**step 6** in this Subheading 4).
3. This mix is going to be used later in the protocol to elute mRNA captured on the oligo dT beads. It must be kept on ice.
4. Prepare one 1.5 mL tube for each ten samples to process.
5. Depending on the magnetic rack strength, it can take up to 5 min for the solution to clear.
6. 50 min is the fragmentation time for RNA inserts of 150–300 bp.
7. This part can be performed in 1 or 2 days. If it is going to be carried out in 2 days, the first safe stopping point is after the purification of the adaptor-ligated DNA.
8. The NEBNext Ultra II End Prep reaction buffer and Enzyme mix are very viscous. Prepare the given volumes plus 10%.
9. The adaptor and USER enzyme are included in the Oligos kit with the indexes.

10. The input of total RNA must be considered in this step of the process. Likewise, the number of PCR cycles in the enrichment step depends on the amount of total RNA used as input. The NEBNext adaptor for Illumina must be diluted according to the protocol instructions.
11. The NEBNext Ultra II Ligation reaction master mix is very viscous. Prepare the given volumes plus 10%.
12. This kit's protocol recommends eluting the dsDNA in 17 μL of 0.1× TE, and to transfer 15 μL of supernatant. Increasing the elution volume to 25 μL and transferring 22 μL, has no negative impact on the DNA output. This also allows for a decrease in the volume of the index primers in the enrichment reaction.
13. Quality control is necessary to normalize the libraries before sequencing. Instruments like TapeStation, will indicate the average fragment size for each library. Additionally, concentration should be assessed with fluorometric methods such as Qubit.
14. Check a few samples in a High Sensitivity RNA TapeStation to verify rRNA depletion.
15. The cDNA amplification obtained with the Takara kit results in outputs that work well with the NEBNext Ultra II FS DNA library prep. Since the cDNA was not previously fragmented, the fragmentation step is necessary.
16. This protocol can be completed in 2 or 3 days. The depletion of ribosomal cDNA with ZapR v3 and R-Probes v3 must be performed in one step. Then, depending on the number of samples, it is recommended to assign 1 day for this step.
17. The prechilled PCR tubes are going to be used to prepare the R-probes v3 in the thermal cycler. After incubation in the thermal cycler, the R-probes v3 are added to the ZapR master mix. Prepare enough volume for all samples plus 10%.

References

1. O'Neil D, Glowatz H, Schlumpberger M (2013) Ribosomal RNA depletion for efficient use of RNA-seq capacity. Curr Protoc Mol Biol Chapter 4:Unit 4.19.1–4.19.8
2. Hrdlickova R, Toloue M, Tian B (2017) RNA-seq methods for transcriptomic analysis WIREs RNA. Wiley Interdiscip Rev RNA 8:1–17
3. Bostick M, Bolduc N, Farmer A (2016) Strand-specific transcriptome sequencing using SMART technology. Curr Protoc Mol Biol 4(27): 1–4.27.18
4. van Dijk E, Jaszczyszyn Y, Themes C (2014) Library preparation methods for next-generation sequencing: tone down the bias. Exp Cell Res 322:12–20
5. Podnar J, Deiderick H, Huerta G, Hunicke-Smith S (2014) Next-generation sequencing RNA-seq library construction. Curr Protoc Mol Biol 4.21.1–4.21.19
6. Zhao S, Zhang Y, Gordon W, Quan J, Xi H, Du S, von Schack D, Zhang B (2015) Comparison of stranded and non-stranded RNA-seq transcriptome profiling and investigation of gene overlap. BMC Genomics 16:675
7. Zhu YY, Machleder EM, Chenchick A, Li R, Siebert PD (2001) Reverse transcriptase reverse switching: a SMART™ approach for full-length

cDNA library construction. BioTechniques 30: 892–897

8. Ura H, Togi S, Niida Y (2022) A comparison of mRNA sequencing (RNA-seq) library preparation methods for transcriptome analysis. BMC Genomics 23:303

9. Shroeder A, Mueller O, Stocker S et al (2006) The RIN: an RNA integrity number for assigning integrity values to RNA measurements. BMC Mol Biol 7:3

Chapter 9

Dissection of Gene Expression at the Single-Cell Level: scRNA-seq

Samar Nasser Chehimi

Abstract

Sequencing approaches that allowed for a better resolution of the transcriptome have been a major goal in the transcriptomics field since the development of RNA-seq techniques. While RNA-seq provides gene expression data from one entire sample in bulk, single-cell analysis allows for a better characterization of gene expression associated to specific cell types. Single-cell RNA-seq (scRNA-seq) is a reliable technique to unravel transcriptomic features of the tissues of interest dissociated at a single-cell level. The main feature of the single-cell technique is its ability to generate barcoded individual cells that allow for tracking the origin of thousands to millions of transcripts and reveal new cell types associated to diseases and different cell types and states. In this chapter, we discuss how scRNA-seq has become the gold standard to deepen the understanding of the gene expression with single-cell resolution.

Key words Single-cell, Gene expression, Transcriptome, scRNA-seq

1 Introduction

1.1 What Is the Single-Cell Assay?

Understanding cell behavior has always been a challenge, especially in a disease context. Several questions were asked, such as: what controls cell behavior in a normal state? What changes occur in the cell circuitry to shift to a disease state? What internal factors control different cell responses? What transcription factors are associated with the outcome of a disease? These questions and many more associated with gene expression found in bulk RNA-seq require an alternative for the answers.

Bulk RNA-seq was a revolutionary advance, which enabled the detection of transcripts and differentially expressed genes (DEGs), and a clearer understanding of how different isoforms could impact functional gene expression [1]. Previous studies that investigated the genome from a microarray perspective achieved a more comprehensive genomic understanding with bulk RNA-seq. The impact of RNA-seq was noticeable in cancer, intellectual disability,

Valerie J. Carabetta and Olaitan Akintunde (eds.), *High Throughput Gene Screening: Methods and Protocols*,
Methods in Molecular Biology, vol. 2866, https://doi.org/10.1007/978-1-0716-4192-7_9,

psychiatric diseases, and many other diseases, especially undiagnosed cases, and ones with inconclusive results [2–5]. In bulk RNA-seq, all of the transcriptomic content of a sample, which can be either enriched for mRNA or evaluated for all RNA species, is extracted to obtain a quantification of the average gene expression from all cells in the original population [1]. This quantification is particularly helpful when comparing samples under different natural or experimental conditions to evaluate global gene expression dysregulation [6]. Quantification of the transcriptome provides overall gene expression analysis, but it lacks the resolution to understand the impact and the role of different expression patterns among different cell populations. It became necessary to characterize which types of cells were related to the different transcription patterns that were identified with bulk RNA-seq. This more detailed cell characterization came with the innovative single-cell techniques developed between 2010–2020.

Single-cell approaches provide a unique opportunity to isolate individual cells or nuclei in nanoliter-size droplets or microwell-plates, and detect all the transcripts generated or features present, such as chromatin states in individual cells or nuclei. One of the most used single-cell approaches currently is provided by 10x Genomics. In these protocols for single-cell RNA-seq (scRNA-seq), the initial starting material is isolated cells or nuclei from tissues or cell culture. When working with tissue, isolated cells and nuclei can be obtained through physical and enzymatic dissociation of the tissue of interest, which allows for the release of individual cells. After washing to remove debris and remaining tissue, the cells or nuclei are pelleted and then resuspended in an optimized buffer. After obtaining a clean cell or nuclei suspension, the sample is subjected to barcoding. This barcoding step is done with the use of gel bead-in-emulsion (GEM), which contains thousands of unique barcodes, which are unique oligonucleotide sequences for each cell. By using these GEMs provided with the single-cell kit, it is possible to track the cell origin of each one of the transcripts. For gene expression investigation, during the GEM incubation and reverse transcription steps, the cell membranes are lysed, and full-length cDNA is generated from poly-adenylated mRNA. All genomic content receives an individual barcode within the GEM, which allows the grouping of all transcripts from the same cell in the downstream analysis. After this process, all GEMs are dissolved to allow for cDNA and library construction. This same method can be executed in isolated nuclei (snRNA-seq) to decipher transcripts pertinent to the cell nucleus. Recently, in 2022, many authors adapted current single-cell assays for long-read sequencing, aiming for the detection of isoforms and better characterization of transcripts, which increases the likelihood to find variants of clinical significance and evidence of gene fusions [7–9]. In 2023, 10x

Genomics confirmed partnership and compatibility with Nanopore and PacBio, known for their long-read sequencers, increasing the resolution power of scRNA-seq.

One of the first reports to disclose major single-cell findings revealed the capacity for this technique to characterize new cell types in blood, a tissue that had been broadly studied before but now was being analyzed under a new perspective [10]. Therefore, scRNA-seq provides valuable insights about the cell diversity and heterogeneity of a sample, allows for the characterization of diverse and complex cell populations, leads to the discovery of new cell types, and is useful to describe different disease states. In addition, it can be used to unravel cellular processes and different cells involved in the same pathway, profile immune features for vaccine development, and track how stem cell differentiation occurs. This chapter will provide an overview of the single-cell approaches with emphasis on scRNA-seq and snRNA-seq.

1.2 Evolution of Single-Cell Approaches

The ability to process individual cells to characterize the RNA content was considered by *Nature Methods* as the method of the year in 2013, but the true impact and broader application of this method was realized a couple of years later, when it became possible to scale up the experiments, with more automated versions of single-cell individualization that allowed for a higher cell throughput. Cellular composition in a specific sample was initially assessed through flow cytometry, a biomarker-based strategy that can detect up to 30 different cell types through distinct fluorophores. It was especially important to evaluate the efficacy of drugs by detecting depletion or not of a specific targeted cell population after treatment [11]. Even though flow cytometry still presents many advantages and is still being broadly used, it is laborious and requires well-defined biomarker selection for gating approaches [11]. Laser-capture microdissection is also used as an approach to isolate single cells. In this method, a heterogeneous tissue is harvested to select specific cells under microscopic visualization, which provides an enriched sample, mostly containing the cell population of interest [12]. Later, the transcriptome of the collected samples can be evaluated to determine the molecular signatures of the cells, and evolution or regression of a disease [12]. However, the drawbacks of this technique are like the ones described for flow cytometry; it is a time-consuming technique and requires morphology and histology expertise for tissue evaluation and dissection.

With the development of single-cell suspension approaches, a new era to investigate individualized cells began. With this technique, better accuracy could be achieved in transcriptome measurements for hundreds to thousands of single cells in one experiment. To process individual cells, two main technical approaches are predominant: microwell-based methods and droplet-based methods (Fig. 1). The microwell-based method is efficient in

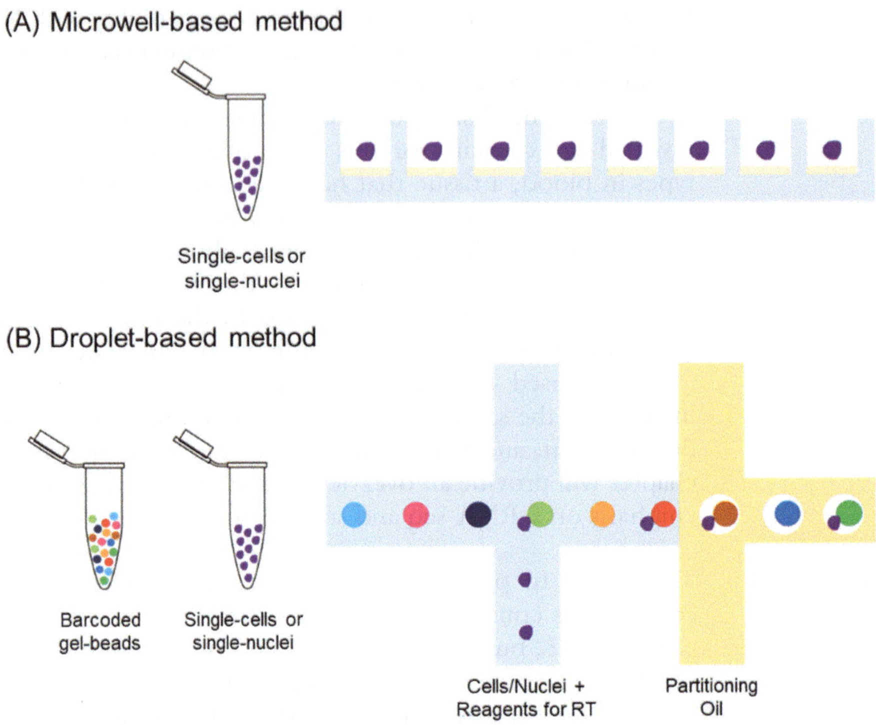

Fig. 1 The most common technical methods for scRNA-seq. (**a**) Microwell-based method. Single cells or nuclei are split in a micro or nanowell-plate in specific wells containing the reagents needed for reverse transcription. (**b**) Droplet-based method. Single cells or nuclei go through a microfluidics system to generate hydrophobic droplets containing one cell or nucleus, one gel-bead with a unique barcode and reagents needed for reverse transcription (RT). The shown method is based on 10x Genomics technology and platform, but other droplet generators are available and based in a similar methodology

individualizing cells and provides the ability to visualize if cells were captured by this method but has low throughput capacity. Each sample goes through a manual separation of cells in wells, which can be performed using laser-capture microdissection. By utilizing a microscope, it is possible to optically identify and isolate specific cells or cell clusters within a tissue sample. This process involves employing a laser beam to delineate the desired cells, which are then excised from the slide, and transferred into separate wells, in either 96- or 384-well plates [12, 13]. Each well contains one cell and the reagents needed to generate cDNA from mRNA to amplify the transcriptome [14]. With the advances in technology, a higher number of cells, thousands or even millions, can be processed using a microfluidics system that individualizes each cell from the single-cell suspension in an automated fashion, generating thousands of individual cell-containing droplets, with a barcoding method that allows for cells to be multiplexed in a single reaction [15]. The main differences between microwell-based and droplet-based methods is the throughput capacity of the method, the hands-on time, and the ability to visually inspect captured cells [14].

2 Samples

Most of the single-cell approaches can be used with either live cells or nuclei, a choice that will be driven by the experimental design, which depends on the type of sample that is available and the hypothesis to be tested. For fresh samples or cell culture, live cells are the best option if the single-cell suspension can be generated in a reasonable time, with a good concentration and a viability higher than 90%. If the goal is to analyze nuclear transcripts, samples can be stored at −80 °C or in liquid nitrogen, the latter option being the best. For samples stored for a long time, postmortem tissues, and sensible cell-types like neurons, snRNA-seq becomes the only feasible option. Even though, nuclei datasets might miss cytosolic transcripts that would be present in scRNA-seq samples, previous studies have shown that snRNA-seq recapitulates results obtained for scRNA-seq in the kidney [16] and inaccessible tissues, like the brain [17]. Therefore, isolation of nuclei is a valid alternative to be used instead of live cells.

Many factors need to be considered when processing a sample that will undergo single-cell assays. Cells or nuclei suspensions can be obtained through different methods depending on the tissue that is being used, such as commercially available nuclei isolation kits, and in-house prepared buffers with sucrose and iodixanol gradients, among other options. Each tissue and condition will have specific properties and therefore, isolation of cells or nuclei must be optimized for each type of sample [18]. The final suspension must follow quality requirements that consider isolated cells without clumped cells, the absence of background debris, fibers, or artifacts that could clog the microfluidics system, and the highest possible viability when handling live cells. Following these requirements ensures that a good-quality sample will be processed to generate reliable data with few ambient RNAs, like degraded RNAs or free-floating RNAs, released in the cell suspension during sample preparation that are not associated to a specific cell, that could impact the analysis [18].

There are available alternatives to improve the quality of the samples when the initial microscopic evaluation of the samples does not meet the optimal quality criteria. Clumped cells and background debris can be removed by filtering the final cell suspension through cell strainers (40 mM filters) or through fluorescence-activated cell sorting (FACS). FACS is an efficient strategy to improve the quality of the sample by removing debris and removing dead cells to increase the viability of a sample [18]. However, multiple considerations must be made when sorting samples. First, the collection buffer should not contain EDTA or excessive amounts of Mg^{2+} since this can interfere with the single-cell workflow. In addition, whenever sorting cells, propidium iodide can be used, but when sorting nuclei, Aminoactinomycin D (7-AAD) dye must be used instead. Although FACS allows for a well-defined labeling, it is a process that can be time consuming, leading to cell stress or increased cell death if not processed quickly enough. If FACS is unavailable to increase cell viability or the quality of a sample, a bead-based approach can be used to select specific cells or to remove dead cells, therefore enhancing the concentration of live cells, and increasing overall viability [19, 20].

Besides good sample quality, accurate cell counting is important for the cell loading and analysis steps. Counting can be performed with automated cell counters or manual counting with a hemacytometer. The downside of an automated method is that the number might be overestimated if debris is present in similar size and shape as the nuclei. If using a manual method, debris and cells are more easily distinguishable, but extra care should be considered when counting nuclei stained with trypan blue, since debris can also stain in a similar pattern. Fluorescence staining, such as acridine orange and propidium iodide, is another alternative and is a reliable method since only dead cells or nuclei will be stained, with an obvious difference with live cells and debris. Assuring a precise cell counting will guarantee the correct initial cell input to achieve the desired cell recovery target.

3 Methods

3.1 Single-Cell Processing

After assuring sample quality and concentration, it is recommended to start the single-cell workflow immediately when working with live cells to maintain viability and membrane integrity. The number of cells that are initially loaded for droplet generation, currently the most common method for the single-cell workflow, will be based on the targeted cell recovery. Although the highest recovery rate is usually the preferred choice to take full advantage of the method, this results in a higher number of multiplets, which are droplets containing more than one cell or one nucleus, that need to be identified and removed bioinformatically to prevent a misguided analysis.

There are multiple ways to process samples for single-cell approaches, but the process is similar if using any droplet-based method. For gene expression investigations, cells or nuclei are initially encapsulated to individualize them and receive a cell barcode. This barcode will be copied throughout the process and all transcripts containing this barcode to be able to be tracked to the cell of origin. Once the cells are labeled and reverse transcribed, cDNA is constructed followed by library preparation. During library construction, amplified cDNA goes under enzymatic fragmentation, followed by end-repair, A-tailing, and adaptor ligation to amplify library content to be sequenced. The initial cDNA contains about 1500 bp and the final libraries are between 400–600 bp (the size of the cDNA and library fragments varies broadly depending on the sample type and method used).

If samples will be used for other applications rather than gene expression, additional steps might be needed. If chromatin analysis is one of the goals, there is a transposition step, which labels all the accessible portions of the chromosome with transposase and then the DNA in open chromatin regions is fragmented. For spatial methodologies, an extra tissue fixation step is required. To maximize the utilization of the droplet capacity in a particular single-cell assay, it is possible to pool cells from different species such as humans, mice, and rats. This can be achieved by incorporating barcodes or tags into the cells, enabling their identification during subsequent analysis stages [21, 22].

3.2 Types of Single-Cell Assays

One of the main focuses of single-cell studies is gene expression through transcriptional changes. However, it is possible to explore many other biological features with this technology, which will be briefly discussed in the following sections.

3.2.1 Single-Cell Gene Expression

Gene expression is the most widely used application for single-cell assays. As previously mentioned, it allows the investigation of different patterns of gene expression at a single-cell level with the capacity to individualize the transcriptome from each cell. The most used method captures the 3′ end of the RNA molecule, but recent publications have overcome the technical challenges implicit in sequencing full-length cDNAs for single-cell gene expression [7, 8, 23]. Capturing the full RNA molecule might become a more common method choice for long-read approaches in the near future.

During the generation of GEMs, it is crucial to use a high dilution for the sample input. The ideal initial concentration for the sample is approximately 1000 cells/μL. This dilution is essential to ensure cell separation and guarantee the individual encapsulation of each cell within the microfluidic-generated droplets. When a sample is processed with a higher cell concentration, there is a risk of generating multiplets that contain more than one cell, thereby compromising the single-cell nature of the sample.

After individualization of cells or nuclei, reverse transcription occurs and then all cDNA products contain the specific barcode associated to the cell of origin. The hydrophobic droplets are then dissolved, and the library generated in one reaction for all the droplets from the same sample. When libraries are being constructed, different samples can receive different indices, which allows for samples to be pooled together for sequencing purposes, which significantly reduces sequencing costs. The presence of a sample index and cell barcodes allow for data to be deconvoluted during the bioinformatics analysis.

The advantages of single-cell gene expression are that it allows for the investigation of specific timepoints of a disease and can be used to describe the evolutionary trajectory for single cells. It also allows for the inference of changes in the cell population constitution or expression patterns. The major limitations of this approach are associated with the lack of standardization in the analysis of single-cell data and sub-clustering of different cell populations, which hinders multiple datasets integration [24].

3.2.2 Single-Cell ATAC-seq

To investigate epigenetic signatures about chromatin accessibility, ATAC-seq (assay for transposase-accessible chromatin using sequencing) was developed in 2013 [25]. This method allows for the identification of epigenomic features by evaluating the structure of chromatin, identifying regions of open chromatin and nucleosome state of regulatory regions in isolated nuclei. Later, this method was adapted to be used in a complementary way with gene expression assays, as it will be later discussed in this chapter. The principle behind the original ATAC-seq technique is the use of a transposase that recognizes and inserts into accessible regions in the chromatin, creating a tag, and followed by cleavage of these regions to identify regions responsible for epigenetic modulation. Once these regions are cleaved, adapters are added to the nucleosome-free regions of the DNA to create fragmented DNA with adapters.

To evaluate ATAC-seq at a single-cell level (snATAC-seq), a unique barcode is added to each nucleus allowing for higher resolution of the accessible chromatin profile, even in mitochondrial DNA [26]. snATAC-seq can only be used for nuclei since it targets chromatin, which is a nuclear structure. The recommended nuclear suspension must contain <5% of live cells if preparing the sample from a fresh tissue or cell culture. Therefore, if the nuclei suspension was obtained from cells, live cells should be filtered out to increase the concentration of nuclei in the suspension. FACS can be an option to sort nuclei if 7-AAD is used as the sorting dye, which maintains nuclear integrity, has high sensitivity, and low background debris. Sequencing of the snATAC-seq libraries generates reads and data for the transposed DNA, including individual cell markers to track back to the origin of each transcript. The detection of chromatin changes provides further insights associated with nucleosome position and identification of regulatory mechanisms, such as enhancers and promoters, which might directly affect epigenetic regulation [27].

3.2.3 Single-Cell Immune Profiles

Another feature that can be observed through single-cell methods is the immune profiling of T- and B-cells, through the classification of different receptors (BCRs and TCRs). CRISPR (Clustered Regularly Interspaced Short Palindromic Repeats) modifications can also be evaluated at the single-cell level or changes in cell-surface proteins, when labeled with specific antibodies, increasing the power to evaluate immune cells [28–30]. In this method, the 5' UTR end is targeted to detect full-length transcripts. The method to generate droplet-based single immune cells is very similar to the single cell gene expression method. The cDNA will be generated depending on what specifically is being evaluated. For example, if evaluating CRISPR modifications, the cDNA will be generated from single-guide RNAs and when evaluating BCRs and TCRs, the V(D)J polyadenylated mRNA transcripts are studied. All cells will contain specific cell barcodes and individual indices for each sample. This method can be used to develop immunotherapies and vaccines, investigate the diversity of the immune repertoire, track specific clones over time or between samples, and detect shared and different immune receptors [29].

3.2.4 Multiomic Single-Cell Sequencing

Single-cell techniques have led to many important discoveries [10, 21, 31]. However, a higher impact can be achieved with multiomic techniques that explore multiple features, such as gene expression and chromatin conformation, in the same nucleus, when compared to standalone assays. Initially, data from single-cell gene expression assays were integrated with data from ATAC-seq and protein assays [32, 33]. However, using the same samples does not mean the exact same nuclei are analyzed, since microfluidics techniques are prone to cell selection bias and since they are always done with highly diluted samples, not all nuclei are processed, thus generating analysis bias. For this reason, there was an effort to design assays able to obtain ATAC-seq and snRNA-seq in the same experiment [21, 34]. In this approach, the same nuclei are tracked for concomitant chromatin and gene expression features. The first step consists in nuclei transposition to detect accessible chromatin regions and labeling with specific nuclei barcodes and unique molecular identifiers (UMIs), like the ATAC-seq method. After this step, part of the sample is used to generate cDNA and libraries for gene expression analysis and the remaining part is used to generate samples for ATAC-seq analysis. For the analysis, there are specific strategies able to combine both sets of data and define an array of characteristics intrinsic to each nucleus and unique cell population [21, 35]. Briefly, in these multimodal approaches, by using the same cell-type markers and the individual nuclei barcodes, datasets for snRNA-seq and snATAC-seq are integrated and gene expression is evaluated simultaneously with chromatin accessibility [21, 35].

3.2.5 Spatial Transcriptomics

More recently, still thinking about exploring integrated genomic aspects of the same sample, spatial technologies were developed to fill the gap about the precise location of individual cells among a population of cells within a tissue. This allowed for a better understanding of complex tissues and cells in different disease states that overlap in functions, as determined by scRNA-seq studies [36, 37]. The advantage of using spatial technologies is the ability to use fresh frozen and formalin-fixed paraffin-embedded (FFPE) samples, that usually present high levels of nucleic acid degradation. This assay requires histology, pathology, and molecular biology expertise, since the initial sample consists of a piece of tissue placed on a glass slide that already contains labeled probes in known locations. The correct selection of the region or tissue to be studied directly impacts the results and analysis. A thin section of the area of interest is mounted and fixed on a slide to facilitate in situ hybridization of the probes with the mRNA. The probes are specific oligonucleotides sequences designed to hybridize with targeted transcripts of genes of interest; the probes can either be selected from commercially available panel options or customized for specific targets. Later, the tissue is detached from the slide to continue library preparation. Some commercially available options also consist of a predefined set of mRNA probes, which are similar to a multiplexed and high-resolution RNAscope method. This method, RNAscope, also allows for the visualization of RNA in a FFPE tissue through in situ hybridization with high sensitivity but for few genes at a time, which is useful to investigate rare cell populations of a specific transcript or genes expressed locally at a spatial resolution [36, 38].

4 Sequencing

When initially developed, single-cell approaches relied on short-read sequencing. Short reads are already an established and reliable sequencing method that generates powerful data with considerable depth and well-developed alignment methods. The most common single-cell methods utilize the Illumina platform, which uses short-reads sequencing and therefore require the addition of specific Illumina adaptors, which hybridize to complementary sequences on the flow cell [39]. However, aiming to increase the power of unraveling small variations, the single-cell approach was recently adapted for analysis using long reads. The advantages of long reads in single-cell samples is the ability to detect full transcripts and isoforms that can be associated with different disease outcomes and individual responses [40]. Each type of single-cell assay will have specific run parameters based on the length of the final library, which will always include barcodes and UMIs, but might contain additional portions depending on the type of assay that is being prepared.

Usually, for sequencing purposes, it is recommended to proceed with a paired-end sequencing method. Although paired-end sequencing has a higher cost than single-end sequencing, it generates data for the 5′ and 3′ directions, ensuring higher quality of sequencing data. In addition, it provides better characterization of cells sub-clustering and enables the detection of rare and novel transcripts in gene expression assays [41]. The recommended number for sequencing depth and targeted reads per nuclei will depend on the quality of the sample being processed, the expected number of single cells to be recovered, the complexity of the cell populations, the detection of rare transcripts and the type of single-cell assay [31]. Therefore, the number of required reads per cell might vary between 10,000 to 50,000 reads per cell [31]. By aligning the fragments and identifying the positions of the cell barcode and UMI, it is possible to assess the sequences that correspond to specific transcripts.

4.1 Single-Cell Data Analysis

After sequencing is completed, all scRNA-seq data is analyzed through specific pipelines. One key step for the analysis is the identification of the distinct cell populations that constitute the sample being analyzed. This cell annotation is done after defining specific and canonical cell type-specific marker genes for each cell type in the sample. The cell patterns identified during the analysis are based on a principal components analysis, a computational approach which evaluates and integrates all variation findings to define individual patterns for each cluster. Some challenges may be involved in the cell annotation, such as some clusters may contain several cell types or unknown cell types that lack a well-defined marker, which cannot be classified. Since the cell clusters are defined based on pre-selected gene markers, gene activity might be a relevant parameter to characterize different cell clusters. For example, two different clusters may not represent two different cell types, but in fact are the same cell population in two different activation states. Once marker genes are defined, it is possible to separate clusters for each cell type and proceed with annotation and dimensionality reduction techniques.

One resource that has been frequently used in single-cell visualization are UMAPs [42]. UMAP stands for uniform manifold approximation and projection and allows for a 2D visualization of the cell clusters by transforming very dense and complex high-dimensional objects into a lower dimensional graph, in a "flat format," for a more interactive and easier format of data visualization and interpretation [42]. Multiple pipelines, tools, and software packages are available in R for snRNA-seq analysis and data visualization, such as seurat [35], scater [43], and sceasy [44], among others.

For data produced with 10x Genomics' platform, Cell Ranger and Loupe are offered by the company and are used to perform demultiplexing, read alignment, generation of feature barcode matrices, cell counting, recovery estimation, performing clustering, and further analyses. Other companies that also develop single-cell applications, such as Parse Biosciences, also developed their own data analysis package that provides summary statistics and allows to assess data quality, identify different cell clusters and gene expression.

5 Future Directions

Single-cell sequencing has been a suitable alternative to characterize different tissues and organoid models [45]. The future of single-cell gene expression involves increasing the number of cells under investigation, compare sex-specific molecular features, and decreasing the cost of the experiment to make it more affordable and accessible. Also, besides technical improvements and the implementation of high-throughput platforms, more integrated modalities should be developed to investigate multiple features of the same cell. Multimodal approaches include spatial technologies, proteomics, and DNA accessibility to track all the changes that occur in a cell during specific and controlled conditions for better perspectives on cell biology and behavior. The field is developing at such a fascinating speed that long-read sequencing is already being adapted at a spatial resolution [23].

One current challenge that needs to be surpassed refers to bioinformatic tools. The fact that the numbers of cell clusters and different cell types varies broadly in the same tissue among different studies reveals the need of a standardization of the analysis programs and scripts. Currently, many atlases are published, but more data, clear analysis, and more samples batched together will provide a more precise cell description and characterization. Considering the power of single-cell genomic techniques, constant improvements are expected, aiming for a complete cellular profiling that will give better directions to targeted treatments.

References

1. Wang Z, Gerstein M, Snyder M (2009) RNA-Seq: a revolutionary tool for transcriptomics. Nat Rev Genet 10(1):57–63. https://doi.org/10.1038/nrg2484
2. Li X, Teng S (2015) RNA sequencing in schizophrenia. Bioinf Biol Insights 9s1:BBI.S28992. https://doi.org/10.4137/bbi.s28992
3. Gonorazky HD, Naumenko S, Ramani AK, Nelakuditi V, Mashouri P, Wang P, Kao D, Ohri K, Viththiyapaskaran S, Tarnopolsky MA, Mathews KD, Moore SA, Osorio AN, Villanova D, Kemaladewi DU, Cohn RD, Brudno M, Dowling JJ (2019) Expanding the boundaries of RNA sequencing as a diagnostic tool for rare Mendelian disease. Am J Hum Genet 104(3):466–483. https://doi.org/10.1016/j.ajhg.2019.01.012
4. Rahman MR, Petralia MC, Ciurleo R, Bramanti A, Fagone P, Shahjaman M, Wu L, Sun Y, Turanli B, Arga KY, Islam MR, Islam T, Nicoletti F (2020) Comprehensive analysis of RNA-Seq gene expression profiling of brain transcriptomes reveals novel genes, regulators, and pathways in autism Spectrum disorder. Brain Sci 10(10):747. https://doi.org/10.3390/brainsci10100747
5. Hong M, Tao S, Zhang L, Diao L-T, Huang X, Huang S, Xie S-J, Xiao Z-D, Zhang H (2020) RNA sequencing: new technologies and

applications in cancer research. J Hematol Oncol 13(1). https://doi.org/10.1186/s13045-020-01005-x

6. Voineagu I, Wang X, Johnston P, Lowe JK, Tian Y, Horvath S, Mill J, Cantor RM, Blencowe BJ, Geschwind DH (2011) Transcriptomic analysis of autistic brain reveals convergent molecular pathology. Nature 474(7351):380–384. https://doi.org/10.1038/nature10110
7. Tian L, Jabbari JS, Thijssen R, Gouil Q, Amarasinghe SL, Voogd O, Kariyawasam H, Du MRM, Schuster J, Wang C, Su S, Dong X, Law CW, Lucattini A, Prawer YDJ, Collar-Fernández C, Chung JD, Naim T, Chan A, Ly CH, Lynch GS, Ryall JG, Anttila CJA, Peng H, Anderson MA, Flensburg C, Majewski I, Roberts AW, Huang DCS, Clark MB, Ritchie ME (2021) Comprehensive characterization of single-cell full-length isoforms in human and mouse with long-read sequencing. Genome Biol 22(1). https://doi.org/10.1186/s13059-021-02525-6
8. Wang Q, Bönigk S, Böhm V, Gehring N, Altmüller J, Dieterich C (2021) Single-cell transcriptome sequencing on the Nanopore platform with ScNapBar. RNA 27(7):763–770. https://doi.org/10.1261/rna.078154.120
9. Ahmed YW, Alemu BA, Bekele SA, Gizaw ST, Zerihun MF, Wabalo EK, Teklemariam MD, Mihrete TK, Hanurry EY, Amogne TG, Gebrehiwot AD, Berga TN, Haile EA, Edo DO, Alemu BD (2022) Epigenetic tumor heterogeneity in the era of single-cell profiling with nanopore sequencing. Clin Epigenetics 14(1). https://doi.org/10.1186/s13148-022-01323-6
10. Villani A-C, Satija R, Reynolds G, Sarkizova S, Shekhar K, Fletcher J, Griesbeck M, Butler A, Zheng S, Lazo S, Jardine L, Dixon D, Stephenson E, Nilsson E, Grundberg I, Mcdonald D, Filby A, Li W, De Jager PL, Rozenblatt-Rosen O, Lane AA, Haniffa M, Regev A, Hacohen N (2017) Single-cell RNA-seq reveals new types of human blood dendritic cells, monocytes, and progenitors. Science 356(6335):eaah4573. https://doi.org/10.1126/science.aah4573
11. Mitra-Kaushik S, Mehta-Damani A, Stewart JJ, Green C, Litwin V, Gonneau C (2021) The evolution of single-cell analysis and utility in drug development. AAPS J 23(5). https://doi.org/10.1208/s12248-021-00633-6
12. Espina V, Wulfkuhle JD, Calvert VS, Vanmeter A, Zhou W, Coukos G, Geho DH, Petricoin EF, Liotta LA (2006) Laser-capture microdissection. Nat Protoc 1(2):586–603. https://doi.org/10.1038/nprot.2006.85
13. Potter SS, Brunskill EW (2014) Building an atlas of gene expression driving kidney development: pushing the limits of resolution. Pediatr Nephrol 29(4):581–588. https://doi.org/10.1007/s00467-013-2602-9
14. Zhou W-M, Yan Y-Y, Guo Q-R, Ji H, Wang H, Xu T-T, Makabel B, Pilarsky C, He G, Yu X-Y, Zhang J-Y (2021) Microfluidics applications for high-throughput single cell sequencing. J Nanobiotechnol 19(1). https://doi.org/10.1186/s12951-021-01045-6
15. Datlinger P, Rendeiro AF, Boenke T, Senekowitsch M, Krausgruber T, Barreca D, Bock C (2021) Ultra-high-throughput single-cell RNA sequencing and perturbation screening with combinatorial fluidic indexing. Nat Methods 18(6):635–642. https://doi.org/10.1038/s41592-021-01153-z
16. Wu H, Kirita Y, Donnelly EL, Humphreys BD (2019) Advantages of single-nucleus over single-cell RNA sequencing of adult kidney: rare cell types and novel cell states revealed in fibrosis. J Am Soc Nephrol 30(1):23–32. https://doi.org/10.1681/asn.2018090912
17. Lake BB, Codeluppi S, Yung YC, Gao D, Chun J, Kharchenko PV, Linnarsson S, Zhang K (2017) A comparative strategy for single-nucleus and single-cell transcriptomes confirms accuracy in predicted cell-type expression from nuclear RNA. Sci Rep 7(1). https://doi.org/10.1038/s41598-017-04426-w
18. Nadelmann ER, Gorham JM, Reichart D, Delaughter DM, Wakimoto H, Lindberg EL, Litviňukova M, Maatz H, Curran JJ, Ischiu Gutierrez D, Hübner N, Seidman CE, Seidman JG (2021) Isolation of nuclei from mammalian cells and tissues for single-nucleus molecular profiling. Curr Protocol 1(5). https://doi.org/10.1002/cpz1.132
19. Lee MY, Lufkin T (2012) Development of the "Three-step MACS": a novel strategy for isolating rare cell populations in the absence of known cell surface markers from complex animal tissue. J Biomol Tech: Official J Assoc Biomol Res Fac 23(2):69–77. https://doi.org/10.7171/jbt.12-2302-003
20. Pezzi HM, Niles DJ, Schehr JL, Beebe DJ, Lang JM (2018) Integration of magnetic bead-based cell selection into complex isolations. ACS Omega 3(4):3908–3917. https://doi.org/10.1021/acsomega.7b01427
21. Mimitou EP, Lareau CA, Chen KY, Zorzetto-Fernandes AL, Hao Y, Takeshima Y, Luo W, Huang T-S, Yeung BZ, Papalexi E, Thakore PI, Kibayashi T, Wing JB, Hata M, Satija R, Nazor KL, Sakaguchi S, Ludwig LS, Sankaran VG, Regev A, Smibert P (2021) Scalable, multimodal profiling of chromatin accessibility, gene expression and protein levels in single

cells. Nat Biotechnol 39(10):1246–1258. https://doi.org/10.1038/s41587-021-00927-2
22. Cheng J, Liao J, Shao X, Lu X, Fan X (2021) Multiplexing methods for simultaneous large-scale transcriptomic profiling of samples at single-cell resolution. Adv Sci 8(17):2101229. https://doi.org/10.1002/advs.202101229
23. Fu Y, Kim H, Adams JI, Grimes SM, Huang S, Lau BT, Sathe A, Hess P, Ji HP, Zhang NR (2023) Single cell and spatial alternative splicing analysis with long read sequencing. Cold Spring Harbor Laboratory. https://doi.org/10.1101/2023.02.23.529769
24. Chen G, Ning B, Shi T (2019) Single-cell RNA-Seq technologies and related computational data analysis. Front Genet 10. https://doi.org/10.3389/fgene.2019.00317
25. Buenrostro JD, Giresi PG, Zaba LC, Chang HY, Greenleaf WJ (2013) Transposition of native chromatin for fast and sensitive epigenomic profiling of open chromatin, DNA-binding proteins and nucleosome position. Nat Methods 10(12):1213–1218. https://doi.org/10.1038/nmeth.2688
26. Lareau CA, Ludwig LS, Muus C, Gohil SH, Zhao T, Chiang Z, Pelka K, Verboon JM, Luo W, Christian E, Rosebrock D, Getz G, Boland GM, Chen F, Buenrostro JD, Hacohen N, Wu CJ, Aryee MJ, Regev A, Sankaran VG (2021) Massively parallel single-cell mitochondrial DNA genotyping and chromatin profiling. Nat Biotechnol 39(4):451–461. https://doi.org/10.1038/s41587-020-0645-6
27. Li Z, Kuppe C, Ziegler S, Cheng M, Kabgani N, Menzel S, Zenke M, Kramann R, Costa IG (2021) Chromatin-accessibility estimation from single-cell ATAC-seq data with scOpen. Nat Commun 12(1). https://doi.org/10.1038/s41467-021-26530-2
28. Hanamsagar R, Reizis T, Chamberlain M, Marcus R, Nestle FO, De Rinaldis E, Savova V (2020) An optimized workflow for single-cell transcriptomics and repertoire profiling of purified lymphocytes from clinical samples. Sci Rep 10(1). https://doi.org/10.1038/s41598-020-58939-y
29. Gilchuk P, Bombardi RG, Erasmus JH, Tan Q, Nargi R, Soto C, Abbink P, Suscovich TJ, Durnell LA, Khandhar A, Archer J, Liang J, Fouch ME, Davidson E, Doranz BJ, Jones T, Larson E, Ertel S, Granger B, Fuerte-Stone J, Roy V, Broge T, Linnekin TC, Linde CH, Gorman MJ, Nkolola J, Alter G, Reed SG, Barouch DH, Diamond MS, Crowe JE, Van Hoeven N, Thackray LB, Carnahan RH (2020) Integrated pipeline for the accelerated discovery of antiviral antibody therapeutics. Nat Biomed Eng 4(11):1030–1043. https://doi.org/10.1038/s41551-020-0594-x
30. Stadtmauer EA, Fraietta JA, Davis MM, Cohen AD, Weber KL, Lancaster E, Mangan PA, Kulikovskaya I, Gupta M, Chen F, Tian L, Gonzalez VE, Xu J, Jung I-Y, Melenhorst JJ, Plesa G, Shea J, Matlawski T, Cervini A, Gaymon AL, Desjardins S, Lamontagne A, Salas-Mckee J, Fesnak A, Siegel DL, Levine BL, Jadlowsky JK, Young RM, Chew A, Hwang W-T, Hexner EO, Carreno BM, Nobles CL, Bushman FD, Parker KR, Qi Y, Satpathy AT, Chang HY, Zhao Y, Lacey SF, June CH (2020) CRISPR-engineered T cells in patients with refractory cancer. Science 367(6481): eaba7365. https://doi.org/10.1126/science.aba7365
31. Haque A, Engel J, Teichmann SA, Lönnberg T (2017) A practical guide to single-cell RNA--sequencing for biomedical research and clinical applications. Genome Med 9(1). https://doi.org/10.1186/s13073-017-0467-4
32. Granja JM, Klemm S, Mcginnis LM, Kathiria AS, Mezger A, Corces MR, Parks B, Gars E, Liedtke M, Zheng GXY, Chang HY, Majeti R, Greenleaf WJ (2019) Single-cell multiomic analysis identifies regulatory programs in mixed-phenotype acute leukemia. Nat Biotechnol 37(12):1458–1465. https://doi.org/10.1038/s41587-019-0332-7
33. Pervolarakis N, Nguyen QH, Williams J, Gong Y, Gutierrez G, Sun P, Jhutty D, Zheng GXY, Nemec CM, Dai X, Watanabe K, Kessenbrock K (2020) Integrated single-cell Transcriptomics and chromatin accessibility analysis reveals regulators of mammary epithelial cell identity. Cell Rep 33(3):108273. https://doi.org/10.1016/j.celrep.2020.108273
34. Tani S, Okada H, Onodera S, Chijimatsu R, Seki M, Suzuki Y, Xin X, Rowe DW, Saito T, Tanaka S, Chung U-i, Ohba S, Hojo H (2023) Stem cell-based modeling and single-cell multiomics reveal gene-regulatory mechanisms underlying human skeletal development. Cell Rep 42(4):112276. https://doi.org/10.1016/j.celrep.2023.112276
35. Hao Y, Hao S, Andersen-Nissen E, Mauck WM, Zheng S, Butler A, Lee MJ, Wilk AJ, Darby C, Zager M, Hoffman P, Stoeckius M, Papalexi E, Mimitou EP, Jain J, Srivastava A, Stuart T, Fleming LM, Yeung B, Rogers AJ, Mcelrath JM, Blish CA, Gottardo R, Smibert P, Satija R (2021) Integrated analysis of multimodal single-cell data. Cell 184(13): 3573–3587.e3529. https://doi.org/10.1016/j.cell.2021.04.048

36. Bassiouni R, Gibbs LD, Craig DW, Carpten JD, Mceachron TA (2021) Applicability of spatial transcriptional profiling to cancer research. Mol Cell 81(8):1631–1639. https://doi.org/10.1016/j.molcel.2021.03.016
37. Ortiz C, Carlén M, Meletis K (2021) Spatial Transcriptomics: molecular maps of the mammalian brain. Annu Rev Neurosci 44(1): 547–562. https://doi.org/10.1146/annurev-neuro-100520-082639
38. Moses L, Pachter L (2022) Museum of spatial transcriptomics. Nat Methods 19(5):534–546. https://doi.org/10.1038/s41592-022-01409-2
39. Ding C, Yan X, Xu M, Zhou R, Zhao Y, Zhang D, Huang Z, Pan Z, Xiao P, Li H, Chen L, Wang Y (2022) Short-read and long-read full-length transcriptome of mouse neural stem cells across neurodevelopmental stages. Sci Data 9(1). https://doi.org/10.1038/s41597-022-01165-0
40. Lebrigand K, Magnone V, Barbry P, Waldmann R (2020) High throughput error corrected Nanopore single cell transcriptome sequencing. Nat Commun 11(1). https://doi.org/10.1038/s41467-020-17800-6
41. Corley SM, Mackenzie KL, Beverdam A, Roddam LF, Wilkins MR (2017) Differentially expressed genes from RNA-Seq and functional enrichment results are affected by the choice of single-end versus paired-end reads and stranded versus non-stranded protocols. BMC Genomics 18(1). https://doi.org/10.1186/s12864-017-3797-0
42. Becht E, Mcinnes L, Healy J, Dutertre C-A, Kwok IWH, Ng LG, Ginhoux F, Newell EW (2019) Dimensionality reduction for visualizing single-cell data using UMAP. Nat Biotechnol 37(1):38–44. https://doi.org/10.1038/nbt.4314
43. McCarthy DJ, Campbell KR, Lun ATL, Wills QF, Hofacker I (2017) Scater: pre-processing, quality control, normalization and visualization of single-cell RNA-seq data in R. Bioinformatics 33(8):1179–1186. https://doi.org/10.1093/bioinformatics/btw777
44. Cakir B, Prete M, Huang N, van Dongen S, Pir P, Kiselev VY (2020) Comparison of visualization tools for single-cell RNAseq data. NAR Genomics Bioinf 2(3). https://doi.org/10.1093/nargab/lqaa052
45. Brady MV, Mariani J, Koca Y, Szekely A, King RA, Bloch MH, Landeros-Weisenberger A, Leckman JF, Vaccarino FM (2022) Characterization of human basal ganglia organoids. Mol Psychiatry 27(12):4823–4823. https://doi.org/10.1038/s41380-022-01914-y

Chapter 10

The Study of Small RNA Sequencing from Biological Samples (miRNA-seq)

Weam Othman Elbezanti, Kishore B. Challagundla, and Manoj K. Pandey

Abstract

Small RNA sequencing (sRNA-seq) has greatly transformed the study of molecular biology by allowing thorough analysis of several types of short RNA molecules, such as microRNAs (miRNAs), small interfering RNAs (siRNAs), and piwi-interacting RNAs (piRNAs). This chapter presents a comprehensive methodology for conducting miRNA -seq on biological materials. The text introduces the principles and uses of miRNA-seq, emphasizing its importance in gene regulation, disease processes, and therapeutic interventions. The chapter subsequently explores the essential stages of miRNA-seq, encompassing sample acquisition, RNA isolation, library construction, sequencing, and data interpretation. Every part contains pragmatic suggestions, guidance for resolving issues, and optimal methods to guarantee excellent outcomes. Upon completion of this chapter, readers will possess a comprehensive comprehension of the miRNA-seq workflow and will be adequately prepared to apply this potent approach in their own research endeavors.

Key words Small RNA, microRNA, RNA-induced silencing complex (RISC), SYBR stain, DNA polymerase

1 Introduction

MicroRNAs (miRNAs) are short, non-coding single-stranded RNAs (18 to 25 nucleotides) that play an important role in gene expression [1, 2]. miRNAs are naturally created, small, single-stranded hairpin RNAs that are complementary to their target mRNAs. Unlike synthesized short interfering RNA (siRNA) that require total complementarity to silence a target gene, miRNA works with partial complementarity [3]. miRNAs can be developed as potential therapeutic agents, as they are more stable in vivo and have long-acting ability over siRNAs [3]. However, dose optimization and pharmacokinetic studies are needed for artificial miRNAs

The original version of the chapter has been revised. A correction to this chapter can be found at https://doi.org/10.1007/978-1-0716-4192-7_17

Valerie J. Carabetta and Olaitan Akintunde (eds.), *High Throughput Gene Screening: Methods and Protocols*, Methods in Molecular Biology, vol. 2866, https://doi.org/10.1007/978-1-0716-4192-7_10,

to determine and avoid potential toxicities. miRNA can be used for gene therapy, evaluation of gene function, and generation of transgenic animal models. In addition, they can be used as biomarkers for diagnosis and act as modulators of drug resistance [4].

miRNAs are transcribed from non-protein coding genes to regulate the function of other genes. It was previously thought that miRNAs act only as repressors through targeting mRNAs for degradation or translational repression through the recruitment of the Argonaute (AGO) protein complex. However, it has been shown that miRNAs can also upregulate translation [5, 6]. There are exonic and intronic-derived miRNAs. Intronic miRNA are synthesized during pre-messenger RNA (pre-mRNA) splicing and intron processing [7]. On the other hand, exonic miRNAs are transcribed inside the nucleus by RNA polymerases II (Pol-II) or III (Pol-III) as a large oligo-nucleotide precursor, 100–120 nucleotides in length, called pri-miRNA. Pri-miRNA is processed by Drosha RNase to produce pre-mRNA (65 to 70 nucleotides) that is ready to be exported to the cytoplasm through Exportin-5. Then, pre-mRNA will be further cleaved by Dicer endoribonuclease to produce 20 nucleotide long miRNA duplex, which binds to AGO and other proteins, such as trinucleotide repeat containing protein 6 (TRNC6), to make RNA-induced silencing complex (RISC) [7]. One strand of RNA, the guide strand, will stay attached to RISC, which then binds to the target and leads to mRNA degradation or repression of translation. The other strand, the passenger strand, is released and degraded [3]. There are miRNA subsets that are produced by non-canonical biogenesis such as tRNAs, introns, and snoRNAs [8]. The function of these subsets remains unknown; however, it has been shown to play an important role in the immune system and stem cell proliferation [2, 8, 9].

Studying miRNA is important as their deregulation has been shown to play important roles in cancer, HIV, and other diseases [10, 11]. Next-generation sequencing (NGS) of miRNAs is a useful technique to determine miRNA sequence and quantity of known and novel miRNAs, which were discovered to be involved in the pathophysiology of several diseases. We aim to provide a detailed protocol on how to extract and prepare miRNA samples for NGS. The general workflow for miRNA sequencing consists of three steps: (1) RNA isolation, (2) construction of cDNA library, and (3) sequencing and bioinformatic analysis (Fig. 1, [5]). RNA isolation depends on using guanidinium thiocyanate, a chaotropic salt, followed by column purification. We use the QIAseq kit (QIAGEN) for library construction, which uses unique molecular indices (UMIs). UMIs are incorporated into each cDNA after adding 3′ and 5′ adapters. The advantage of incorporation of UMIs is to decrease PCR bias [12]. There are different isolation methods that have been used to extract miRNA from serum or plasma. Here, we describe the most efficient, widely used protocol, which can be used for either serum or plasma samples [13].

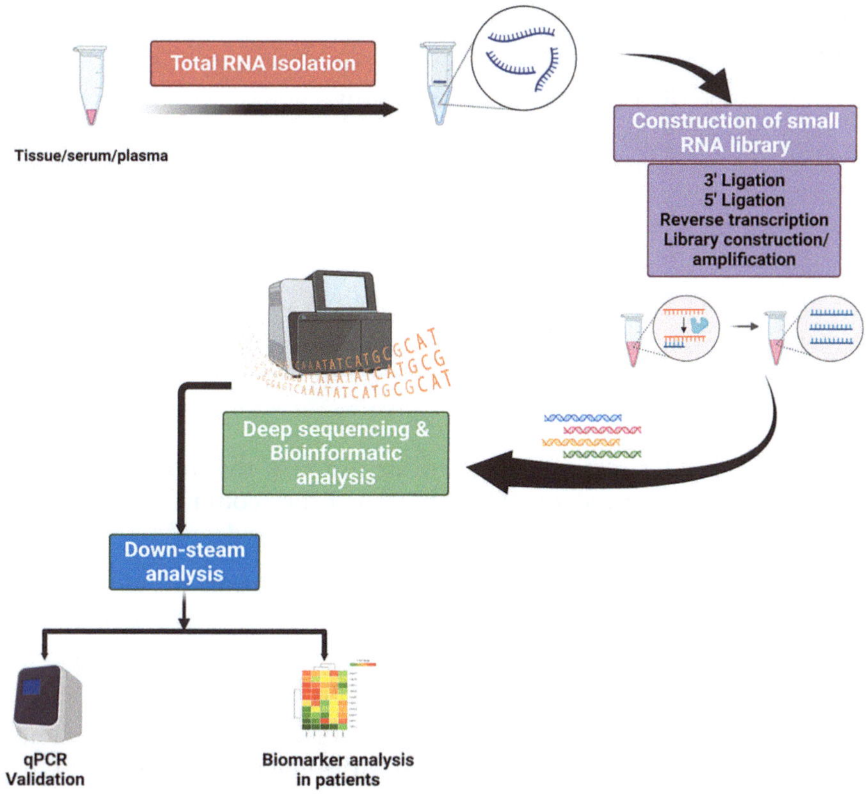

Fig. 1 General steps in miRNA sequencing. RNA is isolated from a sample followed by library preparation and amplification. Created by Biorender.com

2 Materials

2.1 miRNA Extraction

1. Qiagen miRNeasy Serum/Plasma Advanced Kit.
2. 100% isopropanol.
3. 80% ethanol.
4. RNase-free pipet tips.
5. RNase-free water.
6. Microcentrifuge tubes.

2.2 Construction of cDNA Library

1. Qiagen QIAseq miRNA Library Kit.
2. QIAseq miRNA NGS 12 index TF.
3. Plate magnet for 8-well strips.
4. Thermocycler.
5. Magnet.
6. Nuclease-free filter tips.
7. 80% ethanol.

2.3 Gel Electrophoresis

1. Agarose.
2. TBE buffer.
3. Electrophoresis apparatus.
4. DNA stain, such as ethidium bromide or SYBR Safe stain (ThermoFisher Scientific).
5. 6X DNA loading dye.
6. 50 bp ladder.
7. Gel imaging system, such as Gel Doc (BioRad).

3 Methods

3.1 miRNA Extraction

This section is written for use with the miRNeasy Serum/Plasma Advanced Kit (Qiagen). Any other commercially available kit that can isolate miRNAs could also be used.

1. Prepare the RWT working solution by adding 2 volumes of ethanol (96–100%) to RWT buffer that is supplied with the kit.
2. Prepare the RPE working solution by adding 4 volumes of ethanol (96–100%) to RPE buffer that is supplied in the kit.
3. Add 60 μL lysis buffer (RPL) to 200 μL of your sample (plasma or serum) in a microcentrifuge tube (*see* **Note 1**).
4. Vortex for 5 s and incubate for 3 min at room temperature (RT).
5. Add 20 μL buffer RPP buffer and vortex for 20 s. Incubate for 3 min at RT.
6. Centrifuge at 12,000× *g* at RT for 3 min (*see* **Note 2**).
7. Add 200–230 μL of the supernatant in a new microcentrifuge tube, then add one volume of isopropanol and vortex to mix.
8. Transfer the sample to a RNeasy UCP MinElute column and centrifuge at ≥8000× *g* for 15 s. Discard the flowthrough.
9. Wash with 700 μL of RWT buffer and centrifuge for 15 s at ≥8000× *g*. Discard the flowthrough.
10. Wash with 500 μL of RPE buffer and centrifuge at ≥8000× *g* for 15 s. Discard the flowthrough.
11. Wash with 500 μL of 80% ethanol and centrifuge at ≥8000× *g* for 2 min. Discard the flowthrough.
12. Centrifuge the column at ≥8000× *g* for 2 additional min to remove residual ethanol.
13. Place the RNeasy UCP MinElute column into a new collection tube. To dry the membrane, centrifuge for 5 min at full speed. Discard the flowthrough.

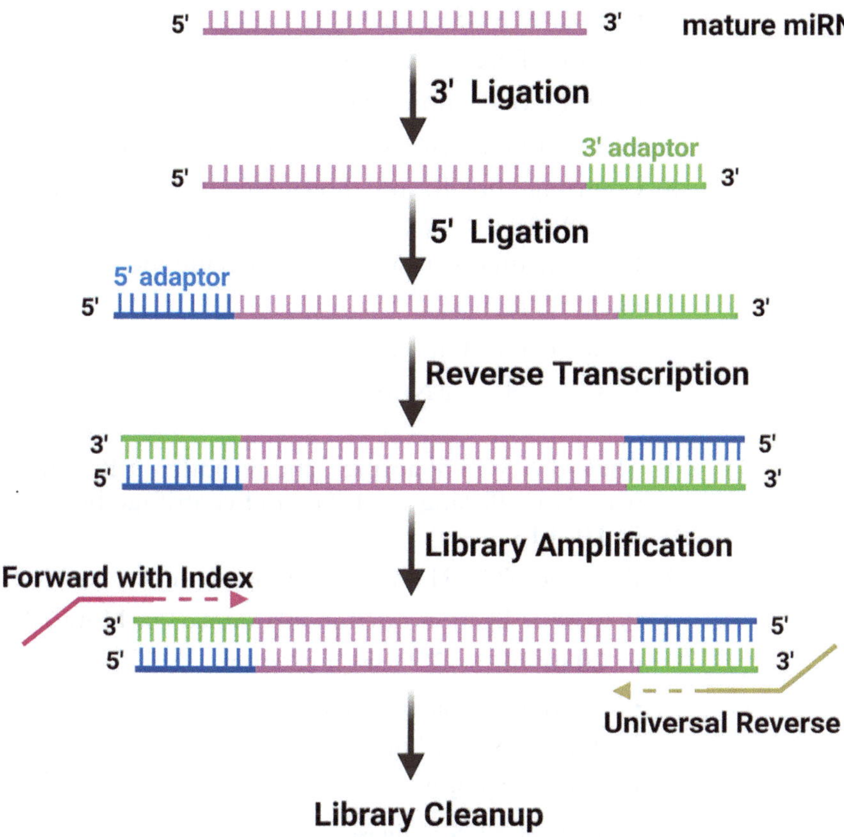

Fig. 2 Computational workflow for the analysis of miRNA profiling data. Created by Biorender.com

14. Place the spin column in a new 1.5 mL microcentrifuge tube. To elute RNA, 20 μL of RNase-free water should be added directly to the center of the spin column membrane and incubated for 1 min. Centrifuge at full speed for 1 min (*see* **Note 3**).

3.2 Construction of cDNA Library

3.2.1 3′ Ligation

This protocol is written for use with the QIAseq miRNA Library Kit (Qiagen). For other available kits, follow manufacturer's instructions (see Fig. 2).

1. Thaw template RNA on ice and centrifuge briefly.
2. Thaw the NGS 3′ adapter, 3′ buffer, and ligation activator at RT, then mix each reagent by flicking the tubes, and centrifuge briefly (*see* **Notes 4–6**).
3. Keep the NGS RI and NGS 3′ ligase on ice for the entire protocol (*see* **Note 7**).
4. Prepare the 3′ ligation reaction mix on ice (*see* **Note 8**):
   - 1 μL of NGS 3′ adapter
   - 1 μL of NGS RI

- 1 μL of NGS 3′ ligase
- 2 μL of NGS 3′ buffer
- 10 μL of ligation activator
- 5 μL of template RNA (total volume: 20 μL)

5. Centrifuge briefly, mix thoroughly by pipetting up and down (12 times), then centrifuge briefly again.
6. Place the tube(s) in a thermocycler and incubate:
   - 28 °C for 1 h.
   - 65 °C for 20 min.
   - 4 °C hold (*see* **Note 9**).

3.2.2 5′ Ligation

1. Thaw the NGS 5′ adapter, and 5′ buffer at RT, then mix each reagent by flicking the tubes and centrifuge briefly (*see* **Notes 5, 10–11**).
2. Keep the NGS RI and NGS 5′ ligase on ice (*see* **Note** 7).
3. Prepare 5′ ligation reaction mix on ice (*see* **Note 12**):
   - 20 μL of the 3′ ligation reaction
   - 15 μL of nuclease-free water
   - 2 μL of NGS 5′ buffer
   - 1 μL of NGS RI
   - 1 μL of NGS 5′ ligase
   - 1 μL of NGS 5′ adapter (total volume: 40 μL)
4. Centrifuge briefly, mix thoroughly by pipetting up and down (12 times), then centrifuge briefly again.
5. Place the tube(s) in the thermocycler and incubate:
   - 28 °C for 30 min.
   - 65 °C for 20 min.
   - 4 °C hold (*see* **Note 13**).

3.2.3 RT Reaction

1. Thaw the NGS RT initiator, RT primer, and RT buffer at RT, then mix by flicking the tubes and centrifuge briefly (*see* **Notes 5, 14–15**).
2. Keep the NGS RI and NGS RT enzymes on ice (*see* **Note** 7).
3. Add 2 μL of NGS RT initiator to the 3′, 5'ligation product (*see* **Note 16**).
4. Centrifuge briefly, mix by pipetting up and down, and centrifuge briefly again.
5. Place the tube(s) in the thermocycler:

- 57 °C for 2 min.
- 70 °C for 2 min.
- 65 °C for 2 min.
- 60 °C for 2 min.
- 55 °C for 2 min.
- 37 °C for 5 min.
- 25 °C for 5 min.
- 4 °C hold.

6. Prepare the RT reaction mix on ice (*see* **Note 17**).
   - 42 µL (the 3′, 5′ ligation product)
   - 2 µL NGS RT primer
   - 2 µL of nuclease-free water
   - 12 µL of NGS RT buffer
   - 1 µL of NGS RI
   - 1 µL of NGS RT Enzyme (total volume: 60 µL)
7. Centrifuge briefly, mix thoroughly by pipetting up and down (12 times), then briefly centrifuge again.
8. Place the tube(s) in the thermocycler and incubate:
   - 50 °C for 1 h.
   - 70 °C for 15 min.
   - 4 °C hold for at least 5 min (*see* **Notes 18–19**).

3.2.4 Preparation of the Beads

1. For each sample, add 400 µL of QIA seq beads to a microcentrifuge tube after vortexing, and centrifuge briefly. Then, immediately isolate beads on a magnetic stand (*see* **Notes 20–21**).
2. Carefully remove the supernatant without disturbing the beads after complete isolation on the magnet (*see* **Notes 22–23**).
3. Wash the beads with the binding buffer by adding 150 µL of vortexed QIA seq miRNA NGS bead binding buffer to the beads. Remove the tube from the magnetic stand and vortex to fully resuspend the beads. Centrifuge briefly and isolate the beads on the magnetic stand (*see* **Note 24**).
4. Carefully, remove supernatant without disturbing the beads (*see* **Note 25**).
5. To combine the beads with the binding buffer, add 400 µL of vortexed QIA seq miRNA NGS bead binding buffer to the beads. Remove the tube from the magnetic stand and vortex to fully resuspend the beads (*see* **Note 26**).
6. The QMN beads are now ready and you can proceed to the cDNA cleanup (*see* **Note 27**).

3.2.5 cDNA Cleanup

1. Briefly centrifuge the 60 μL cDNA product before use.
2. Add 143 μL of vortexed QMN beads to the cDNA product. Vortex for 3 s, briefly centrifuge, and incubate at RT for 5 min (*see* **Note 28**).
3. Place the tubes on the magnetic stand and leave until all the beads entirely migrate (*see* **Note 29**).
4. Without disturbing the beads, discard the supernatant.
5. Add 200 μL of 80% ethanol to the beads while they are still isolated on the magnet to remove non-specific products and immediately discard the ethanol.
6. Repeat the ethanol wash by adding 200 μL of 80% ethanol to the beads on the magnet and discard the ethanol immediately (*see* **Note 30**).
7. Air dry the beads on the magnet for 10-15 min at RT (*see* **Note 31**).
8. To elute the cDNA, add 17 μL of nuclease-free water to the beads after air drying, while still isolated on the magnet (*see* **Note 32**).
9. Remove the tube from the magnet and resuspend the beads in the nuclease-free water by pipetting up and down.
10. Centrifuge briefly and incubate for 2 min at RT.
11. Return the tube to the magnetic stand and wait for 2 min to ensure that all beads are isolated on the magnet.
12. Remove 15 μL of supernatant containing cDNA and transfer to a clean tube.

3.2.6 Library Amplification (See Notes 33 and 34)

1. Thaw the following reagents: QIAseq miRNA NGS library buffer, the required index primer(s) from the QIAseq miRNA 12 Index TF, and QMI TF library rev primer. Mix each kit reagent by flicking tubes.
2. To collect residual liquid from the edges of each tube, centrifuge briefly.
3. Prepare library amplification reaction on ice (*see* **Notes 35–37**):
   - 15 μL of cDNA
   - 16 μL of QIAseq miRNA NGS library buffer
   - 3 μL of HotStart Taq DNA polymerase
   - 2 μL of QMI TF IP1 through IP12 (index primers)
   - 2 μL of QMI TF lib rev primer
   - 42 μL of nuclease-free water (total volume: 80 μL)
4. Centrifuge briefly, mix through pipetting up and down (12 times). Then, centrifuge briefly again.
5. Place the tube(s) in the thermocycler and incubate:

- Initial denaturation, 95 °C for 15 min.
- Denaturation, 95 °C for 15 s.
- Annealing, 60 °C for 30 s.
- Extension, 72 °C for 15 s.

The number of cycles depends on the initial amount of RNA. Use 22 cycles if starting amount was ≤1 ng from serum or plasma, 19 cycles for 10 ng, 16 cycles for 100 ng, or 13 cycles for 500 ng.

- Final extension, 72 °C for 2 min.
- Hold at 4 °C for at least for 5 min.

6. Take out the prepared QMN beads from the refrigerator and vortex. Briefly centrifuge the library amplification reaction.
7. In a new tube, mix 75 μL of the QMN beads and 75 μL of the amplification product.
8. Vortex for 3 s, centrifuge briefly, and incubate at RT for 5 min.
9. Place tube(s) on the magnetic stand and leave for 4 min or until all the beads are isolated.
10. Remove the supernatant and transfer 145 μL to a new tube (*see* **Note 38**).
11. Add 130 μL of vortexed QMN beads to the 145 μL of the supernatant, vortex for 3 s, centrifuge briefly, and incubate at RT for 5 min.
12. Place the tube(s) on a magnetic stand until ensuring all the beads are isolated.
13. Without disturbing the beads, discard supernatant (*see* **Note 39**).
14. For washing, add 200 μL of 80% ethanol to the beads while on the magnetic rack and immediately discard the first ethanol wash.
15. Repeat the ethanol wash by adding 200 μL of 80% ethanol to the beads on the magnetic rack and discard the ethanol immediately.
16. Air dry the beads on the magnetic rack for 10-15 min at RT (*see* **Note 30**).
17. For DNA elution, add 17 μL of nuclease-free water to the tube (s) after air drying, while still on the magnetic rack and cover the tube(s).
18. Remove the tube(s) from the magnet and resuspend the beads in the nuclease-free water by pipetting up and down carefully until thoroughly mixed.

19. Centrifuge briefly and incubate for 2 min at RT.
20. Return the tube(s) to the magnet and wait for 2 min to ensure that all are isolated.
21. Aliquot 15 μL of eluted DNA (miRNA sequencing library) into three new tubes of 5 μL each (*see* **Notes 40** and **41**).

3.2.7 miRNA Library Pre-sequencing QC

Using gel electrophoresis (*see* **Note 42**):

1. To prepare a 6% agarose gel, add 6.0 grams of agarose to 100 mL of 1X tris-borate EDTA buffer (TBE) in a flask.
2. Add 10 μL of SYBR Safe DNA stain to the agarose solution and mix well.
3. Microwave for 1–3 min with swirling the flask occasionally until the agarose is completely dissolved.
4. Let agarose solution cool down to about 50 °C, about 5 min.
5. Pour the agarose into a gel tray with the comb in place.
6. Let the gel sit at RT for 20–30 min, until it has completely solidified.
7. Add DNA loading dye to each of your samples.
8. Once solidified, place the agarose gel into the electrophoresis chamber and fill with 1X TBE to cover the gel.
9. Add 3 μL of a molecular weight ladder into the first well of the gel.
10. Load the gel with 3 μL of each sample and run for 1 h at 120 V.
11. Take an image picture using a gel imaging system, like the Gel-Doc (BioRad), in which you should see a band of an approximate size of 173 bp, which corresponds to the size of the miRNA library (*see* **Note 43**).

3.2.8 Sending for Sequencing and Bioinformatic Analysis

1. Determine the concentration of the miRNA library by using a Qubit fluorimeter (ThermoFisher Scientific) and dilute each library to a final concentration of 4 nM.
2. For data analysis, use geneglobe.qiagen.com and click analyze. Log in or create an account.
3. Select NGS and QIAseq miRNA library kit—primary quantification and then click start analysis.
4. Upload the sequencing files from your computer by selecting "upload new," which is under the upload tab that is found under read files and add UBAM files.
5. For each file that you want to map, select by checking the box and select analyze.
6. Define the sample group by going to the "sample manager" under analysis setup.

7. Select "Normalization method under analysis" setup to choose the normalization option.
8. Observe the fold-regulation and fold-change results under Analysis. Also, observe the visual representations of the data under "plots & charts."
9. Export your data.

4 Notes

1. Please note that this protocol can be scaled up. The increase of sample volume needs to be adjusted.
2. The supernatant should be colorless.
3. The resulting volume of the eluate is 18 μL if you elute with 20 μL, because the dead volume of the column is 2 μL. Aliquot into two tubes: 8 μL (to be used) and 10 μL (for backup) and store at −20 °C until use.
4. Do not vortex the template RNA, NGS RI, NGS 3′ ligase, or the 3′ ligation reaction.
5. Ensure full thawing with no precipitates.
6. Determine the RNA concentration using the nanodrop, or equivalent. If your total RNA quantity in 5 μL is less than 100 ng, dilute the NGS 3′ adapter 1:5. If the concentration is less than 10 ng, dilute the adapter 1:10.
7. Return immediately to −20 °C freezer after use.
8. It is important to prepare the 3′ ligation reaction on ice. If you have more than one sample, prepare a master mix for all samples with an extra 10% of what is needed. Pipet slowly when mixing because the 3′ ligation reaction is viscus.
9. Once you have completed the 3′ ligation reaction, immediately begin the 5′ ligation.
10. Do not vortex the NGS RI, NGS 5′ ligase, or 5′ ligation reaction.
11. It is important to prepare the 5′ ligation reaction on ice. If you have more than one sample, prepare a master mix for all samples with an extra 10% of what is needed. Pipet up slowly for mixing because the 5′ ligation reaction is viscus.
12. If your total RNA is less than 100 ng, dilute the NGS 5′ adapter 1:2.5. If the concentration in 5 μL is less than 10 ng, dilute the adapter to 1:5.
13. Once you completed the 5′ ligation reaction, immediately begin the RT reaction.
14. Do not vortex the RT reaction.

15. It is important to prepare the RT reaction on ice. If you have more than one sample, prepare a master mix for all samples with an extra 10% of what is needed.
16. Add RT components directly to the entire 5′ ligation product (40 μL).
17. If your total RNA is less than 100 ng, dilute the NGS RT primer 1:5. If the concentration in 5 μL is less than 10 ng, dilute the adapter to 1:10.
18. It is important to keep the samples at 4 °C for at least 5 min to terminate the reaction. Because cDNA is stable, it can be kept for longer at 4 °C.
19. Once the RT reaction is completed, proceed immediately with the NGS beads.
20. Keep the beads on ice. Be sure to vortex beads and bead binding buffer thoroughly and immediately before use (it is important to use a homogeneous bead suspension).
21. If you have more than one sample, prepare the beads for all samples with an extra 10% of what is needed.
22. Isolation of beads on a magnet takes about 15 min because of high viscosity.
23. It is acceptable if residual supernatant remains.
24. Both the beads storage buffer and bead binding buffer are viscus, so pipette carefully.
25. At this step completely remove the supernatant, leaving no residual liquid, without disturbing the beads.
26. The QIAseq beads in addition to the QIAseq miRNA bead binding buffer makes QMN beads.
27. You can store the QMN beads at 2–8 °C after preparation for 1 week. The prepared beads can now be used for the cDNA cleanup and in library amplification steps.
28. Keep QMN beads on ice and vortex them thoroughly and immediately before use.
29. The original protocol indicates that the time of bead isolation takes about 4 min. However, in our hands it took about 20 min.
30. To remove all ethanol, briefly centrifuge and remove ethanol using a 200 μL pipette. Then, use a 10 μL pipette to remove any residual ethanol.
31. It is important to keep the tubes on the magnet because the beads would be disrupted when removed from the magnet.
32. After finishing the cDNA cleanup, you can proceed immediately to library amplification or store in a – 20 °C freezer.

33. This protocol is compatible with ThermoFisher Scientific NGS systems.
34. Here, we describe the protocol for tube indices in library amplification that have up to 12 indices. However, if you are using a plate, you can use HT plate indices, such as the QIAseq miRNA NGS 48 index TF [13].
35. Prepare the library amplification reactions on ice. If you have more than one sample, prepare a master mix for all samples with an extra 10% of what is needed.
36. Do not vortex the library amplification reactions or the Hot-Start Taq DNA polymerase.
37. Remove HotStart Taq DNA polymerase from the −20 °C freezer before use and keep on ice. Return to the freezer immediately after use.
38. At this step, do not discard the supernatant. However, discard the beads.
39. At this step, do not discard the beads.
40. The three aliquots are used as follows: 5 μL for quality control, 5 μL for sequencing, and 5 μL for backup.
41. After finishing the library amplification steps, you can proceed immediately to miRNA library pre-sequencing QC or store in a − 20 °C freezer until further use.
42. There are two options for QC: using the Agilent Bioanalyzer 2100 or gel electrophoresis. We routinely use gel electrophoresis because it is easier and available for use.
43. If you find a band at approximately 181 bp, that means that your preparation was contaminated with piRNA. If you find a band at approximately 150 bp, that means your preparation was contaminated with primer dimer. To prevent primer dimerization, use 1 ng of RNA. In either case, you will need to repurify your cDNA libraries by running them on gel and excising the appropriate size.

References

1. Bartel DP (2004) MicroRNAs: genomics, biogenesis, mechanism, and function. Cell 116(2): 281–297. https://doi.org/10.1016/s0092-8674(04)00045-5
2. Zhang N, Hu G, Myers TG, Williamson PR (2019) Protocols for the analysis of microRNA expression, biogenesis, and function in immune cells. Curr Protoc Immunol 126(1): e78. https://doi.org/10.1002/cpim.78
3. Ying S-Y, Chang DC, Miller JD, Lin S-L (2006) MicroRNA protocols. Perspectives. Methods Mol Biol 342:351–358. https://doi.org/10.1385/1-59745-123-1:351
4. Hanna J, Hossain GS, Kocerha J (2019) The potential for microRNA therapeutics and clinical research. Front Genet 10. https://doi.org/10.3389/fgene.2019.00478
5. Benesova S, Kubista M, Valihrach L (2021) Small RNA-sequencing: approaches and considerations for miRNA analysis. Diagnostics (Basel) 11(6). https://doi.org/10.3390/diagnostics11060964

6. Vasudevan S, Tong Y, Steitz JA (2007) Switching from repression to activation: microRNAs can up-regulate translation. Science 318(5858):1931–1934. https://doi.org/10.1126/science.1149460
7. Crocker OJ, Trigg NA, Conine CC (2022) Cloning and sequencing eukaryotic small RNAs. Curr Protocol 2(8):e495. https://doi.org/10.1002/cpz1.495
8. Abdelfattah AM, Park C, Choi MY (2014) Update on non-canonical microRNAs. Biomol Concepts 5(4):275–287. https://doi.org/10.1515/bmc-2014-0012
9. Moore S, Järvelin AI, Davis I, Bond GL, Castello A (2018) Expanding horizons: new roles for non-canonical RNA-binding proteins in cancer. Curr Opin Genet Dev 48:112–120. https://doi.org/10.1016/j.gde.2017.11.006
10. Peng Y, Croce CM (2016) The role of MicroRNAs in human cancer. Signal Transduct Target Ther 1:15004. https://doi.org/10.1038/sigtrans.2015.4
11. Sardo L, Vakil PR, Elbezanti W, El-Sayed A, Klase Z (2016) The inhibition of microRNAs by HIV-1 tat suppresses beta catenin activity in astrocytes. Retrovirology 13:25. https://doi.org/10.1186/s12977-016-0256-y
12. Wong RKY, MacMahon M, Woodside JV, Simpson DA (2019) A comparison of RNA extraction and sequencing protocols for detection of small RNAs in plasma. BMC Genomics 20(1):446. https://doi.org/10.1186/s12864-019-5826-7
13. Ali SA, Gandhi R, Potla P, Keshavarzi S, Espin-Garcia O, Shestopaloff K, Pastrello C, Bethune-Waddell D, Lively S, Perruccio AV, Rampersaud YR, Veillette C, Rockel JS, Jurisica I, Appleton CT, Kapoor M (2020) Sequencing identifies a distinct signature of circulating microRNAs in early radiographic knee osteoarthritis. Osteoarthr Cartil 28(11):1471–1481. https://doi.org/10.1016/j.joca.2020.07.003

Chapter 11

Biological Insights from RNA–RNA Interactomes in Bacteria, as Revealed by RIL-seq

Aviezer Silverman and Sahar Melamed

Abstract

Bacteria reside in constantly changing environments and require rapid and precise adjustments of gene expression to ensure survival. Small regulatory RNAs (sRNAs) are a crucial element that bacteria utilize to achieve this. sRNAs are short RNA molecules that modulate gene expression usually through base-pairing interactions with target RNAs, primarily mRNAs. These interactions can lead to either negative outcomes such as mRNA degradation or translational repression or positive outcomes such as mRNA stabilization or translation enhancement. In recent years, high-throughput approaches such as RIL-seq (RNA interaction by ligation and sequencing) revolutionized the sRNA field by enabling the identification of sRNA targets on a global scale, unveiling intricate sRNA–RNA networks. In this review, we discuss the insights gained from investigating sRNA–RNA networks in well-studied bacterial species as well as in understudied bacterial species. Having a complete understanding of sRNA-mediated regulation is critical for the development of new strategies for controlling bacterial growth and combating bacterial infections.

Key words Small RNA, Hfq, Regulatory networks, RNA-seq, Regulatory RNAs, RIL-seq, Machine learning

1 Introduction

Bacteria are known for responding and adjusting to environmental changes quickly and efficiently. This ability is facilitated by regulatory networks which tune the identity and quantity of the components needed for the bacterial response. Regulation in these networks is facilitated mainly at two levels, the transcriptional and the post-transcriptional level. Transcription factors that can act as activators or repressors control gene expression at the transcriptional level. The key players at the post-transcriptional level are regulatory small RNAs (sRNAs). These are mostly 50–400 nucleotide long RNA molecules that use sequence complementarity to bind to their RNA targets and regulate their stability and/or translation (reviewed in [1]). Most of the studied sRNAs regulate their targets *in trans* where the base-pairing is facilitated by an RNA

Valerie J. Carabetta and Olaitan Akintunde (eds.), *High Throughput Gene Screening: Methods and Protocols*, Methods in Molecular Biology, vol. 2866, https://doi.org/10.1007/978-1-0716-4192-7_11,

chaperone. The most investigated RNA chaperone, primarily in Gram-negative bacteria, is the hexameric, ring-shaped Hfq protein [2]. While transcription factor networks have been studied thoroughly for decades [3], the mapping of sRNAs regulatory networks was lagging behind due to limited tools for finding sRNA targets (reviewed in [4]). In recent years, several approaches were developed to provide a global map of sRNA–RNA interactions in bacteria (reviewed in [5]). Herein, we focus on the broadly used RIL-seq (RNA interaction by ligation and sequencing) methodology that integrates experimental and computational tools [6]. RIL-seq provides a comprehensive map of RNA–RNA interactions that take place on an RNA chaperone under a certain growth condition. In the experimental protocol, RNAs are in vivo crosslinked to the RNA chaperone (e.g., Hfq) by exposure to UV, the protein is immunoprecipitated with its bound RNAs, and neighboring RNA ends are then ligated. This gives rise to RNA chimeras that can be sequenced from both ends for the identification of RNA pairs. In the computational pipeline, reads are mapped to the genome and chimeric fragments are filtered for fragments that are statistically over-represented in the data to eliminate spurious interactions. RIL-seq was initially applied to *Escherichia coli* K12 [7–9], but later it was used to reveal the sRNA interactome in multiple Gram-negative [10–13] and Gram-positive bacteria [14]. In the current review, we focus on the growth of knowledge in the regulation of gene expression in bacteria, gained by RIL-seq. We describe the expansion of sRNA networks, discuss a few examples of novel RNA sponges, elaborate on the advantage of using RIL-seq data for machine learning, highlight new properties of RNA-binding proteins, and re-visit the roles sRNAs have in bacterial pathogenicity.

2 Revealing Intricate RNA–RNA Networks

sRNAs are a diverse group of gene expression regulators. While they act by different mechanisms [15], they also differ in the number of targets they have. In *E. coli* for example, RyhB is an iron metabolism regulator that has dozens of validated targets [16], while other sRNAs, like ChiX, have only a handful of known targets [17]. Moreover, some targets can be regulated by several sRNAs, adding another level of complication, and establishing a connection between different pathways in bacteria. The sRNA–RNA networks drawn based on these interactions were useful but somewhat limited (e.g., resulting in 150 interactions in *E. coli* pre-RIL-seq; Table 1). RIL-seq technology serves as a tool for a significant expansion of these networks, uncovers sRNA–RNA networks in bacterial species where no network was known, and enables learning about the dynamics of the sRNA interactome. To date, RIL-seq dramatically expanded our knowledge in multiple Gram-

Table 1
Summary of the number of RNA–RNA interactions identified by RIL-seq in different studies. "Unique interactions" represents the number of unique interactions that were found in each study. Each pair was counted only once except for cases where the same pair appeared in a different order in the chimera, and then it was counted twice

Bacterial species	Pre RIL-seq estimated sRNA network	RIL-seq network				Reference
		Growth conditions	RNA chaperone	Number of interactions	Unique interactions	
E. coli	*~150*	Log phase	Hfq	1027	3138	[9]
		Stationary phase	Hfq	1844		
		Iron limitation	Hfq	1947		
		Rich media	Hfq	2069	2413	[8]
		Minimal media	Hfq	1109		
		Rich media	ProQ	322	320	
		Minimal media	ProQ	33		
EPEC	*~155*	Activating	Hfq	810	1555	[12]
		Non-activating	Hfq	1050		
P. aeruginosa	*~300*	Exponential phase	Hfq	997	1446	[13]
		Stationary phase	Hfq	702		
Salmonella	*<100*	Early stationary 1	Hfq	883	1240	[11]
		Early stationary 2	Hfq	891		
V. cholerae	*<100*	Low cell density	Hfq	3193	4274	[10]
		High cell density	Hfq	1860		
C. difficile	*<5*	Sporulation conditions	Hfq	1569	1186	[14]

negative bacteria and in one Gram-positive bacteria (Table 1). Valuable online platforms to visualize the RNA–RNA networks and RIL-seq data were generated and are summarized in Table 2.

The extensive networks drawn by RIL-seq opened an uncharted world of sRNA-mediated regulation. This includes the identification of novel sRNAs from previously unimagined regions, such as internal to the coding sequence (CDS) and from other regions, such as the 3′ UTR or 5′ UTR, expansion of the target sets of known sRNAs, discovery of new sRNA–sRNA interactions, interactions involving tRNAs, and mRNA–mRNA interactions. While a portion of these interactions represent canonical sRNA mechanisms of action (reviewed in [15]), some require more

Table 2
Summary of online platforms for visualization and analysis of RIL-seq data

Platform	Bacterial species	Description	Data from	Link
UCSC genome browser	*E. coli*	Hfq RIL-seq single fragments data Hfq RIL-seq chimeric fragments data	[9]	Single fragments Chimeric fragments
		Hfq and ProQ RIL-seq data Hfq and ProQ RIL-seq data in *hfq* and *proQ* deletion backgrounds	[8]	Hfqand ProQ *hfq*and*proQ*deletion backgrounds
	P. aeruginosa	Hfq RIL-seq data in exponential growth phase Hfq RIL-seq data in stationary growth phase	[13]	Exponential phase data Stationary phase data
RILseqDB	*E. coli* and EPEC	A user-friendly interface to query and view the results of RIL-seq experiments. Interactions of specific RNA can be queried	[9, 12]	http://rilseqdb.cs.huji.ac.il/
RNA-RNA interactome browser	*Salmonella*	The browser allows for an interactive search on sRNAs and coding genes, thereby enabling an intuitive visualization of sRNA regulons, and offering users the option to upload and visualize additional RIL-seq datasets	[11]	http://resources.helmholtz-hiri.de/rilseqset/
RILSeqExplorer V. cholerae	*V. cholerae*	A dynamic and searchable web interface that provides a network view of RIL-seq interactions	[10]	http://rnaseqtools.vmguest.uni-jena.de/
RNA–RNA interactome browser	*C. difficile*	*C. difficile* RIL-seq network	[14]	https://resources.helmholtz-hiri.de/rilseqcd/
sInterBase	*E. coli*	Platform for mining sRNA-mRNA interactions in *E. coli*	[62]	https://sinterbase.cs.bgu.ac.il/

thorough deciphering. For others, we do not have sufficient knowledge to understand their meaning. A realization resulting from these valuable datasets is that sRNA-mediated regulation is as abundant as transcription factor-mediated regulation, emphasizing the complexity of regulation of gene expression in bacteria. In different bacterial species, further insights can be gained. Enteropathogenic *E. coli* (EPEC) can be grown in conditions that resemble those in the host environment and induce the expression of virulence genes (activating conditions) or in conditions that do not resemble the host environment (non-activating conditions) [18]. Mizrahi et al. carried out RIL-seq in non-activating and in activating conditions. Under non-activating conditions, ~16% of the RIL-seq chimeras involved genes from accessory regions (e.g., regions that encode pathogenicity traits) whereas under activating conditions, this percentage increased to ~44%, implying a more significant involvement of sRNAs in EPEC pathogenicity than previously thought [12]. In *Salmonella*, a total of ~1240 significant chimers were found suggesting 74 sRNAs, 25 of which were previously characterized as Hfq-dependent sRNAs, multiplying by three the number of Hfq-dependent sRNAs [11]. RIL-seq performed in *Pseudomonas aeruginosa* uncovered that a single sRNA called PhrS dominates the RNA–RNA interaction network by pairing with ~800 targets [13]. In *E. coli*, the application of RIL-seq to two different RNA chaperones revealed that the RNA chaperones compete for the same RNA pairs, playing opposite roles [8]. As more RNA–RNA networks are expected to be published, computational analysis followed by experimental ones will be able to give us new insights about the evolution of sRNAs and their targets.

3 Identification of RNA Sponges

Traditionally, sRNA functions and their ability to regulate gene expression were studied through the mRNA targets to which they were binding. However, an increasing number of independent studies in recent years documented the binding of two sRNAs, where one regulates the activity of the other, a mechanism that we now call "sponging." Since the focus has been on documenting effects of a sRNA on its target, a target that was not affected by the sRNA was ignored and therefore RNA sponges have been overlooked. The concept of RNA sponges was first described for microRNAs (miRNAs) by Albert et al. in 2007 and was called "target mimicry" [19]. In this study, they showed that by engineering a synthetic RNA that carried repeats of specific sequences, it could base pair with other known miRNAs and compete with their natural targets, subsequently causing the removal of this miRNA from the regulatory pool and depleting its action. The first RNA sponge to be identified in bacteria was an intergenic region between *chbB*-

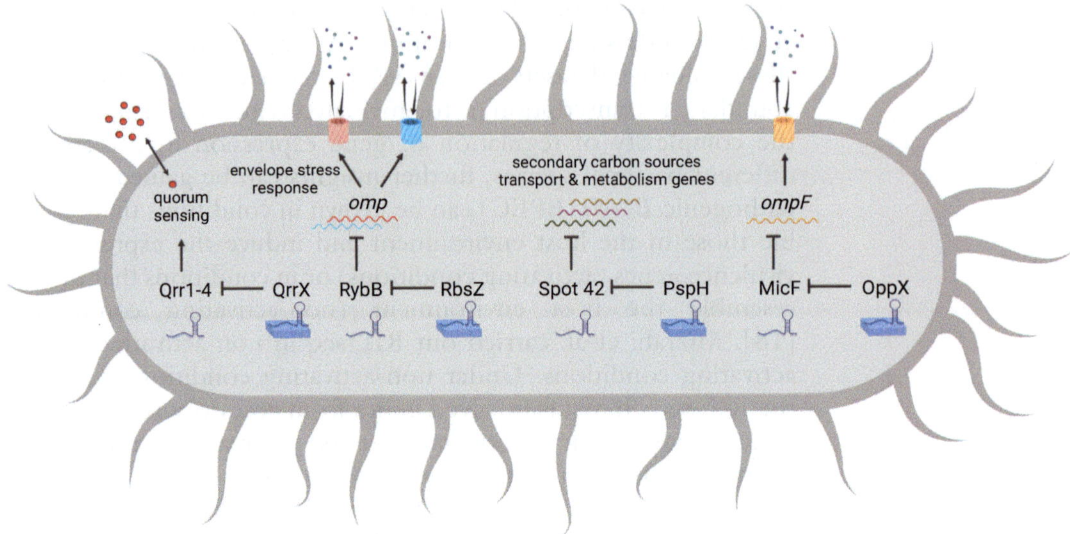

Fig. 1 RNA sponges regulate the activity of sRNAs, affecting diverse pathways in a range of bacterial species. QrrX base-pairs with Qrr1–4 sRNAs in *V. cholerae*, leading to their degradation and impairing quorum sensing regulation. In *E. coli*, RbsZ binding to RybB leads to its degradation, enhancing the synthesis of outer membrane porins. Downregulation of Spot 42 by PspH enables the expression of transport and secondary carbon sources utilization genes. The OppX sRNA positively affects the synthesis of the key porin OmpF in *Salmonella* by sponging MicF. The schematic drawing of the bacterial cell does not represent any specific species

chbC, two genes that are part of the chitobiose operon. It was shown to sponge a sRNA called ChiX, allowing for the expression of ChiX targets that were previously repressed by it [20, 21]. This discovery was followed by the identification of two other RNA sponges in *E. coli* and *Salmonella*, SroC, which sponges the global regulator GcvB [22], and 3′ETS^{leuZ}, an RNA fragment processed from a tRNA precursor, that regulates the sigma E-dependent, iron homeostasis-related sRNA RyhB [23]. The mechanism by which RNA sponges act involves base pairing with a sRNA, depleting its availability in the cell and possibly causing its degradation.

The development of various high-throughput RNA-sequencing-based approaches allows unbiased identification of RNA sponges. Herein, we discuss the identification of RNA sponges by RIL-seq (summarized in Fig. 1) while deliberation on sponges found by other approaches can be found elsewhere [24]. RIL-seq analysis typically yields between a few hundred to thousands of statistically significant RNA pairs. Of which, a significant subset are sRNA–sRNA interactions, providing a large pool of potential RNA sponges. Matera et al. applied the RIL-seq to *Salmonella enterica* during early stationary phase and they identified a new potential target of the classic porin repressor MicF [25]. As expected, MicF's primary target, the outer membrane porin *ompF*, was found with MicF in many chimers. However, the analysis recovered nearly twice as many MicF chimeras with the major inner membrane transporter mRNA, *oppA*. This observation was surprising because previous work on MicF targets did not find any change in the translation of the *oppABCDF* operon [26, 27]. The base pairing between MicF and *oppA* was actually taking place upstream of the start codon, in a region from which a separate transcript was generated, and they called this transcript OppX. Further research classified this 5′ UTR as a major sRNA sponge-binding MicF and depleting its availability in the cell without causing its degradation, hence acting as a "soaking"sponge. 5′ UTR sRNAs can be easily missed due to limited transcriptome annotation, whereas RIL-seq analysis can point to potential 5′ UTR-originated sRNAs. The identification of OppX adds another level of complexity to the regulatory network of MicF and helps to better understand this regulatory circuit. In another study, Huber et al. applied RIL-seq to *Vibrio cholerae* and found 81 sRNA–sRNA interactions [10]. Out of all those sRNA pairs, a novel sRNA drew the authors' attention as the data suggested that it base pairs exclusively with the four Qrr sRNAs known to modulate the quorum sensing pathway in *V. cholera*. The sRNA was termed QrrX and the characterization of its interaction with the Qrrs revealed that it functions as an RNA sponge, affecting the levels and activity of the Qrr sRNAs. Specificity of RNA sponges was also observed in the case of PspH in *E. coli* [9]. PspH is a 3′ UTR-derived sRNA, interacting solely with Spot 42 sRNA, a key player in the regulation of carbon transport and metabolism. Overexpression of PspH led to an almost five-fold reduction in Spot 42 levels and to increased levels of several Spot 42 targets. RbsZ is another RNA sponge in *E. coli* identified by RIL-seq and its characterization uncovered an additional level of complexity to sRNA-mediated regulation [8]. In this case, Melamed et al. demonstrated that RIL-seq could be applied to other RNA chaperones than Hfq and they applied it to both Hfq and the understudied ProQ. A comparison of the RNA pairs found on Hfq and ProQ revealed that about a third of the RNA pairs captured on ProQ were also found on Hfq, suggesting overlapping and/or competing roles for the two RNA-binding proteins. One of these overlapping RNA pairs was between RybB, a sRNA known to be active under cell envelope stress, and a novel sRNA, RbsZ. RbsZ corresponds to the 3′ UTR of *rbsB*, which is part of the ribose catabolism operon. Characterization of the RybB–RbsZ interaction revealed a fascinating interplay between two RNA-binding proteins. It was found that RbsZ downregulates RybB by base-pairing, leading to RybB degradation, and relieving the RybB-dependent regulation on its targets. RbsZ-dependent regulation was mediated by Hfq and blocked by ProQ, resulting in a continuous competition between the two RNA-binding proteins on this RNA pair.

The increasing number of RNA sponges found in recent years enhances our understanding of sRNA mechanisms of action. While previously the regulation of sRNA activity was thought to be achieved mainly by controlling its expression, it is now acknowledged that post-transcriptional regulation by RNA sponges is just as important. Most of the studied RNA sponges to date share two main traits in the RIL-seq data: (1) An RNA is found exclusively in chimeras with another sRNA (e.g., does not have any other RNA partners), and (2) It is ranked high in that sRNA target set based on the number of chimeras. The examples above were taken from the studies in which RIL-seq was performed. Later studies used the published RIL-seq datasets to uncover additional RNA sponges [28], demonstrating how RIL-seq is an extremely valuable tool for identifying RNA sponges in regions that were not previously thought to encode sRNAs.

4 Identification of Novel sRNAs Using Machine Learning

Initially, bacterial-encoded sRNAs such as Spot 42 [29, 30], MicF [31], and OxyS [32] were found sporadically and thought to be encoded only in intergenic regions. Today, it is well documented that sRNAs can be encoded from diverse genomic loci including portions of mRNAs and tRNAs (reviewed in [33]), making it harder to detect them systematically. The development of high-throughput techniques, like the microarray at the beginning [34] and RNA sequencing a decade later (reviewed in [35]), allowed for the identification of an increasing number of sRNAs. The development of RNA-seq-based techniques that define RNA boundaries, such as differential RNA-seq [36] or Term-seq [28, 37], gave another boost to the identification of sRNAs. However, while all these methods could point on the existence of sRNAs, they could not provide any data on the functionality of these short RNA fragments expressed from the genome.

The RIL-seq data withholds valuable information for the interacting RNAs that can be interpreted to better understand the function of the interacting RNAs (e.g., target or regulatory RNA). Bar et al. have nicely developed a machine learning algorithm to identify sRNAs from RIL-seq data based on several characteristics of the RNAs in the data [38]. The authors used 26 to 29 known sRNAs as a training dataset to detect significant features of sRNAs and applied the machine learning algorithm on the remaining ~1000 RNAs found in the *E. coli* Hfq RIL-seq data, in each growth condition [9]. By doing so, they were able to identify in high probability 69 sRNAs, 28 of which served as the training set, 18 were experimentally confirmed by the authors or by other groups recently, and 23 were defined as promising sRNA candidates (Fig. 2). These findings contribute to the scientific community in

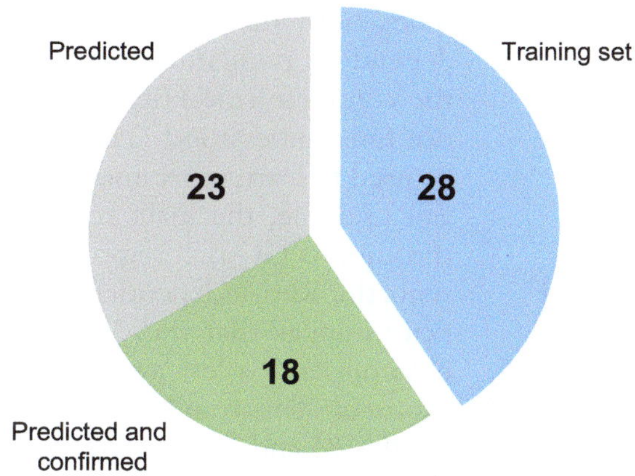

Fig. 2 Visualization of the results of the machine learning algorithm based on RIL-seq data, developed by Bar et al. [38]. The graph shows the number of known sRNAs that were used as a training set for the algorithm (blue), predicted sRNAs via machine learning that were confirmed by the authors and others (green), and predicted sRNAs via machine learning that are yet to be experimentally verified (gray)

multiple ways. First, this study demonstrates that there are specific characteristics of sRNAs in the RIL-seq data that can be used to predict novel sRNAs. This will be extremely useful in less-studied bacteria where only a handful of sRNAs are known, and the characteristics of these sRNAs in the RIL-seq data can be used to train the algorithm for the prediction of novel sRNAs. Second, RIL-seq analysis can be overwhelming due to the amount of data it generates, especially when most of the findings are yet to be understood and are not well documented. The ability to study and analyze all of this information experimentally is limited and combining computational tools with experimental ones can be a game changer for successful research.

5 Expansion of RNA-Binding Proteins Properties

RNA molecules are naturally unstable and thus need to be synthesized in large numbers, form stable secondary structures, or be protected by other proteins. For example, the bacterial mRNA is transcribed and simultaneously translated into a protein. This way, the mRNA is protected from degradation by RNases, and once the translation is complete, the molecule is most likely to be degraded and recycled [39]. In contrast, sRNAs are not translated and therefore may be exposed to RNases and quickly turned over. RNA-binding proteins bind to specific RNA molecules at specific sites based on sequence or structural recognition (reviewed in [40]). These proteins are often referred to as chaperones because they mediate the function of the RNAs. As mentioned earlier, Hfq binds both the sRNA and its target mRNA, facilitating sRNA-dependent regulation, and is the most studied one [2, 41]. Yet, the complete role Hfq plays in sRNA-mediated regulation is still not fully understood. The extensive data achieved by RIL-seq can be used to learn more about the characteristics of RNA chaperones. For example, the main region by which a sRNA binds to Hfq is through its GC-rich region followed by a poly-U tail, which represents the Rho-independent transcriptional terminators. This specificity implies that the 3′ end of the sRNA is buried in the Hfq hexamer and is less available for the ligation step in the RIL-seq protocol. This resulted in a strong enrichment for sRNAs located on the 3′ end of the RIL-seq chimeras, recapitulating what was already known. Faigenbaum-Romm et al. used *E. coli* Hfq RIL-seq data to gain new insights about the hierarchy in the chaperone occupancy of sRNA targets [7]. The authors estimated the binding capabilities of two RNA molecules by calculating their free energy using two tools (RNA duplex and RNAup), and the amount of Hfq bound to a single RNA from the read counts (single and chimeric) from RIL-seq experiments. While they observed a weak correlation between the free energy of a bound sRNA to its target and the number of chimeric fragments, they did find a strong correlation between the sRNA-target interaction frequency and the Hfq occupancy of the target in a specific condition. This suggests that the ability of one target to interact with the Hfq protein has a large impact on its level of regulation by its corresponding sRNA. This competition among targets explains the difference between affected and unaffected RNAs by the sRNA Hfq-mediated regulation. They found that sRNAs that bind many Hfq molecules have a more extensive impact on gene regulation as the amount of Hfq in the cell is closely regulated and tends to be maintained within a certain concentration range. Thus, sRNAs with better binding affinity for Hfq are more likely to have a more substantial effect on the cell's physiology. Gaining a better understanding of how RNA chaperones facilitate the regulation of gene expression will also help to develop new ways to manipulate them.

While RIL-seq was initially applied to Hfq, it can be applied to other RNA-binding proteins, as described earlier in this review [8]. The observation that ProQ blocked the Hfq-mediated RbsZ-dependent regulation led to the hypothesis that the protective ability of ProQ might be broader than just affecting one RNA pair. To test this idea, ProQ was briefly overexpressed and the effect on the *E. coli* transcriptome was studied. In support of the hypothesis, it was found that ProQs protected many transcripts. This finding expanded the role of ProQ and suggested that it has a global role in protecting and/or stabilizing RNA transcripts. The interplay between the *E. coli* Hfq and ProQ opens the door for other types of relationships between RNA-binding proteins to be studied in the future (reviewed in [42]).

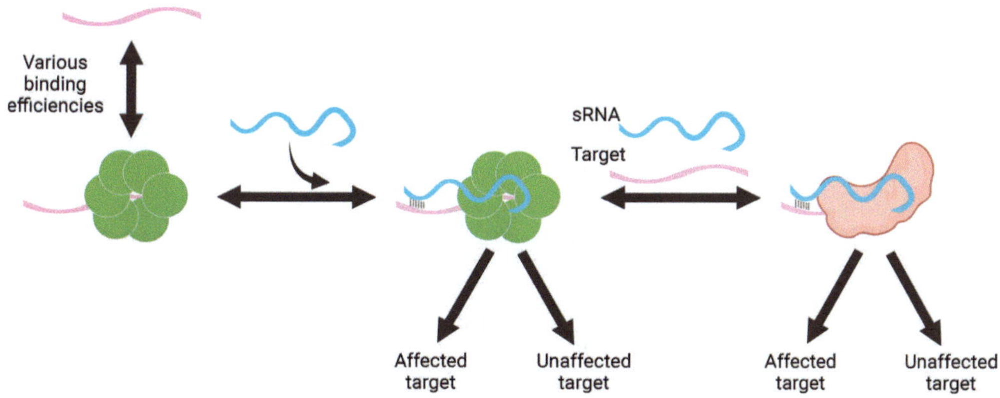

Fig. 3 The binding properties between RNA-binding proteins and the RNAs they bind govern the outcome of sRNA-mediated regulation. The binding efficiency of an mRNA target to Hfq (green) affects its fate. ProQ (red) can bind sRNA–mRNA pairs and thus compete with Hfq

In conclusion, RIL-seq enables an expansion of our knowledge regarding proteins that take part in the regulation of gene expression (Fig. 3), and it can be applied to other RNA-binding proteins of interest such as CsrA [43] or others that were recently found [44]. It suggests that sRNA-dependent regulation is more complex than previously thought and that RNA-binding proteins have multiple means to fine tune sRNA-dependent regulation.

6 Uncovering sRNAs Involved in Pathogenicity

Bacterial pathogens play a crucial role in many environments, including the human body. Bacterial species often acquire pathogenicity genes through mutational events, horizontal gene transfer, and bacteriophage infections [45]. Performing RIL-seq in pathogenic bacteria provides exclusive opportunities to understand better how sRNAs couple the core genes with virulence traits encoded by these additional genes at the post-transcriptional level (Fig. 4). To date, RIL-seq analysis was performed in several common pathogens, such as EPEC [12], *Salmonella* [11], *Clostridioides difficile* [14], and *V. Cholera* [10].

EPEC has a similar genome to that of *E. coli* K-12, genes that are identical are called the core genome, whereas additional genes that are pathogenesis-associated are called "accessory genes" and cluster in pathogenic islands. Pearl Mizrahi et al. applied RIL-seq to EPEC and cross-referenced the RNA–RNA interactions with non-pathogenic *E. coli*. Many interactions were shared between the strains, but a subset of the chimeras involved the accessory genome. They hypothesized that these interactions are important for the pathogenicity of EPEC. In this study, EPEC was grown in

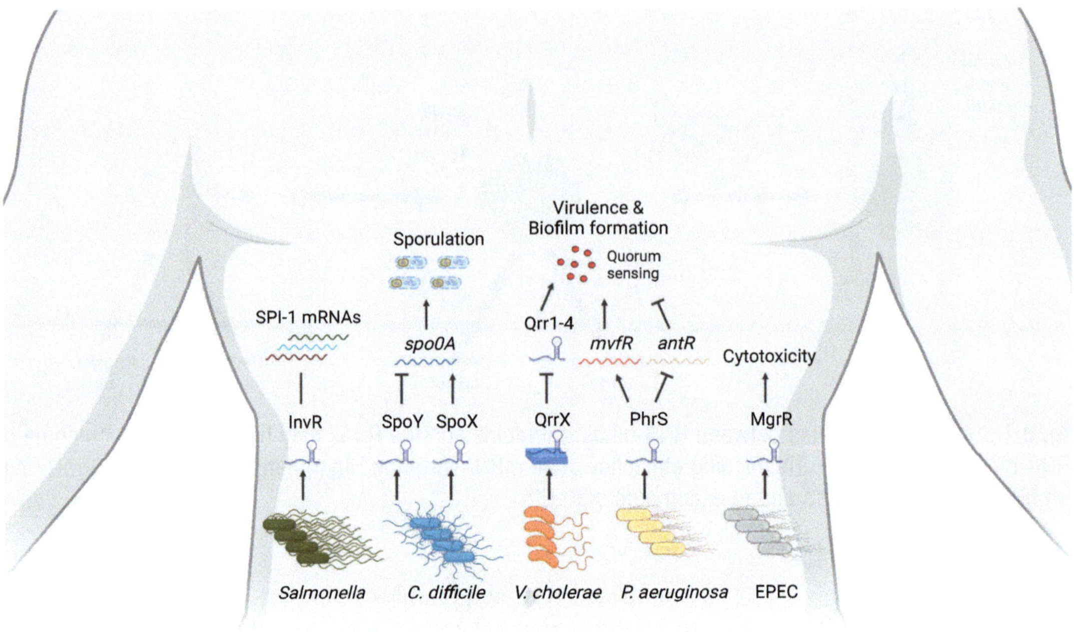

Fig. 4 sRNAs regulate bacterial pathogenicity in multiple species. The *Salmonella* SPI-1-encoded sRNA InvR was found with many SPI-1 mRNA targets in RIL-seq data. Sporulation in *C. difficile* is regulated by two sRNAs, SpoY and SpoX, with opposite outcomes. PhrS promotes the synthesis of quorum sensing signaling molecules in *P. aeruginosa* by regulating two transcription factors, MvfR and AntR. The network of QrrX with the four Qrr sRNAs controls virulence and biofilm formation, among other pathways in *V. cholerae*. EPEC cytotoxicity is regulated by the core genome-encoded sRNA MgrR

two distinct environments, one causing the activation of pathogenic traits, such as quorum sensing and aggregation, and the other non-activating. RIL-seq revealed a significant difference in the RNA interactome between these two growth conditions. Interestingly, they noticed that the abundance of several sRNAs, like Spot 42, RyhB, MgrR, and CpxQ, under the non-activated condition was higher and that the core sRNA network changed under activating conditions, implying sRNAs regulate EPEC pathogenicity. The authors focused on MgrR, a sRNA that had more interactions in the activating state than the inactivated state and had chimeras with accessory genes. The top target of MgrR was *cesT*, encoding the major virulence-associated chaperone. MgrR adjusted the level of EPEC cytotoxicity via regulation of *cesT* expression and by enhancement of the transcription of all the LEE (locus of enterocyte effacement) genes, including that of *cesT*, through the regulation of another target.

Salmonella is one of the leading bacterial foodborne causes of death in low and middle-income countries [46–48]. As Gram-negative bacteria, they have a complicated system of inner membrane and outer membrane proteins that must balance membrane integrity and nutrient uptake. Matera et al. studied the sRNA interactome in *Salmonella* and provided important insights on the fine regulation of membrane permeability controlled by sRNAs [11]. As discussed earlier in this review, OppX is a novel sRNA that acts as an RNA sponge and regulates MicF, a well-studied prototypical porin repressor. In this work, the bacteria were grown in conditions that express the *Salmonella* pathogenicity island 1 (SPI-1) virulence genes, which encodes many virulence proteins and a type III secretion system (T3SS) for their secretion. Interestingly, while previous target searches for the InvR sRNA encoded by SPI-1 excluded effects of this sRNA on SPI-1 virulence gene expression [49], RIL-seq data reveal many chimeras between InvR and SPI-1 encoded mRNAs (Table S2 in [11]). Further studies will be required to determine the nature of these interesting findings and their significance for *Salmonella* pathogenicity.

Quorum sensing is the ability of a bacterium to sense a signal from neighboring cells in the population when they reach a critical concentration and synchronously control processes that are only efficient above this cell concentration. This ability allows pathogenic bacteria to minimize host immune responses by delaying the production of virulence factors until sufficient bacteria are present and prepared to overcome the host defense mechanisms [50]. One of the model bacterial species to study quorum sensing is *Vibrio.* Recently, RIL-seq analysis was applied to *V. cholera* [10], identifying hundreds of sRNA–mRNA interactions and RNA duplexes. One novel sRNA had an interesting interactome, in that it exclusively interacted with all four quorum regulator RNAs (Qrr1–4). As described above, further experiments confirmed the hypothesis that this sRNA, termed QrrX, acts as an RNA sponge and adds another layer of regulation to the quorum sensing pathway in *Vibrio.* In *P. aeruginosa*, the PhrS sRNA regulates quorum sensing by upregulating the transcription factor MvfR required for the synthesis of quorum sensing signaling molecules [51]. A RIL-seq study revealed that PhrS negatively regulates the *antR* transcript that encodes AntR, a regulator that directs a precursor of quorum sensing signaling molecules called anthranilate towards the TCA cycle [13]. Thus, PhrS upregulates the synthesis of quorum sensing signaling molecules through a two-tiered mechanism, the upregulation of *mvfR* and downregulation of *antR*.

Gram-positive pathogens such as *C. difficile* pose a significant challenge to healthcare systems as they form antibiotic-resistant endospores that lead to relapsing and recurrent infections. Understanding the molecular mechanisms and the environmental signals that control spore generation is essential for developing efficient treatments. Fuchs et al. applied RIL-seq, for the first time in Gram-positive bacteria, to *C. difficile* under sporulating conditions [14]. The authors identified sRNA-mediated post-transcriptional regulation of *spo0A*, the master regulator of sporulation, as a new mechanism contributing to sporulation initiation in *C. difficile*. They found two novel sRNAs, SpoY and SpoX, that bind to *spo0A* mRNA, resulting in inhibition or activation of translation, respectively.

The emergence of antibiotic-resistant bacteria is an increasing global concern [52]. It is suggested that by the year of 2050, as many as ten million people could die each year due to resistant microbes [53]. Deciphering the mechanisms of bacterial pathogenicity is crucial for developing novel effective therapeutics. As we discussed above, regulatory RNAs in pathogenic bacteria participate in such mechanisms and a further understanding of their mechanism of action can help develop novel antimicrobial agents [54].

7 Perspectives

While our knowledge of sRNA regulatory networks and modes of action is growing by the day, there is still much left to learn. The growing use of RIL-seq and other RNA–RNA identification methods such as iRIL-seq [55], CLASH [56], or Hi-GRIL-seq [57] provides vast amounts of data that can be mined to deepen our understanding of sRNA-mediated regulation. While this review was in its final stages for publication, several RIL-seq studies revealed the RNA interactome of nitrogen-starved *E. coli* [58], the RNA interactome of the opportunistic human pathogen *Klebsiella pneumoniae* [59, 60], and the interactome of *V. cholerae* infected with a lysogenic phage induced by quorum sensing [61], emphasizing the rapid progress in deciphering RNA–RNA interaction networks. Current and future studies will take advantage of this knowledge, together with the increasing power of artificial intelligence and machine learning tools, to decipher specific mechanisms by which sRNAs act and how they integrate with other regulatory pathways, constituting multiplayer regulation of gene expression. The gained knowledge can be used for development of therapeutics targeting sRNAs or using them for targeting other genes in bacterial pathogens. Furthermore, integration of RNA sequencing-based methods with other multi-omics data, such as genome sequencing and metabolomics, will lead to a more comprehensive understanding of the role sRNAs play in bacteria.

Acknowledgments

We appreciate the useful comments from Philip Adams. We are grateful to the Melamed lab members for helpful discussions and comments on the manuscript. The study was supported by the Israel Science Foundation (grants 826/22 and 2859/22). Figures 1, 3, and 4 were created with BioRender.com.

Disclosure Statement The authors are unaware of any affiliations, memberships, funding, or financial holdings that might be perceived as affecting the objectivity of this review.

References

1. Hör J, Matera G, Vogel J, Gottesman S, Storz G (2020) Trans-acting small RNAs and their effects on gene expression in *Escherichia coli* and *Salmonella enterica*. EcoSal Plus ESP-0030-2019. https://doi.org/10.1128/eocsalplus
2. Updegrove TB, Zhang A, Storz G (2016) Hfq: the flexible RNA matchmaker. Curr Opin Microbiol 30:133–138. https://doi.org/10.1016/j.mib.2016.02.003
3. Tierrafría VH, Rioualen C, Salgado H, Lara P, Gama-Castro S, Lally P et al (2022) RegulonDB 11.0: comprehensive high-throughput datasets on transcriptional regulation in *Escherichia coli* K-12. Microb Genom 8(5). https://doi.org/10.1099/mgen.0.000833
4. Papenfort K, Melamed S (2023) Small RNAs, large networks: post-transcriptional regulons in gram-negative bacteria. Ann Rev Microbiol 77:23–43
5. Saliba AE, Santos SC, Vogel J (2017) New RNA-seq approaches for the study of bacterial pathogens. Curr Opin Microbiol 35:78–87. https://doi.org/10.1016/j.mib.2017.01.001
6. Melamed S, Faigenbaum-Romm R, Peer A, Reiss N, Shechter O, Bar A et al (2018) Mapping the small RNA interactome in bacteria using RIL-seq. Nat Protoc 13(1):1–33. https://doi.org/10.1038/nprot.2017.115
7. Faigenbaum-Romm R, Reich A, Gatt YE, Barsheshet M, Argaman L, Margalit H (2020) Hierarchy in Hfq chaperon occupancy of small RNA targets plays a major role in their regulation. Cell Rep 30(9):3127–38.e6. https://doi.org/10.1016/j.celrep.2020.02.016
8. Melamed S, Adams PP, Zhang A, Zhang H, Storz G (2020) RNA-RNA interactomes of ProQ and Hfq reveal overlapping and competing roles. Mol Cell 77(2):411–25.e7. https://doi.org/10.1016/j.molcel.2019.10.022
9. Melamed S, Peer A, Faigenbaum-Romm R, Gatt YE, Reiss N, Bar A et al (2016) Global mapping of small RNA-target interactions in bacteria. Mol Cell 63(5):884–897. https://doi.org/10.1016/j.molcel.2016.07.026
10. Huber M, Lippegaus A, Melamed S, Siemers M, Wucher BR, Hoyos M et al (2022) An RNA sponge controls quorum sensing dynamics and biofilm formation in vibrio cholerae. Nat Commun 13(1):7585. https://doi.org/10.1038/s41467-022-35261-x
11. Matera G, Altuvia Y, Gerovac M, El Mouali Y, Margalit H, Vogel J (2022) Global RNA interactome of *Salmonella* discovers a 5' UTR sponge for the MicF small RNA that connects membrane permeability to transport capacity. Mol Cell 82(3):629–44.e4. https://doi.org/10.1016/j.molcel.2021.12.030
12. Pearl-Mizrahi S, Elbaz N, Argaman L, Altuvia Y, Katsowich N, Socol Y et al (2021) The impact of Hfq-mediated sRNA-mRNA interactome on the virulence of enteropathogenic *Escherichia coli*. Sci Adv 7(44):eabi8228-eabi. https://doi.org/10.1126/sciadv.abi8228
13. Gebhardt MJ, Farland EA, Basu P, Macareno K, Melamed S, Love SD (2023) Hfq licensed RNA-RNA interactome in Pseudomonas aeruginosa reveals a keystone sRNA. Proc Natl Acad Sci USA 120(21):e2218407120
14. Fuchs M, Lamm-Schmidt V, Lenče T, Sulzer J, Bublitz A, Wackenreuter J et al (2023) A network of small RNAs regulates sporulation initiation in *Clostridioides difficile*. EMBO J 42(12):e112858. https://doi.org/10.15252/embj.2022112858
15. Dutta T, Srivastava S (2018) Small RNA-mediated regulation in bacteria: a growing palette of diverse mechanisms. Gene 656:60–72. https://doi.org/10.1016/j.gene.2018.02.068
16. Chareyre S, Mandin P (2018) Bacterial iron homeostasis regulation by sRNAs. Microbiol Spectr 6(2). https://doi.org/10.1128/microbiolspec.RWR-0010-2017
17. Plumbridge J, Bossi L, Oberto J, Wade JT, Figueroa-Bossi N (2014) Interplay of transcriptional and small RNA-dependent control mechanisms regulates chitosugar uptake in *Escherichia coli* and *salmonella*. Mol Microbiol 92(4):648–658. https://doi.org/10.1111/mmi.12573
18. Ronin I, Katsowich N, Rosenshine I, Balaban NQ (2017) A long-term epigenetic memory switch controls bacterial virulence bimodality. elife 6:e19599. https://doi.org/10.7554/eLife.19599

19. Franco-Zorrilla JM, Valli A, Todesco M, Mateos I, Puga MI, Rubio-Somoza I et al (2007) Target mimicry provides a new mechanism for regulation of microRNA activity. Nat Genet 39(8):1033–1037. https://doi.org/10.1038/ng2079
20. Figueroa-Bossi N, Valentini M, Malleret L, Fiorini F, Bossi L (2009) Caught at its own game: regulatory small RNA inactivated by an inducible transcript mimicking its target. Genes Dev 23(17):2004–2015. https://doi.org/10.1101/gad.541609
21. Overgaard M, Johansen J, Møller-Jensen J, Valentin-Hansen P (2009) Switching off small RNA regulation with trap-mRNA. Mol Microbiol 73(5):790–800. https://doi.org/10.1111/j.1365-2958.2009.06807.x
22. Miyakoshi M, Chao Y, Vogel J (2015) Cross talk between ABC transporter mRNAs via a target mRNA-derived sponge of the GcvB small RNA. EMBO J 34(11):1478–1492. https://doi.org/10.15252/embj.201490546
23. Lalaouna D, Carrier M-C, Semsey S, Brouard J-S, Wang J, Wade Joseph T et al (2015) A 3′ external transcribed spacer in a tRNA transcript acts as a sponge for small RNAs to prevent transcriptional noise. Mol Cell 58(3):393–405. https://doi.org/10.1016/j.molcel.2015.03.013
24. Denham EL (2020) The sponge RNAs of bacteria—how to find them and their role in regulating the post-transcriptional network. Biochim Biophys Acta Gene Regul Mech 1863(8):194565. https://doi.org/10.1016/j.bbagrm.2020.194565
25. Andersen J, Forst SA, Zhao K, Inouye M, Delihas N (1989) The function of micF RNA. micF RNA is a major factor in the thermal regulation of OmpF protein in *Escherichia coli*. J Biol Chem 264(30):17961–17970
26. Corcoran CP, Podkaminski D, Papenfort K, Urban JH, Hinton JC, Vogel J (2012) Superfolder GFP reporters validate diverse new mRNA targets of the classic porin regulator. MicF RNA Mol Microbiol 84(3):428–445. https://doi.org/10.1111/j.1365-2958.2012.08031.x
27. Holmqvist E, Unoson C, Reimegård J, Wagner EG (2012) A mixed double negative feedback loop between the sRNA MicF and the global regulator Lrp. Mol Microbiol 84(3):414–427. https://doi.org/10.1111/j.1365-2958.2012.07994.x
28. Adams PP, Baniulyte G, Esnault C, Chegireddy K, Singh N, Monge M et al (2021) Regulatory roles of *Escherichia coli* 5′ UTR and ORF-internal RNAs detected by 3′ end mapping. elife 10:e62438. https://doi.org/10.7554/eLife.62438
29. Ikemura T, Dahlberg JE (1973) Small ribonucleic acids of *Escherichia coli*. II. Noncoordinate accumulation during stringent control. J Biol Chem 248(14):5033–5041
30. Ikemura T, Dahlberg JE (1973) Small ribonucleic acids of *Escherichia coli*. I. Characterization by polyacrylamide gel electrophoresis and fingerprint analysis. J Biol Chem 248(14):5024–5032
31. Mizuno T, Chou MY, Inouye M (1984) A unique mechanism regulating gene expression: translational inhibition by a complementary RNA transcript (micRNA). Proc Natl Acad Sci USA 81(7):1966–1970. https://doi.org/10.1073/pnas.81.7.1966
32. Altuvia S, Weinstein-Fischer D, Zhang A, Postow L, Storz G (1997) A small, stable RNA induced by oxidative stress: role as a pleiotropic regulator and antimutator. Cell 90(1):43–53. https://doi.org/10.1016/S0092-8674(00)80312-8
33. Adams PP, Storz G (2020) Prevalence of small base-pairing RNAs derived from diverse genomic loci. Biochim Biophys Acta Gene Regul Mech. 1863(7):194524. https://doi.org/10.1016/j.bbagrm.2020.194524
34. Wassarman KM, Repoila F, Rosenow C, Storz G, Gottesman S (2001) Identification of novel small RNAs using comparative genomics and microarrays. Genes Dev 15(13):1637–1651. https://doi.org/10.1101/gad.901001
35. Westermann AJ, Vogel J (2021) Cross-species RNA-seq for deciphering host-microbe interactions. Nat Rev Genet 22(6):361–378. https://doi.org/10.1038/s41576-021-00326-y
36. Sharma CM, Vogel J (2014) Differential RNA-seq: the approach behind and the biological insight gained. Curr Opin Microbiol 19:97–105. https://doi.org/10.1016/j.mib.2014.06.010
37. Dar D, Shamir M, Mellin JR, Koutero M, Stern-Ginossar N, Cossart P et al (2016) Term-seq reveals abundant ribo-regulation of antibiotics resistance in bacteria. Science 352(6282):aad9822. https://doi.org/10.1126/science.aad9822
38. Bar A, Argaman L, Altuvia Y, Margalit H (2021) Prediction of novel bacterial small RNAs from RIL-Seq RNA-RNA interaction data. Front Microbiol 12:635070. https://doi.org/10.3389/fmicb.2021.635070

39. Artsimovitch I (2018) Rebuilding the bridge between transcription and translation. Mol Microbiol 108(5):467–472. https://doi.org/10.1111/mmi.13964
40. Holmqvist E, Vogel J (2018) RNA-binding proteins in bacteria. Nat Rev Microbiol 16(10):601–615. https://doi.org/10.1038/s41579-018-0049-5
41. Woodson SA, Panja S, Santiago-Frangos A (2018) Proteins that chaperone RNA regulation. Microbiol Spectr 6(4):10.1128/microbiolspec.RWR-0026-2018. https://doi.org/10.1128/microbiolspec.RWR-0026-2018
42. Melamed S (2020) New sequencing methodologies reveal interplay between multiple RNA-binding proteins and their RNAs. Curr Genet 66(4):713–717. https://doi.org/10.1007/s00294-020-01066-y
43. Rojano-Nisimura AM, Simmons TR, Leistra AN, Mihailovic MK, Buchser R, Ekdahl AM et al (2023) CsrA shows selective regulation of sRNA-mRNA networks. bioRxiv. https://doi.org/10.1101/2023.03.29.534774
44. Gerovac M, Vogel J, Smirnov A (2021) The world of stable ribonucleoproteins and its mapping with grad-seq and related approaches. Front Mol Biosci 8:661448. https://doi.org/10.3389/fmolb.2021.661448
45. Boerlin P (2022) Evolution of bacterial pathogens. In: Pathogenesis of bacterial infections in animals, pp 15–31
46. Jajere SM (2019) A review of *salmonella enterica* with particular focus on the pathogenicity and virulence factors, host specificity and antimicrobial resistance including multidrug resistance. Vet World 12(4):504–521. https://doi.org/10.14202/vetworld.2019.504-521
47. Urban-Chmiel R, Marek A, Stępień-Pyśniak D, Wieczorek K, Dec M, Nowaczek A et al (2022) Antibiotic resistance in bacteria-a review. Antibiotics (Basel) 11(8). https://doi.org/10.3390/antibiotics11081079
48. Newson JP, Gaissmaier MS, McHugh SC, Hardt WD (2022) Studying antibiotic persistence in vivo using the model organism *Salmonella Typhimurium*. Curr Opin Microbiol 70:102224. https://doi.org/10.1016/j.mib.2022.102224
49. Pfeiffer V, Sittka A, Tomer R, Tedin K, Brinkmann V, Vogel J (2007) A small non-coding RNA of the invasion gene Island (SPI-1) represses outer membrane protein synthesis from the *Salmonella* core genome. Mol Microbiol 66(5):1174–1191. https://doi.org/10.1111/j.1365-2958.2007.05991.x
50. Deep A, Chaudhary U, Gupta V (2011) Quorum sensing and bacterial pathogenicity: from molecules to disease. J Lab Physicians 3(1):4–11. https://doi.org/10.4103/0974-2727.78553
51. Sonnleitner E, Gonzalez N, Sorger-Domenigg T, Heeb S, Richter AS, Backofen R et al (2011) The small RNA PhrS stimulates synthesis of the *Pseudomonas aeruginosa* quinolone signal. Mol Microbiol 80(4):868–885. https://doi.org/10.1111/j.1365-2958.2011.07620.x
52. Ventola CL (2015) The antibiotic resistance crisis: part 1: causes and threats. P T 40(4):277–283
53. O'Neill J (2016) Tackling drug-resistant infections globally: final report and recommendations. Review on Antimicrobial Resistance, London
54. Chan H, Ho J, Liu X, Zhang L, Wong SH, Chan MT et al (2017) Potential and use of bacterial small RNAs to combat drug resistance: a systematic review. Infect Drug Resist 10:521–532. https://doi.org/10.2147/idr.S148444
55. Liu F, Chen Z, Zhang S, Wu K, Bei C, Wang C et al (2023) In vivo RNA interactome profiling reveals 3'UTR-processed small RNA targeting a central regulatory hub. Nat Commun 14(1):8106. https://doi.org/10.1038/s41467-023-43632-1
56. Waters SA, McAteer SP, Kudla G, Pang I, Deshpande NP, Amos TG et al (2017) Small RNA interactome of pathogenic *E. coli* revealed through crosslinking of RNase E. EMBO J 36(3):374–387. https://doi.org/10.15252/embj.201694639
57. Zhang Y-F, Han K, Chandler CE, Tjaden B, Ernst RK, Lory S (2017) Probing the sRNA regulatory landscape of *P. Aeruginosa*: post-transcriptional control of determinants of pathogenicity and antibiotic susceptibility. Mol Microbiol 106(6):919–937. https://doi.org/10.1111/mmi.13857
58. McQuail J, Matera G, Gräfenhan T, Bischler T, Haberkant P, Stein F et al (2024) Global Hfq-mediated RNA interactome of nitrogen starved *Escherichia coli* uncovers a conserved post-transcriptional regulatory axis required for optimal growth recovery. Nucleic Acids Res 52(5):2323–2339. https://doi.org/10.1093/nar/gkad1211
59. Ruhland E, Siemers M, Gerst R, Späth F, Vogt LN, Figge MT et al (2024) The global RNA-RNA interactome of *Klebsiella pneumoniae* unveils a small RNA regulator of cell division. Proc Natl Acad Sci USA 121(9):e2317322121. https://doi.org/10.1073/pnas.2317322121

60. Goh KJ, Altuvia Y, Argaman L, Raz Y, Bar A, Lithgow T, Margalit H, Gan YH (2024) RIL-seq reveals extensive involvement of small RNAs in virulence and capsule regulation in hypervirulent Klebsiella pneumoniae. Nucleic Acids Res 52(15):9119–9138. https://doi.org/10.1093/nar/gkae440
61. Sprenger M, Siemers M, Krautwurst S, Papenfort K (2024) Small RNAs direct attack and defense mechanisms in a quorum sensing phage and its host. Cell Host Microbe 32(5): 727–738.e6. https://doi.org/10.1016/j.chom.2024.03.010
62. Cohen S, Maximof E, Rokach S, Tadeski M, Veksler-Lublinsky I (2023) sInterBase: a comprehensive database of *E. coli* sRNA-mRNA interactions. Bioinformatics. https://doi.org/10.1093/bioinformatics/btad172

Chapter 12

Extraction, Purification, and Next-Generation Sequencing (NGS) Analysis of DNA and RNA from Formalin-Fixed and Paraffin-Embedded (FFPE) Tissue

Mainul Hoque, Veera D'Mello, Seema Husain, and Patricia Soteropoulos

Abstract

Formalin fixed paraffin embedded (FFPE) tissues have long been used for immunohistological analyses. FFPE tissues can be stored at room temperature for several years enabling analyses to be performed later. Ease of storage and transport makes these tissues an attractive source of biological material. However, formalin fixation results in chemical modifications of proteins and nucleic acids that poses a major challenge to any type of analysis. Recovery of nucleic acids for quantitative assays is rendered difficult due to degradation resulting from fixation and long-term storage, producing low usable yields. Extensive efforts in the last 20 years have led to significant improvements in use of FFPE tissues for DNA and RNA analyses and resulted in development of sensitive assays for a wide range of applications, including next-generation sequencing. In this chapter, we describe the optimization of methods for sequential extraction of DNA and RNA from FFPE tissue and subsequent preparation of DNA-seq and RNA-seq libraries for use with the Illumina platform using commercially available reagents/kits.

Key words Formalin fixed paraffin embedded tissue, DNA-seq, RNA-seq, Next-generation sequencing, Sequential extraction, Library preparation, Formaldehyde modifications, Genome, Transcriptome

1 Introduction

The cutting-edge next-generation sequencing (NGS) technology is useful for DNA and RNA sequencing and epigenetic analysis in both research and clinical diagnosis. One of the biggest challenges with patient samples is the collection and storage of such samples. Solid tissue samples can be flash frozen and cryopreserved to extract DNA or RNA later or fixed with formalin for transport and long-term storage. Formalin fixed paraffin embedded (FFPE) tissue

Mainul Hoque and Veera D'Mello have equally contributed to this chapter.

Valerie J. Carabetta and Olaitan Akintunde (eds.), *High Throughput Gene Screening: Methods and Protocols*, Methods in Molecular Biology, vol. 2866, https://doi.org/10.1007/978-1-0716-4192-7_12,

preparation is widely used in clinical and laboratory settings, as it is an efficient and cost-effective method for long-term storage for immunohistological analyses. With rapid advances in the field of molecular biology and NGS technologies, FFPE tissues have become a valuable resource for harnessing global genomic and transcriptomic data from archival samples.

The recovery of DNA and RNA from FFPE tissues for these analyses, however, is a challenging task due to modification and degradation of nucleic acids during fixation steps. Formalin fixation causes significant damage to DNA [1] and RNA [2, 3] that includes fragmentation, base modifications, and cross linking to proteins, resulting in inadequate and erroneous base pairing, leading to poor amplification and sequencing artifacts. Nucleic acid recovery is further rendered difficult by the fact that these tissues are often retrieved for analyses after prolonged periods of storage, which leads to further DNA and RNA degradation. Furthermore, DNA-seq and RNA-seq library yields and sequencing reads are also significantly affected by fixation [4]. It therefore becomes necessary to modify and optimize the standard extraction protocols in order to maximize nucleic acid recovery from FFPE tissues for downstream applications.

Several commercial kits are now available for extraction of nucleic acids from FFPE tissues that result in good yield which can be used for library preparation for downstream NGS applications. A comparison of DNA insert sizes (Fig. 1) and RNA extraction profile (Fig. 2) from fresh frozen (FF) and FFPE tissues using commercial kits shows that both DNA and RNA invariably give rise to shorter fragments. In this chapter, we describe sequential extraction of DNA and RNA from FFPE samples using MagMAX™ FFPE DNA/RNA Ultra Kit combined with NEBNext Ultra II FS DNA Library Prep kit and RNA library preparation using SMARTer® Stranded Total RNASeq Kit—Pico Input Mammalian from Takara Bio for Illumina NGS platform or TruSeq RNA Exome kit from Illumina. We also touch upon strategies to tackle a few alternate scenarios. The sequential approach is of particular advantage in correlating genomic changes to expression profiles, and in differentiating between genomic versus transcriptomic effects.

2 Materials

For each patient sample, paraffin ribbons containing 2–4 sections, or 3 sections mounted on glass slides, each 7–8 μm thick with an approximate area of 1 cm^2 are used. If the area of a tissue section is smaller, the number of sections increased proportionately. For example, if the area of a tissue section is approximately 0.4 cm^2, 5–6 sections are used as available.

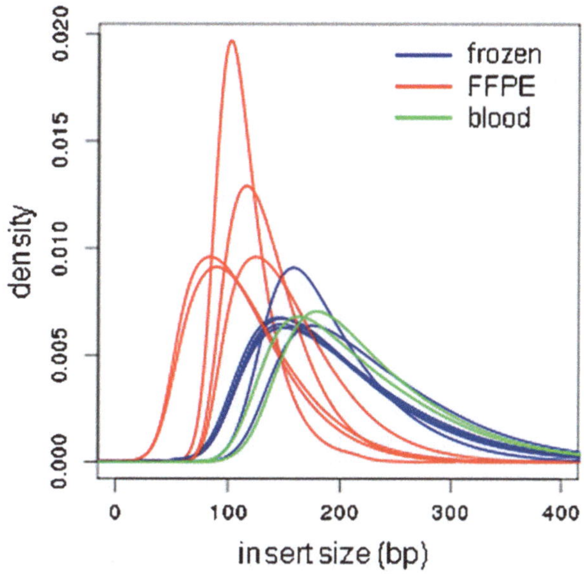

Fig. 1 Comparison of DNA insert sizes from FF and FFPE tissues. The distribution of insert sizes was calculated from properly mapped paired reads. 3 μg of genomic DNA was randomly fragmented using Covaris. Exome capture was performed using NimbleGen exome 2.1 M array or SureSelect. The median insert sizes of the frozen samples (blue lines) were about 150 ~ 200 bp, while those of the matched FFPE samples (red lines) were about 100 bp. Reproduced from [20], with permission

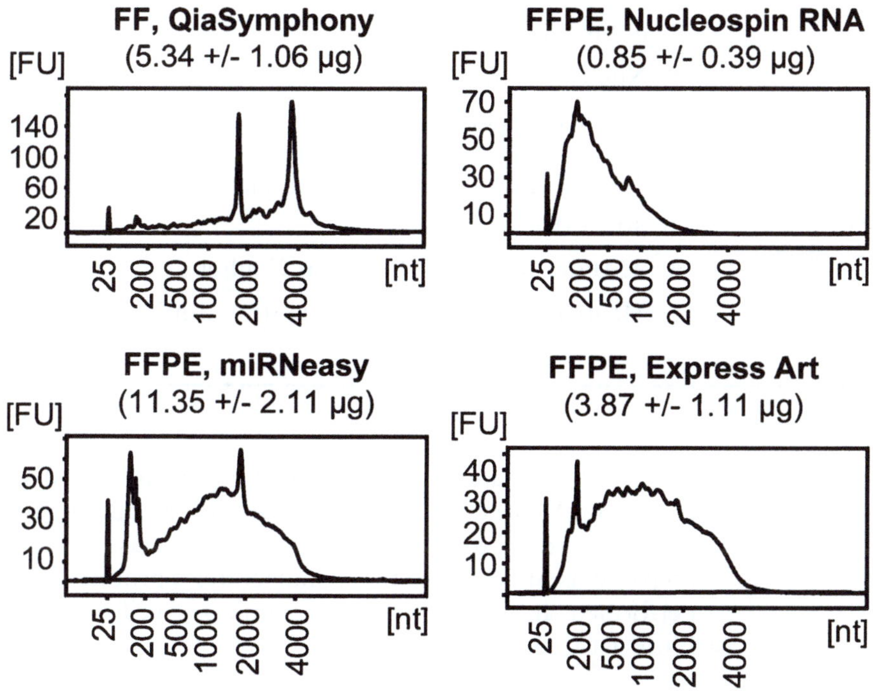

Fig. 2 Comparison of RNA from FF and FFPE tissues. Bioanalyzer profiles and yields of RNA isolated from FF tissue and from sections of the matching FFPE block of a normal tonsil (reactive) sample. RNA was isolated in triplicate (representative profile shown) from sections of the FFPE block using three different purification kits. Yields of RNA and purification kits are stated in the text above each profile. Reproduced [4], with permission

2.1 Sequential Extraction of DNA and RNA

1. MagMAX™ FFPE DNA/RNA Ultra Kit (Applied Biosystems).
2. CitriSolv (Decon Labs) or xylene.
3. 100% ethanol.
4. Isopropanol.
5. Sterile razor blades.
6. Nuclease-free, PCR inhibitor free, low adhesion 1.5 mL microcentrifuge tubes.
7. Nuclease-free filter tips.
8. Heat block with shaker.
9. Magnetic rack for 1.5 mL microcentrifuge tubes.

2.2 Genomic DNA Library Preparation

1. NEBNext Ultra II FS DNA Library Prep kit for Illumina (New England BioLabs).
2. 0.2 ml PCR tube strips.
3. Nuclease-free filter tips.
4. Magnetic bead separator plate for 0.2 mL tubes.
5. SPRIselect beads (Beckman Coulter).

2.3 RNA Library Preparation with rRNA Depletion

1. SMARTer Stranded Total RNA-Seq Kit—Pico Input Mammalian (Takara Bio).
2. 0.2 mL PCR tube strips.
3. Nuclease-free filter tips.
4. Magnetic bead separator plate for 0.2 mL tubes.

2.4 RNA Exome Library Preparation Using Exon-Targeted Probes

1. TruSeq RNA Exome kit (Illumina).
2. 80% Ethanol.
3. AMPure XP beads (Beckman Coulter).
4. Magnetic bead separator plate for 0.2 mL tubes.

3 Methods

3.1 Sequential Extraction of DNA and RNA

DNA and RNA are sequentially extracted using MagMAX™ FFPE DNA/RNA Ultra Kit (Applied Biosystems) by binding to magnetic beads according to manufacturer's instructions. All solutions are prepared as indicated in the manufacturer's protocol. All steps are performed at room temperature unless specified otherwise (*see* **Note 1** for kit alternatives).

3.1.1 Deparaffinization, Tissue Lysis and Reversal of Formaldehyde Modification

1. To deparaffinize tissues, add 1 mL CitriSolv to microcentrifuge tubes containing samples, vortex, and spin to collect all tissue in the solvent (*see* **Notes 2–5**. For sections mounted on slides, *see* **Note 3**).

2. Place tubes in a pre-heated block at 50 °C and incubate for 3 min.
3. Centrifuge tubes at 12,000 × *g* for 2 min to pellet tissue.
4. Carefully discard supernatant using a 1 mL pipette and leave about 100 μL solvent in the tube (*see* **Note 4**).
5. Add 1 mL of 100% ethanol and vortex thoroughly.
6. Centrifuge tubes at 12,000 × *g* for 2 min to pellet tissue.
7. Carefully discard supernatant using a 1 mL pipette. Remove the residual ethanol using a 200 μL pipette.
8. Repeat **steps 5–7** once more to remove all solvent.
9. Air dry the pellet at room temperature for 30 min.
10. To lyse and digest the tissue pellet, add 110 μL of protease solution, included in the kit, and mix the pellet by flicking the side of the tube or quickly vortexing for 2–5 s. If the pellet is large, increase volume to 210 μL (*see* **Note 6**). Ensure that all the pellet is submerged in the protease solution by briefly centrifuging the tube and inspecting the walls for any unsubmerged particles. Push particles in solution using a pipette tip. This step digests tissue and inactivates nucleases (*see* **Note** 7 for additional details).
11. Incubate at 55 °C for 1 h.
12. Centrifuge briefly to collect condensation.
13. Incubate at 90 °C for 1 h to reverse formalin fixation (*see* **Notes 8** and **9**).
14. Centrifuge briefly to collect condensation.
15. Cool the samples to room temperature.

3.1.2 Isolation of DNA and RNA

At this point, it is important to separate tubes containing larger pellets (LP)/higher volume of protease solution from smaller pellets (SP)/lower volume of protease solution (210 μL vs. 110 μL). Prepare a master mix of DNA binding buffer separately for each group according to the recipe in the protocol. A small pellet is equivalent to sections ≤40 μm and large pellet is equivalent to sections >40 μm. Similarly, prepare a master mix of RNA binding buffer separately for each group according to the recipe in the protocol. Proceed with manual isolation of DNA and RNA sequentially as follows:

1. Add 155 μL of DNA binding buffer to each SP tube and 270 μL to each LP tube and mix by pipetting up and down gently 1–2 times.
2. Shake SP tubes at 1000 rpm and LP tubes at 1150 rpm in an orbital shaker for 5 min (*see* **Note 10**).

3. Place tubes in magnetic rack until the solution clears and the beads are collected against the magnet (~2–3 min).
4. *Collect the RNA-containing supernatant carefully and add to a new labeled tube.* Separate the SP-RNA tubes and LP-RNA tubes for further processing. Do not discard. Place RNA-containing tubes on ice or in the refrigerator.
5. Add 200 μL of DNA wash buffer to the DNA-bound magnetic beads in each SP tube and 400 μL to each LP tube. Mix by pipetting up and down gently 1–2 times.
6. Shake for 2 min at 1100 rpm.
7. Place on magnetic rack until the solution clears (2–3 min). Discard the supernatant.
8. Repeat **steps 5–7**.
9. Wash the DNA-bound beads with Wash solution 2. Add 200 μL to each SP tube and 500 μL to each LP tube.
10. Shake at 1150 rpm for 1 min.
11. Place the tubes on a magnetic rack until the solution clears. Discard supernatant.
12. Repeat **steps 9–11**.
13. Shake beads at 1150 rpm for 2 min with the lids open to dry the beads. *Do not over-dry in order to avoid loss of yield* (*see* **Note 11**).
14. Add 50 μL of Elution solution to the beads and resuspend the beads by gently pipetting up and down 2–3 times.
15. Shake the tubes at 1150 rpm for 5 min.
16. Place the tubes on a magnetic rack for 2–3 min until the solution is completely clear. *The DNA is now contained in the supernatant.*
17. Collect the DNA-containing supernatant in a new labeled tube and store at −20 °C.
18. Equilibrate RNA-containing tubes to room temperature for about 10–15 min.
19. Add 20 μL of Nucleic acid binding beads to the RNA-containing tubes.
20. Add 400 μL RNA binding buffer to SP-RNA tubes and 580 μL buffer to LP-RNA tubes.
21. Shake at 1150 rpm for 5 min.
22. Place tubes on a magnetic rack until the solution clears (2–3 min). Discard the supernatant. *The RNA is now bound to the beads.*

23. Add 500 μL of RNA wash buffer and mix gently to resuspend beads.
24. Shake at 1150 rpm for 1 min.
25. Place tubes on a magnetic rack until the solution clears (2–3 min). Discard the supernatant.
26. Resuspend beads with 500 μL of Wash solution 2.
27. Shake at 1150 rpm for 1 min.
28. Place tubes on a magnetic rack until the solution clears (2–3 min). Discard the supernatant.
29. Shake beads at 1150 rpm for 2 min with the lids open to dry the beads. Do not over-dry in order to avoid loss of yield.
30. Resuspend the beads gently using 100 μL of DNase solution.
31. Shake at 1000 rpm at 37 °C (or room temperature) for 20 min.
32. Add 450 μL of RNA re-binding buffer.
33. Shake at 1150 rpm for 5 min.
34. Place tubes on magnetic stand until solution clears (~5 min). Discard supernatant.
35. Gently resuspend beads in 500 μL RNA wash buffer.
36. Shake at 1150 rpm for 1 min.
37. Place tubes on magnetic stand until solution clears (2–3 min). Discard supernatant.
38. Gently resuspend beads with 500 μL of Wash solution 2.
39. Shake at 1150 rpm for 1 min.
40. Place tubes on magnetic rack until solution clears (2–3 min). Discard supernatant.
41. Repeat **steps 38–40**.
42. Shake beads at 1150 rpm for 2 min with the lids open to dry the beads. Do not over-dry in order to avoid loss of yield.
43. Add 50 μL of Elution solution to the beads and resuspend the beads by gently pipetting up and down 2–3 times.
44. Shake the tubes at 1150 rpm for 5 min.
45. Place the tubes on a magnetic rack for 2–3 min until the solution is completely clear. The supernatant now contains purified RNA.
46. Collect the RNA-containing supernatant in a new labeled tube and store at −80 °C.
47. The quality of extracted genomic DNA and RNA is analyzed on TapeStation (Figs. 3 and 4). Alternatively, Bioanalyzer or similar gel electrophoresis system for NGS can be used for quality check.

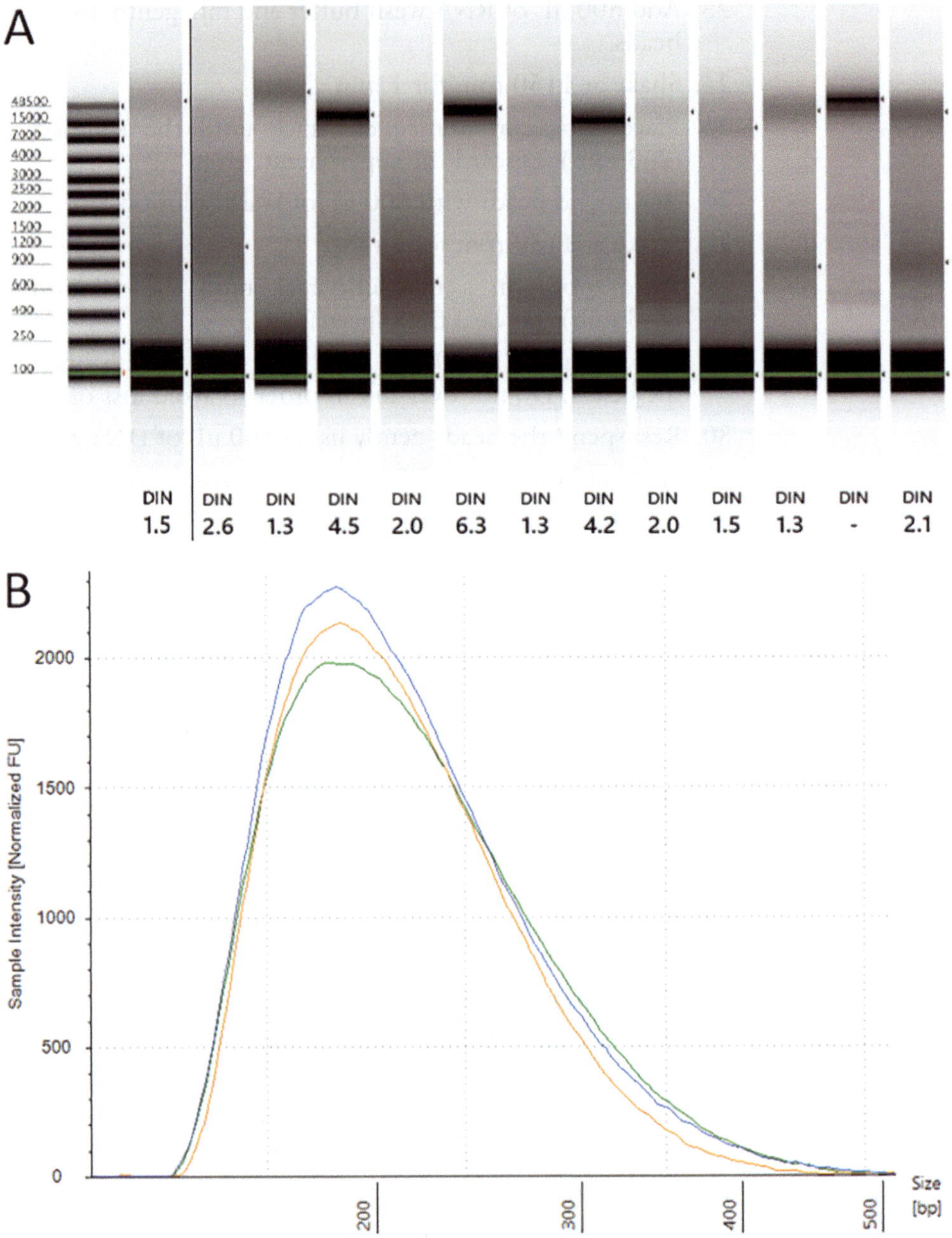

Fig. 3 Analysis of DNA from FFPE samples and purified gDNA libraries. (**a**). Genomic DNA was extracted using MagMAX FFPE DNA/RNA Ultra kit (Applied Biosystems). The concentration of extracted DNA varied between samples. Smearing pattern in lanes is typical of DNA degradation in FFPE tissues. (**b**). The gDNA library was prepared with NEB Ultra II FS DNA library prep kit. With proper selection and optimized fragmentation condition, library sizes obtained were similar, with an average insert length of approximately 200 bp. Library concentrations differed between samples. Insert lengths of three representative libraries from the pool that fall within similar range of concentration are shown

Due to the degraded nature of nucleic acids in FFPE tissues, genomic DNA appears as a smear with few samples showing the expected 48 kb bands. Extracted total RNA similarly shows a smeared pattern. For FFPE samples, RNA integrity number (RIN) is not a recommended measure of quality. Instead, a DV_{200} index is used which calculates % of RNA fragments >200 bp. Although DV_{200} of at least 55% or more is preferred for FFPE samples, samples with lower DV_{200} value can also be used to generate RNA libraries [5]. However, the possibility of obtaining an RNA-seq library decreases with decreasing DV_{200}. Figure 4 shows that most samples had RNA fragment sizes less than 200 bp. By carefully selecting and optimizing commercially available RNA-seq protocols, we were able to generate libraries from shorter RNA fragments that could be sequenced successfully.

3.2 Preparation of Genomic DNA Libraries

The NEBNext Ultra II FS DNA Library Prep kit for Illumina is used for whole-genome sequencing. The protocol for use with inputs ≤100 ng is followed exactly according to the manual, except for the fragmentation step and number of PCR amplification cycles. The recommended gDNA input range for this kit is 100 pg to 100 ng. We use 100 ng of gDNA as the maximum available input. Volumes of samples are diluted accordingly to a final volume of 26 μL (*see* **Notes 12–14**).

1. Dilute 100 ng of purified gDNA to 26 μL and combine with 9 μL of FS buffer-enzyme master mix (7 μL of FS reaction buffer +2 μl of FS enzyme mix per sample), vortex, and centrifuge briefly.
2. Add samples to a thermocycler with the lid temperature set to 75 °C, with the following conditions:

   5 min at 37 °C.

   30 min at 65 °C.

   Hold at 4 °C.
3. Prepare the adapter ligation mixture by adding 30 μL of ligation master mix, 1 μL of ligation enhancer, and 2.5 μL of undiluted adapter to the reaction mixture, according to manufacturer's recommendation for adapter dilution.
4. Vortex the samples briefly three times to ensure uniform mixing and centrifuge briefly to collect all liquid to the bottom of tubes.
5. Incubate at 20 °C for 15 min in a thermocycler with the heated lid off.
6. Add 3 μL of USER enzyme to the reaction product from **step 5**, vortex, and centrifuge briefly (*see* **Note 15**).

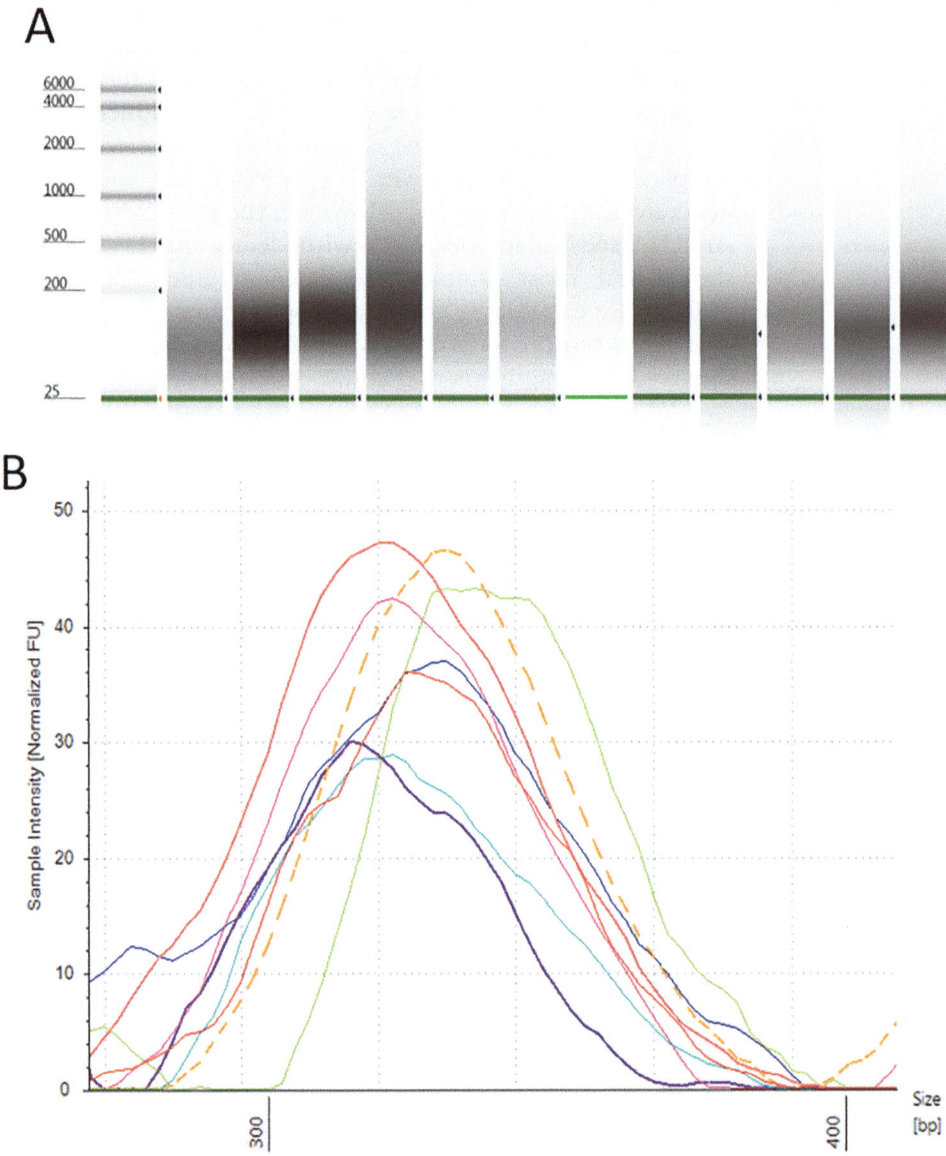

Fig. 4 Analysis of RNA and purified cDNA libraries. (**a**). Total RNA was extracted using MagMAX FFPE DNA/RNA Ultra kit (Applied Biosystems) and analyzed on TapeStation system using high-sensitivity tape. A variable range of RNA concentrations was observed for different samples. 18S (~2 kb) and 28S (~5 kb) rRNA bands were not detected. All samples contain degraded RNA with a fragment size of 200 bp or less. (**b**). cDNA library prepared using Takara Bio ribo-depletion protocol. With proper selection and optimized conditions, library sizes obtained were similar with an average insert length between 325 and 350 bp

7. Incubate samples in a thermocycler at 37 °C for 15 min with heated lid set to 50 °C.
8. Proceed to next step or store samples overnight at −20 °C.
9. Clean dsDNA fragments using SPRIselect beads as outlined in the protocol (*see* **Note 16** for avoiding DNA loss during ethanol washes).
10. Elute DNA in 17 μL of nuclease-free water instead of 0.1× TE.
11. Assemble the PCR mix as follows (*see* **Note 17**):

    15 μL of adapter-ligated and cleaned DNA fragments from **step 10**.

    25 μL of Q5 master mix.

    1.5 μL of i7 primer (instead of 5 μL).

    1.5 μL of i5 primer (instead of 5 μL).

    PCR cycling parameters:

Step	Temperature	Time	No. of cycles
Initial denaturation	98 °C	30 s	1
Denaturation Annealing/extension	98 °C 65 °C	10 s 75 s	12[a]
Final extension	65 °C	5 min	1
Hold	4 °C	∞	

[a]Increase the number of cycles because of low DNA library output from FFPE tissue. The recommended number of cycles for 100 ng of intact input gDNA is 3–4 cycles

12. Clean the resulting PCR products using SPRIselect beads according to the manufacturer's protocol.
13. Elute the DNA library in 33 μL of nuclease-free water and store at −20 °C.
14. Measure the concentration of each library by qPCR (recommended method) or by Qubit fluorometric quantification (ThermoFisher Scientific). For qPCR, NGS Library Quantification Kit (Agilent Technologies) can be used to accurately measure library concentrations or follow the protocol described in Illumina support guide [6].
15. Analyze the quality of DNA-seq libraries on the TapeStation system (Agilent Technologies, Inc.) using D1000 ScreenTape or High Sensitivity D1000 ScreenTape (Agilent Technologies, Inc.) following manufacturer's instructions depending on concentration measured by Qubit. Other similar gel electrophoresis systems developed for NGS such as QIAxcel Connect

System by Qiagen or Bioanalyzer can also be used. Note the average base pair size for each library and calculate the molarity of each library.

nM concentration of library = concentration in ng/μL × 10^6 /660 / Av. base pair size of library.

16. Pool the libraries and sequence on the NovaSeq Illumina platform according to Illumina library dilution, pooling, and loading protocols.

3.3 Preparation of RNA-seq Libraries Using the rRNA Depletion Method

As RNA yields obtained for most samples will likely be in the picogram range, RNA-seq libraries are prepared using SMARTer Stranded Total RNASeq Kit-Pico Input Mammalian (Takara Bio). The recommended total RNA input range for this kit is 250 pg to 10 ng. Our samples ranged from 24 pg to 5 ng. The advantages of using this kit is that ribosomal RNA (rRNA), which accounts for >90% of total RNA, is digested after cDNA synthesis and adapter ligation, using probes specific to mammalian rRNA. This allows for amplification of lower abundance RNA molecules thereby minimizing loss. Both coding and rRNA depleted non-coding RNAs are amplified by this method. RNA-seq libraries are prepared according to manufacturer's protocol without fragmentation of RNA, as recommended for FFPE tissues. A constant volume of 8 μL of purified RNA supernatant is used as input for each sample, irrespective of concentration.

1. Add 1 μL of SMART Pico Oligos Mix to 8 μL of undiluted purified RNA supernatant and incubate at 72 °C in a preheated, hot-lid thermal cycler for 3 min.
2. Immediately place tubes on ice.
3. Add 11 μL of First-strand master mix (containing the SMART UMI-template switching oligo, RNase inhibitor, and reverse transcriptase) to each tube.
4. Incubate tubes in a pre-heated thermocycler as follows:

   42 °C for 180 min.

   70 °C for 10 min.

   Hold at 4 °C.
5. Add 28 μL of PCR 1 master mix (containing DNA polymerase) to each tube.
6. Add 2 μL of the premixed SMARTer RNA Unique Dual Index primers to each tube, taking the utmost care that a different index was added to each sample.

7. Perform PCR as follows:

Temperature	Time	No. of cycles
94 °C	1 min	1
98 °C 55 °C 68 °C	15 s 15 s 30 s	10
68 °C	2 min	1
4 °C	∞	

8. Proceed to next step or store at −20 °C for up to 2 weeks.
9. Purify the amplified RNA-seq library using NucleoMag beads according to manufacturer's protocol (*see* **Note 18**).
10. Resuspend the magnetic beads in 22 μL of ZapR master mix and incubate at room temperature for 5 min.
11. Place the tubes on the magnetic stand until the solution is clear. The supernatant contains RNA.
12. Remove 20 μL of the supernatant to a new labeled PCR tube (*see* **Note 19**).
13. Add 80 μL of PCR 2 master mix (containing a single pair of primers and DNA polymerase) to each tube containing purified RNA.
14. Perform PCR as follows:

Temperature	Time	No. of cycles
94 °C	1 min	1
98 °C 55 °C 68 °C	15 s 15 s 30 s	16
68 °C	2 min	1
4 °C	∞	

15. Proceed to next step or store at −20 °C for up to 2 weeks.
16. Purify the RNA-seq library using NucleoMag beads according to manufacturer's protocol and analyze on TapeStation. Re-purify any samples where primer dimer bands are observed. Dimer bands are usually seen in samples that have low RNA concentration and are between 150 and 200 bp, whereas RNA library bands are between 300 and 400 bp in size.

17. Determine the quality of the RNA-seq libraries using the TapeStation. Note the average base pair size for each library.
18. Determine the concentration of each library using Qubit and calculate the molarity.
19. Pool libraries and sequence on a NovaSeq Illumina platform according to Illumina library dilution, pooling, and loading protocols.

3.4 Preparation of RNA Exome Libraries

Another method for generating usable RNA-seq libraries from a limited amount of total RNA is by using the TruSeq RNA Exome kit by Illumina. This method of RNA-seq library preparation selectively enriches for coding sequences and is independent of polyA+ selection bias, resulting in uniform amplification of polyA+ and alternate transcript forms, such as polyA-, bimorphic transcripts, and transcripts with short polyA tails. Although polyA+ transcripts are the most abundant, sequencing studies show that the latter transcripts constitute a significant proportion of the transcriptome and a few are critical to cellular function [7, 8]. Briefly, total RNA is primed using random hexamers and transcribed into cDNA. As before, fragmentation is not performed for FFPE tissues. Probes directed against coding RNAs are hybridized to cDNA and captured onto streptavidin beads. As selection probes are targeted to coding RNAs, non-coding RNAs are excluded by default. A minimum of 20 ng of total RNA input is recommended for FFPE tissues. For this experiment, all steps were followed exactly according to manufacturer's protocol.

1. Dilute total RNA in nuclease-free water to a final volume of 8.5 μL.
2. Prepare a master mix of reverse transcriptase and first strand synthesis actinomycin D mix (FSA mix). Add 8 μL of this master mix to 8.5 μL of diluted RNA, agitate at 1600 rpm on a shaker for 20 s and incubate in a thermocycler using following program with the pre-heated lid set to 100 °C:

   25 °C for 10 min.

   42 °C for 15 min.

   70 °C for 15 min.

   Hold at 4 °C.
3. Add 5 μL of resuspension buffer to each sample, 20 μL of second strand master mix, agitate at 1600 rpm for 20 s on a shaker, and incubate at 16 °C for 1 h.
4. Purify cDNA synthesized in this reaction using AMPure XP beads as detailed in the manufacturer's protocol.

5. Add 2.5 μL of resuspension buffer to 15 μL of the bead eluate from the previous step. Add 12.5 μL of A-tailing mix to each sample and agitate at 1800 rpm on a shaker for 2 min (*see* **Note 20**).
6. Centrifuge briefly and incubate at 37 °C for 30 min.
7. Incubate at 70 °C for 5 min and cool the reaction by placing on ice for 1 min.
8. Add 2.5 μL of resuspension buffer to the sample, followed by 2.5 μL of ligation mix and 2.5 μL of RNA adapters. Agitate at 1800 rpm on a shaker for 2 min and centrifuge briefly.
9. Incubate samples at 30 °C for 10 min and place on ice.
10. Add 5 μL of stop buffer (STL) to the reaction, agitate at 1800 rpm on a shaker for 2 min and centrifuge briefly.
11. Purify the adapter-ligated cDNA using AMPure XP beads as detailed in manufacturer's protocol.
12. Add 5 μL of PCR primer cocktail (PPP) to 20 μL of the bead eluate from the previous step on ice.
13. Add 25 μL PCR master mix (PMM) to each sample, agitate at 1600 rpm on a shaker for 20 s, and centrifuge briefly.
14. Incubated the reaction in the thermocycler using the following PCR program, with a pre-heated lid set to 100 °C:

Temperature	Time	No. of cycles
98 °C	30 s	1
98 °C 60 °C 72 °C	10 s 30 s 30 s	15
72 °C	5 min	1
4 °C	∞	

15. Purify the amplified DNA as detailed in the manufacturer's protocol.
16. Quantify the libraries using Qubit and determine the quality of each library using the TapeStation system. The expected cDNA library size is approximately 250 bp.
17. *First Hybridization and Capture of Probes*:
    (a) Pool 200 ng of each library in a total volume of 45 μL, combine with 50 μL of capture buffer (CT3) and 5 μL of coding exome oligos (CEX), agitate at 1200 rpm on a shaker for 1 min, and centrifuge briefly.
    (b) Incubate the samples in a thermocycler using the following program, with a pre-heated lid set to 100 °C:

95 °C for 10 min.

18 cycles of 1 min each, starting at 94 °C, then decreasing 2 °C per cycle. Set up thermocycler program manually if touchdown PCR machine is not available.

58 °C for 90 min.

(c) Combine 100 μL of pooled libraries with 250 μL of streptavidin magnetic beads (SMB) and agitate at 1200 rpm on a shaker for 5 min.

(d) Incubate at room temperature for 25 min.

(e) Separate the beads from the supernatant on a magnetic rack and discard the supernatant.

(f) Wash the captured DNA with wash solution (EWS) and elute with elution pre-mix (EE1 + HP3) according to manufacturer's protocol. Transfer 21 μL to a new plate well or tube and add 4 μL of ET2 buffer. Agitate at 1200 rpm for 1 min and centrifuge at 280 × *g* for 1 min. The samples can be sealed and stored at −20 °C for up to a week at this point.

18. *Second Hybridization and Capture of Probes*:

(a) Add 20 μL of resuspension buffer, 50 μL CT3, and 5 μL CEX to 25 μL of sample from **Step 17f**.

(b) Repeat **Steps 17b–17f**.

(c) After the second hybridization, capture, washing and elution of library pool, clean the library as detailed in the protocol using the AMPure XP beads.

19. *Amplification of Enriched Library*:

(a) Add 5 μL of PCR primer cocktail (PPC) and 20 μL of PCR mix (EPM) to 25 μL of the cleaned library from **Step 18c**, agitate at 1200 rpm on a shaker for 1 min, and briefly centrifuge.

(b) Incubate in a thermocycler with the pre-heated lid set at 100 °C as follows:

Temperature	Time	No. of cycles
98 °C	30 s	1
98 °C 60 °C 72 °C	10 s 30 s 30 s	10
72 °C	5 min	1
10 °C	Until next step[a]	

[a]Or store at 2–8 °C for up to 2 days

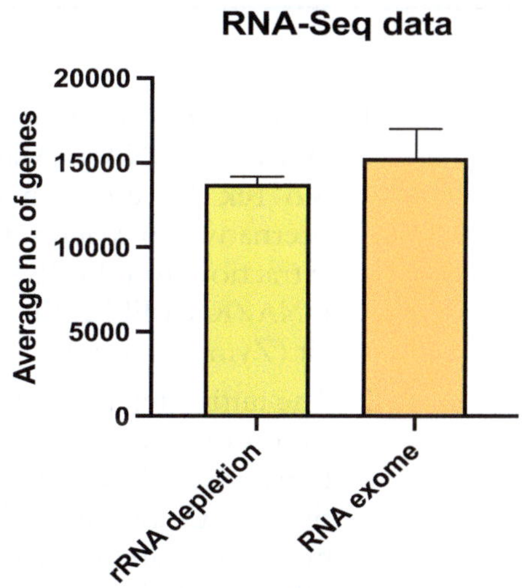

Fig. 5 Comparison of RNA-Seq libraries generated using rRNA depletion vs. RNA exome methods. RNA-seq libraries were prepared and sequenced on Illumina NovaSeq 6000 sequencer. The average number of genes identified across all samples for the rRNA depletion method was 13,737 $\pm$ 452 (n $=$ 51 pooled libraries). The average number of genes identified across all samples for the RNA exome method was 15,291 $\pm$ 1716 (n $=$ 77 pooled libraries)

20. Clean the amplified DNA library as detailed in the manufacturer's protocol.
21. Quantify libraries using Qubit (or kit-based qPCR methods for NGS library quantification) and analyze the quality of each library using the TapeStation.
22. Pool libraries and sequence on the NovaSeq Illumina platform according to Illumina library dilution, pooling, and loading protocols.

A comparison of RNA-seq libraries prepared by the rRNA depletion method and the RNA exome capture method is depicted in Fig. 5. RNA-seq libraries prepared by both methods identified a similar number of genes. We have found that rRNA depletion protocols reduces the time and overall cost for RNA-seq library construction, such as the one used here for FFPE tissues without loss of exome diversity. Comparable exome coverage by the rRNA depletion method can be attributed to the fact that non-coding non-ribosomal RNAs constitute a small fraction of the RNA pool [9], and size selection would further eliminate some species, which would largely favor amplification of coding sequences.

4 Notes

1. Another kit that is widely used for sequential extraction of DNA and RNA is Mag-Bind FFPE DNA/RNA by Omega Bio-Tek that uses mineral oil de-paraffinization as a non-toxic alternative to xylene. Other kits for sequential nucleic acid extraction from FFPE samples are also available such as AllPrep DNA/RNA FFPE Kit (Qiagen) and Quick-DNA/RNA FFPE kit (Zymo Research).
2. The initial step preceding nucleic acid extraction is deparaffinization of sections mounted on slides or obtained as wax ribbons or scrolls. Deparaffinization is achieved by treating FFPE tissue with xylene or the less toxic alternative CitriSolv. This procedure can be performed either at room temperature or at 50 °C for 2–5 min to dissolve paraffin wax. The organic solvent is then thoroughly washed using 100% ethanol and the tissue is air dried.
3. If sections are mounted on slides, deparaffinize by placing slides in glass jars containing xylene or CitriSolv in a fume hood on an orbital shaker for 5 min at room temperature (RT), followed by a wash in 100% ethanol for 5 min on the orbital shaker at RT. Air dry slides completely for 15–30 min. Apply 2–4 μL of protease digestion buffer, provided in the kit, directly to the section, then quickly scrape off using a sterile razor blade and collect in a labeled microcentrifuge tube for further processing. Do not allow section to dry after applying protease digestion buffer.
4. Some solvent will be left in the tube so that the pellet is not lost. Subsequent wash steps will get rid of the solvent.
5. There arise scenarios when there is need for dissecting a defined area of tissue from a larger section, such as collecting an injured or inflamed core from surrounding tissue or cancer cells within normal tissue. The Pinpoint solution (Zymo Research) has been successfully employed by several groups to obtain targeted areas from FFPE specimens, including those stained with hematoxylin-eosin, that resulted in usable DNA yields [10, 11]. This is a simple and inexpensive method. The Pinpoint solution is applied to a region of interest on the section, allowed to dry, and then scraped off for processing. Other methods such as laser capture microdissection require additional elaborate equipment [12, 13]. Recently, companies such as 10X Genomics, NanoString, and Vizgen have developed FFPE-compatible spatial transcriptomics platforms but also require additional equipment [14–16].

6. If the tissue pellet in the protease solution occupies more than half of the solvent volume, add 100 μL more of protease solution to bring final volume to 210 μL.
7. Most commercial kits employ a combination of chemical and enzymatic methods for lysing tissue. Lysis buffer typically constitutes a tris-EDTA solution containing a detergent that destabilizes cell membranes and proteinase K, a broad-spectrum serine protease. The primary purpose of protease treatment is to destroy endogenous nucleases and protect nucleic acids from degradation. The sample is incubated with lysis buffer at 50–55 °C for at least 1 h or in some cases overnight for optimal lysis of tissue [17, 18]. Ultrasonication of tissues in lysis buffer can be included to increase yield; taking care that the quality of nucleic acids is not compromised by shortening incubation time in lysis buffer [18].
8. Formalin fixation can be partially reversed by heating the sample at different temperatures for different lengths of time in buffer solutions and should be optimized to experimental conditions [2, 19]. Buffer solutions, temperature, and time conditions vary with kits and are usually proprietary information. Here, we followed manufacturer-recommended temperature and time conditions.
9. If only one heat block is available, it will not be possible to pre-heat the block to 90 °C. If this is the case, place tubes in the heat block immediately after centrifugation and measure the time that it took to increase the temperature from 55 °C to 90 °C. Increase the incubation time accordingly. For example, if it took 10 min for the temperature to reach 90 °C, incubate the samples for 1 h and 10 min.
10. If using the same orbital shaker, shake the SP samples first and place on magnetic rack. Then, put the LP samples on the shaker for 5 min. Meanwhile, collect RNA-containing supernatant from SP samples.
11. Beads should not be left to dry for more than 5 min. Beads should appear dark brown and glossy. Overdried beads appear light brown and lack glossy appearance.
12. Although NEBNext Ultra II FS DNA Library Prep kit for Illumina is not specified for FFPE tissues, we have successfully used this protocol for an input of 10 ng gDNA with FFPE tissues.
13. An important consideration when using NEBNext Ultra II FS DNA Library Prep kit for Illumina with FFPE tissues is fragmentation time. Considering the degraded nature of nucleic acids obtained from FFPE tissues, we decrease fragmentation time to 5 min.

14. The fragmentation/End Prep step combines fragmentation of purified gDNA using dsDNA fragmentase and end repair by 5′ phosphorylation and 3′ dA-tailing of DNA fragments.
15. Addition of an adapter creates closed loops at the ends of dsDNA. USER (Uracil-specific excision reagent) enzyme removes uracil base from the adapter generating free open ends on the adapter-ligated dsDNA fragments.
16. Be extremely careful not to disturb the beads during the ethanol wash. Ethanol should be slowly added along the opposite wall of the tube from the magnetic beads and aspirated out by maintaining the pipette tip close to the opposite side. Do not resuspend the beads in ethanol. This will cause loss of DNA. It is helpful to leave 20 μL ethanol at the bottom of the tubes during washes. Aspirate out remaining 20 μL after the last wash using a 10 or 20 μL pipette.
17. The PCR step uses Illumina specific i5/i7 index primers to generate uniquely barcoded libraries. It is important to use a unique combination of i5/i7 primers for each sample. To conserve primers, a 1.5 μL volume per primer per sample was optimized for low input material instead of the recommended 5 μL.
18. Preparation for the rRNA depletion step is carried out during the incubation times of the purification step.
19. It is best to use a10 μL pipette and collect the supernatant in two rounds.
20. Adenylation of 3′ ends involves the addition of a single "A" nucleotide to the 3′ end of each strand in the duplex and thus prevents concatenation of cDNA product.

References

1. Do H, Dobrovic A (2015) Sequence artifacts in DNA from formalin-fixed tissues: causes and strategies for minimization. Clin Chem 61(1): 64–71. https://doi.org/10.1373/clinchem.2014.223040
2. Masuda N, Ohnishi T, Kawamoto S, Monden M, Okubo K (1999) Analysis of chemical modification of RNA from formalin-fixed samples and optimization of molecular biology applications for such samples. Nucleic Acids Res 27(22):4436–4443. https://doi.org/10.1093/nar/27.22.4436
3. Farragher SM, Tanney A, Kennedy RD, Paul Harkin D (2008) RNA expression analysis from formalin fixed paraffin embedded tissues. Histochem Cell Biol 130(3):435–445. https://doi.org/10.1007/s00418-008-0479-7
4. Hedegaard J, Thorsen K, Lund MK, Hein AM, Hamilton-Dutoit SJ, Vang S, Nordentoft I, Birkenkamp-Demtroder K, Kruhoffer M, Hager H, Knudsen B, Andersen CL, Sorensen KD, Pedersen JS, Orntoft TF, Dyrskjot L (2014) Next-generation sequencing of RNA and DNA isolated from paired fresh-frozen and formalin-fixed paraffin-embedded samples of human cancer and normal tissue. PLoS One 9(5):e98187. https://doi.org/10.1371/journal.pone.0098187
5. Matsubara T, Soh J, Morita M, Uwabo T, Tomida S, Fujiwara T, Kanazawa S, Toyooka S, Hirasawa A (2020) DV200 index for assessing RNA integrity in next-generation sequencing. Biomed Res Int 2020:9349132. https://doi.org/10.1155/2020/9349132

6. Sequencing Library qPCR Quantification Guide. Illumina. https://support.illumina.com/content/dam/illumina-support/documents/documentation/chemistry_documentation/qpcr/sequencing-library-qpcr-quantification-guide-11322363-c.pdf
7. Yang L, Duff MO, Graveley BR, Carmichael GG, Chen LL (2011) Genomewide characterization of non-polyadenylated RNAs. Genome Biol 12(2):R16. https://doi.org/10.1186/gb-2011-12-2-r16
8. Marzluff WF, Wagner EJ, Duronio RJ (2008) Metabolism and regulation of canonical histone mRNAs: life without a poly(A) tail. Nat Rev Genet 9(11):843–854. https://doi.org/10.1038/nrg2438
9. Palazzo AF, Lee ES (2015) Non-coding RNA: what is functional and what is junk? Front Genet 6:2. https://doi.org/10.3389/fgene.2015.00002
10. Snow AN, Stence AA, Pruessner JA, Bossler AD, Ma D (2014) A simple and cost-effective method of DNA extraction from small formalin-fixed paraffin-embedded tissue for molecular oncologic testing. BMC Clin Pathol 14:30. https://doi.org/10.1186/1472-6890-14-30
11. Verdon TJ, Ballantyne KN, John Mitchell R, van Oorschot RAH (2011) The pinpoint DNA isolation system as a novel DNA sampling method in forensic biology. Forensic Sci Int Genet Suppl Ser 3(1):e91–e92
12. Ong CJ, Tan QX, Lim HJ, Shannon NB, Lim WK, Hendrikson J, Ng WH, Tan JWS, Koh KKN, Wasudevan SD, Ng CCY, Rajasegaran V, Lim TKH, Ong CK, Kon OL, Teh BT, Tan GHC, Chia CS, Soo KC, Teo MCC (2020) An optimised protocol harnessing laser capture microdissection for transcriptomic analysis on matched primary and metastatic colorectal tumours. Sci Rep 10(1):682. https://doi.org/10.1038/s41598-019-55146-2
13. Fend F, Raffeld M (2000) Laser capture microdissection in pathology. J Clin Pathol 53(9):666–672. https://doi.org/10.1136/jcp.53.9.666
14. Meylan M, Petitprez F, Becht E, Bougouin A, Pupier G, Calvez A, Giglioli I, Verkarre V, Lacroix G, Verneau J, Sun CM, Laurent-Puig P, Vano YA, Elaidi R, Mejean A, Sanchez-Salas R, Barret E, Cathelineau X, Oudard S, Reynaud CA, de Reynies A, Sautes-Fridman C, Fridman WH (2022) Tertiary lymphoid structures generate and propagate anti-tumor antibody-producing plasma cells in renal cell cancer. Immunity 55(3):527–541 e525. https://doi.org/10.1016/j.immuni.2022.02.001
15. Chagani S, De Macedo MP, Carapeto F, Wang F, Marzese DM, Wani K, Haydu LE, Peng W, Ong GT, Warren SE, Beechem JM, Hoon DSB, Mills GB, Tetzlaff MT, Lazar AJ, Kwong LN, Davies MA (2023) Multiplatform analysis of intratumoral Pten heterogeneity in melanoma. J Invest Dermatol. https://doi.org/10.1016/j.jid.2023.01.034
16. Beal DaA B (2022) Vizgen launches MERSCOPE™ formalin-fixed paraffin-embedded (FFPE) solution kit. Business Wire, Cambridge
17. McDonough SJ, Bhagwate A, Sun Z, Wang C, Zschunke M, Gorman JA, Kopp KJ, Cunningham JM (2019) Use of FFPE-derived DNA in next generation sequencing: DNA extraction methods. PLoS One 14(4):e0211400. https://doi.org/10.1371/journal.pone.0211400
18. Amini P, Ettlin J, Opitz L, Clementi E, Malbon A, Markkanen E (2017) An optimised protocol for isolation of RNA from small sections of laser-capture microdissected FFPE tissue amenable for next-generation sequencing. BMC Mol Biol 18(1):22. https://doi.org/10.1186/s12867-017-0099-7
19. Kennedy-Darling J, Smith LM (2014) Measuring the formaldehyde protein-DNA cross-link reversal rate. Anal Chem 86(12):5678–5681. https://doi.org/10.1021/ac501354y
20. Oh E, Choi YL, Kwon MJ, Kim RN, Kim YJ, Song JY, Jung KS, Shin YK (2015) Comparison of accuracy of whole-exome sequencing with formalin-fixed paraffin-embedded and fresh frozen tissue samples. PLoS One 10(12):e0144162. https://doi.org/10.1371/journal.pone.0144162

Part IV

Sequencing Technologies for DNA Structure

Chapter 13

Using Chromosome Conformation Capture Combined with Deep Sequencing (Hi-C) to Study Genome Organization in Bacteria

Qin Liao and Xindan Wang

Abstract

Genome organization is fundamental to all living organisms. Long DNA molecules are organized in hierarchical orders to be accommodated into eukaryotic nuclei or bacterial cells, which are thousands of folds shorter. Over the past two decades, chromosome conformation capture (3C) techniques substantially advanced our understanding of genome folding inside cells. 3C involves crosslinking and proximity ligation, and quantifies the physical contacts between two DNA regions within the genome. Coupled with high-throughput sequencing, 3C-seq and Hi-C techniques detect genome-wide DNA interactions, providing a comprehensive view of global genome organization. Here, we describe a detailed method to prepare Hi-C libraries using *Bacillus subtilis*, which includes procedures of crosslinking chromatin, digesting the crosslinked genome, labeling DNA ends with biotin, ligating DNA, and preparing the DNA library for sequencing using an Illumina platform.

Key words Hi-C, Chromosome conformation, Capture, Genome organization, Bacteria, *Bacillus subtilis*

1 Introduction

Genome organization plays critical roles in the regulation of gene expression and genome stability. Microscopy-based techniques have documented many architectural features of genomes, such as nucleus compartmentalization, chromatin fibers, and subnuclear positions of various chromosomal loci [1–3]. Complementary methods involving chromosome conformation capture (3C) assays have extend our understanding of genome organization [4]. Through chromatin crosslinking, genome digestion, re-ligation, and quantification of ligation frequency, 3C-based methods measure spatial proximity between DNA sequences and provide an understanding of the organization of a genome.

Valerie J. Carabetta and Olaitan Akintunde (eds.), *High Throughput Gene Screening: Methods and Protocols*, Methods in Molecular Biology, vol. 2866, https://doi.org/10.1007/978-1-0716-4192-7_13,

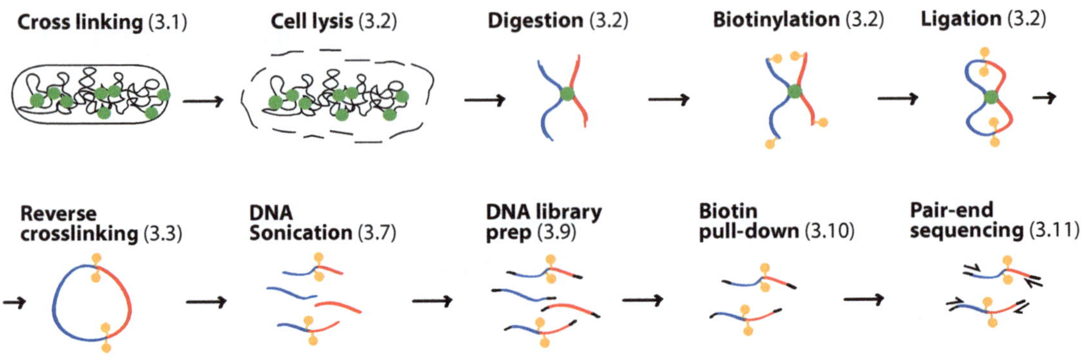

Fig. 1 Overview of the Hi-C experiment. Cells are crosslinked with formaldehyde to maintain protein–DNA and DNA–DNA interactions. After cell lysis, the chromatin is subjected to digestion with a restriction enzyme, resulting in 5′ overhangs which are subsequently filled in with biotinylated nucleotides. Blunt ends are ligated using T4 ligase. After the reversal of crosslinking and protein removal, DNA is purified and further fragmented by sonication. To prepare the DNA library for sequencing, necessary adaptors are ligated to the ends of sheared DNA. Biotinylated junctions are pulled down with streptavidin beads. Finally, the DNA products are amplified by PCR and analyzed by paired-end sequencing. Steps (**3.1–3.11**) in the Methods section are indicated above the schematics

While the original 3C method [5] only quantified the contact between two specific regions, when combined with deep sequencing, 3C-based methods can be used to measure the frequency of genome-wide DNA interactions at high resolution. These methods enable the reconstitution of three-dimensional architecture of an entire genome. While deep sequencing can be costly, one method called Hi-C [6] reduces the cost by enriching the ligation products before sequencing. Essentially, Hi-C involves the following steps: DNA fragments in proximity are covalently crosslinked, usually by formaldehyde, to preserve the DNA interactions; after cell lysis, the crosslinked chromatin is digested using restriction enzymes (either 4-bp or 6-bp cutters) to generate 5′ overhangs; protruding ends of the digestion products are filled with biotinylated nucleotides by the Klenow fragment; the blunt ends are ligated using T4 ligase; after the reversal of crosslinking, DNA is further fragmented by sonication; then the biotinylated fragments are enriched by streptavidin pull-down; and finally, the DNA is prepared for paired-end deep sequencing (Fig. 1).

Hi-C studies have greatly advanced our understanding of genome organization across all domains of life. At the local level, genomes are folded into small domains known as topologically associating domains in eukaryotes [7] and chromosomal interaction domains in bacteria [8, 9]. At a higher level, eukaryotic chromosomes are organized in compartments while many bacteria have juxtaposed chromosome arms [6, 8, 10–13]. Furthermore, both eukaryotes and bacteria with multipartite genomes exhibited a plethora of inter-chromosomal interactions [14–17]. We believe that more secrets of genome biology await Hi-C to uncover. In this chapter, we describe a detailed protocol to prepare a Hi-C library from *Bacillus subtilis*. The method can be adapted to other bacterial species.

2 Materials

2.1 Cell Culture Collection and Fixation

1. Defined rich casein hydrolysate (CH) medium [18].
2. 250 mL 3XD BAF Shake Flask.
3. Waterbath.
4. 50 mL Polypropylene conical centrifuge tube.
5. 15 mL Polypropylene conical centrifuge tube.
6. 37% formaldehyde (*see* **Note 1**).
7. Rocker.
8. 2.5 M glycine.
9. 1× Phosphate-buffered saline (PBS): dissolve 8 g of NaCl, 0.2 g of KCL, 1.44 g of Na_2HPO_4, 0.24 g of KH_2PO_4 in 1 L of ddH_2O. Adjust pH to 7.4 with NaOH. Sterilize by autoclaving.
10. 1.5 mL Eppendorf DNA LoBind microcentrifuge safe-lock tube.

2.2 Cell Lysis, Genome Digestion, Biotin Labeling and Ligation

1. 1× TE buffer: 10 mM Tris, 1 mM EDTA in ddH_2O. Adjust pH to 8.0. Autoclaved.
2. Proteinase inhibitor cocktail.
3. Lysozyme, recommended Ready Lyse (VWR).
4. 10% sodium dodecyl sulfate (SDS).
5. Nuclease-free H_2O.
6. HindIII enzyme.
7. NEB buffer 2.1 (supplied with HindIII).
8. 10% Triton-X.
9. Nuclease-free PCR tubes.
10. 2 mM dGTP.
11. 2 mM dTTP.
12. 2 mM dCTP.
13. 1 mM Biotin-14-dATP (Axxora).
14. DNA Polymerase I, Large (Klenow) Fragment.
15. Thermocycler.
16. 2 mL Eppendorf DNA LoBind safe-lock tube.
17. T4 DNA ligase.

18. T4 DNA ligase reaction buffer.
19. 20 mg/mL Recombinant Albumin (rAlbumin, New England Biolabs).
20. 100 mM ATP.

2.3 Reverse Crosslinking

1. 0.5 M EDTA, pH 8.0.
2. Proteinase K.

2.4 DNA Extraction and RNA Digestion

1. Phenol/Chloroform/Isoamyl Alcohol (PCI, 25:24:1 mixture).
2. 1.7 mL microcentrifuge polypropylene tubes.
3. 3 M NaOAc: dissolve 246.1 g of $NaC_2H_3O_2$ in 800 mL of ddH_2O. Adjust pH to 5.2 with glacial acetic acid and increase volume to 1 L with ddH_2O. Sterilize by autoclaving.
4. 20 mg/mL glycogen.
5. Vortex mixer.
6. 100% ethanol.
7. Vacuum Aspirator.
8. 70% Ethanol.
9. QIAGEN elution buffer (EB).
10. RNaseA.

2.5 Removal of Biotin from Non-ligated Ends

1. 2 mM dATP.
2. T4 DNA polymerase.

2.6 DNA Sonication

1. 0.5 mL sonication tube (BrandTech).
2. Qsonica Q800R2 sonicator (Qsonica).
3. 2.2% agarose gel.

2.7 Cleanup of Fragmented DNA

1. AMPure XP beads (Beckman).
2. 1.5 mL low adhesion microcentrifuge tubes.
3. Magnetic stand.
4. 80% Ethanol.
5. 0.1× TE.

2.8 DNA Library Preparation for Illumina

1. NEBNext Ultra II DNA Library Prep Kit for Illumina (New England Biolabs).
2. Tris/NaCl (10 mM Tris-HCl pH 8.0, 10 mM NaCl).
3. NEBNext Multiplex Oligos for Illumina (New England Biolabs).
4. USER enzyme (New England Biolabs).

5. 2× NTB buffer: 10 mM Tris-HCl (pH 8.0), 1 mM EDTA and 2 M NaCl in ddH_2O.
6. 1× NTB buffer.
7. Dynabeads MyOne Streptavidin C1 beads (Fisher Scientific).
8. Qubit dsDNA HS assay kit (Fisher Scientific).
9. Qubit Assay Tubes (Fisher Scientific).
10. Qubit Fluorometers (Fisher Scientific).

3 Methods

3.1 Cell Culture Collection and Fixation

1. Inoculate cells in 5 mL of CH medium at 22 °C overnight with aeration.
2. The next morning, in a 250 mL flask, set up a subculture of 30 mL by diluting the overnight culture in fresh CH medium to an optical density at 600 nM (OD_{600}) of 0.02.
3. Put the flask in a 37 °C shaking waterbath. Grow cells to mid-exponential-growth phase, to an OD_{600} of ~0.3 (*see* **Note 2**).
4. Transfer 20 mL of the cells to a 50 mL conical centrifuge tube containing 1.8 mL of 37% formaldehyde (3% final concentration). Incubate at room temperature (RT) for 30 min with gentle mixing on a rocker (*see* **Note 3**).
5. Add 1 mL of 2.5 M glycine (125 mM final concentration). Incubate at RT for 5 min with gentle mixing on a rocker.
6. Pellet cells by centrifugation at 12,000 × g for 10 min at 4 °C. Remove the supernatant.
7. Wash cells by resuspending the pellet using 10 mL of ice-cold 1× PBS. Transfer into a 15 mL conical tube. Centrifuge cells again at 12,000 × g for 10 min at 4 °C. Remove the supernatant.
8. Resuspend the pellet in 500 μL of ice-cold 1× PBS.
9. Make a 1:40 dilution (25 μL resuspension in 975 μL of PBS) and measure its OD_{600}. Save aliquots of 1 OD unit of cells (~5×10^8 cells) in 1.5 mL tubes.
10. Freeze cells using liquid nitrogen and store at −80 °C.

3.2 Cell Lysis, Genome Digestion, Biotin Labeling and Ligation

1. Thaw the frozen cells on ice.
2. Centrifuge cells at 21,000 × g for 3 min at 4 °C. Carefully remove the supernatant (*see* **Note 4**).
3. Resuspend the cell pellet in 50 μL of lysis buffer (50 μL of TE + 0.5 μL of proteinase inhibitor cocktail +3 μL of lysozyme) and incubate at RT for 60 min. Pipette the lysate to mix.

4. Add 5 μL of 10% SDS (1% final concentration) and incubate at RT for 30 min (*see* **Note 5**).
5. Transfer 12 μL of the cell lysate into digestion buffer (84 μL of nuclease-free H_2O, 12 μL of NEB buffer 2.1, and 12 μL of 10% triton-X). Mix the reaction well and incubate at RT for 10 min.
6. Add 6 μL of HindIII into the mixture, and incubate at 37 °C for 2 h.
7. Once digestion is finished, cool the reaction down on ice for 5 min (*see* **Note 6**).
8. To label the overhangs of the digested fragments with biotin, in a PCR tube, set up a 120 μL of reaction by mixing the following components: 1.8 μL of 2 mM dGTP, 1.8 μL of 2 mM dTTP, 1.8 μL of 2 mM dCTP, 3.6 μL of 1 mM Biotin-14-dATP, 2.4 μL of DNA Polymerase I, Large (Klenow) Fragment, 11 μL of cold nuclease-free H_2O, and 100 μL of cooled digestion product. Pipette to mix thoroughly.
9. Incubate the reaction at 25 °C for 75 min in a thermocycler.
10. Stop the reaction by adding 6 μL of 10% SDS (0.5% final concentration). Pipette to mix thoroughly. Incubate the reaction at RT for 10 min.
11. Set up a ligation reaction in a 2 mL safe-lock tube on ice. Prepare ligation mixture by mixing the following components: 623 μL of cold nuclease-free H_2O, 80 μL of 10% Triton, 93 μL of ligase buffer, 2.3 μL of 20 mg/mL rAlbumin, and 1.9 μL of 100 mM ATP.
12. Add all SDS-treated digestion products from **Step 10** into the mixture and incubate on ice for 10 min.
13. Add 6 μL of T4 DNA ligase into the mixture and invert to mix.
14. Incubate the tube at 16 °C overnight (*see* **Note 7**).

3.3 Reverse Crosslinking

1. After the overnight ligation, add 20 μL of 0.5 M EDTA to inactivate the enzymes (~10 mM final concentration) and invert the tube to mix.
2. Add 10 μL proteinase K and invert the tube to mix.
3. Add 20 μL of 10% SDS and invert the tube to mix.
4. Incubate the sample at 65 °C overnight (*see* **Note 8**).

3.4 DNA Extraction and RNA Digestion

1. After overnight incubation, add 1 mL of PCI to the 2 mL tube. Vortex vigorously for 30 s. Then, centrifuge at 21,000 × g for 5 min at RT.
2. Carefully transfer the aqueous phase to a new 2 mL tube. Add 1 mL of PCI and repeat vortex and centrifugation as in **Step 1**.

3. Transfer the aqueous phase to a 1.7 mL microcentrifuge tube (*see* **Note 9**).
4. Add 2 μL of glycogen and vortex vigorously for 10 s. Then, add 200 μL of 3 M NaOAc and vortex vigorously for 10 s.
5. Split the mixture in half by transferring ~550 μL of the mixture to 1.7 mL microcentrifuge tubes.
6. Add 1.1 mL of ice-cold 100% EtOH and vortex vigorously for 10 s.
7. Incubate at −80 °C for 1 h.
8. Pellet DNA by centrifugation at 21,000 × g for 30 min at 4 °C. Carefully remove the supernatant by aspiration.
9. Add 1 mL of freshly prepared 70% ethanol (EtOH) to each tube. Vortex vigorously for 10 s.
10. Centrifuge at 21,000 × g for 30 min at 4 °C. Carefully remove the supernatant by aspiration.
11. Air-dry pellets for 3 min.
12. Add 20 μL of EB buffer.
13. Pipette to resuspend. Then, combine the two tubes of the same sample into one tube.
14. Add 2 μL of RNase A and then incubate at 65 °C for 30 min.
15. Cool down the sample on ice for 5 min before setting up the following reaction.

3.5 Removal of Biotins from Non-ligated Ends

1. Set up the reaction in a PCR tube on ice by adding 10.3 μL of nuclease-free H_2O, 6.4 μL of NEBuffer 2.1, 1.6 μL of 2 mM dATP, 1.6 μL T4 DNA polymerase, and all purified DNA.
2. Mix by pipetting followed by a pulse spin to collect all liquid from the side of the tube. Incubate at 20 °C for 4 h in a thermocycler.

3.6 DNA Extraction

1. Transfer the reaction to a 1.7 mL microcentrifuge tube.
2. Add 200 μL of PCI and vortex vigorously for 10 s to inactivate enzymes.
3. Add 150 μL of nuclease-free H_2O to bring up the volume. Vortex vigorously for 30 s.
4. Centrifuge at 21,000 × g for 5 min.
5. Carefully transfer aqueous phase to a new 1.7 mL microcentrifuge tube (*see* **Note 9**).
6. Add 1 μL of glycogen and vortex vigorously for 10 s. Then, add 40 μL of 3 M NaOAc and vortex vigorously for 10 s.
7. Add 1 mL of ice-cold 100% EtOH and vortex vigorously for 10 s.

8. Incubate at −80 °C for 45 min.
9. Pellet the DNA by centrifugation at 21,000 × g for 30 min at 4 °C. Carefully remove the supernatant by aspiration.
10. Add 1 mL of freshly prepared 70% EtOH to the tube. Vortex vigorously for 10 s.
11. Centrifuge at 21,000 × g for 10 min at 4 °C. Carefully remove the supernatant by aspiration.
12. Air-dry the pellet for 3 min. Avoid over-drying.
13. To resuspend DNA, add 105 μL of nuclease-free H_2O. Pipette to mix.

3.7 DNA Sonication

1. Transfer all DNA into a 0.5 mL sonication tube. Cool down the sample on ice for 5 min.
2. Using Qsonica Q800R2 sonicator, sonicate for 6 min without pause at 20% amplitude (*see* **Note 10**).
3. Vortex and quickly spin the tube, and then repeat sonication one more time using the same conditions.
4. Check the sample by DNA electrophoresis using 2.2% agarose gel. A DNA smear below 1 kb is desired. Sonicate more if needed.

3.8 Cleanup PF Fragmented DNA

1. Warm AMPure XP beads up to RT for at least 30 min.
2. Transfer the fragmented DNA into a 1.5 mL low adhesion microcentrifuge tube.
3. Vortex the AMPure XP beads to resuspend. Add 160 μL of AMPure XP beads to the DNA, mix well, and incubate for 5 min at RT (*see* **Note 11**).
4. Quickly spin the tube and place it on an appropriate magnetic stand to separate the beads from the supernatant. After the solution is clear (about 5 min), carefully remove and discard the supernatant (*see* **Note 12**).
5. Add 200 μL of freshly prepared 80% ethanol to the tube while in the magnetic stand. Incubate at RT for 30 s, and then carefully remove and discard the supernatant.
6. Repeat **Step 5** for a total of 2 washes.
7. Air-dry beads for 5 min while the tube is on the magnetic stand with the lid open (*see* **Note 13**).
8. Elute the DNA target from the beads into 53 μL of 0.1× TE. Mix well on a vortex mixer or by pipetting up and down. Let sit at RT for 5 min, and then vortex and quickly spin the tube.
9. Place the tube back on a magnetic stand. After the solution is clear (about 5 min), transfer 50 μL to a PCR tube and leave the tube on ice (*see* **Note 14**).

3.9 DNA Library Preparation for Illumina Sequencing

3.9.1 NEBNext End Prep

1. Add 3 μL of NEBNext Ultra II End Prep Enzyme Mix and 7 μL of 10× NEBNext Ultra II End Repair Reaction Buffer into the above PCR tube containing 50 μL of purified fragmented DNA.
2. Mix by pipetting followed by a pulse spin to collect all liquid from the side of the tube.
3. Place the PCR tube in a thermocycler. Run the following program:
   - 30 min at 20 °C.
   - 30 min at 65 °C.
   - Hold at 4 °C.

3.9.2 Adaptor Ligation

1. Dilute 15 μL of adaptors using 60 μL of a Tris/NaCl solution to achieve 75 μL of a 1:5 dilution of adaptors (*see* **Note 15**).
2. Add 2.5 μL of diluted adaptors directly to the End Prep reaction mixture from Subheading 3.9.1, **Step 3** and pipette to mix thoroughly (*see* **Note 16**).
3. Add 1 μL of NEBNext Ultra II Ligation Enhancer and 30 μL of NEBNext Ultra II Ligation Master Mix into the reaction mixture. Mix by pipetting followed by a pulse spin to collect all liquid from the side of the tube.
4. Incubate at 20 °C for 15 min in a thermocycler.
5. Add 3 μL of USER enzyme to the ligation mixture.
6. Pipette to mix thoroughly and incubate at 37 °C for 30 min in a thermocycler. Put the tube on ice when done.
7. Cleanup the DNA using 87 μL well-resuspended AMPure XP beads for each reaction, following the same procedure as described in Subheading 3.8. Elute the DNA from the beads into 28 μL of 0.1× TE. Use 25 μL of the DNA for the next step.

3.10 Biotin Pull-Down

1. Vortex streptavidin beads in the vial for 30 s to mix. Transfer 5 μL of beads into a 1.5 mL low adhesion microcentrifuge tube (*see* **Note 17**).
2. Add 400 μL of 1× NTB buffer to the beads and pipette to mix. Transfer the resuspension to a new 1.5 mL low adhesion microcentrifuge tube.
3. Place the tube on an appropriate magnetic stand for 3 min to isolate the beads. Carefully remove and discard the supernatant.
4. Repeat **Steps 2** and **3** two more times for a total of 3 washes.
5. Resuspend the beads in 25 μL of 2× NTB buffer per 5 μL of beads used in **Step 1** and transfer to a 1.5 mL low adhesion

microcentrifuge tube (*see* **Note 18**). Place the tube on ice while preparing DNA.

6. Transfer 25 μL of DNA from Subheading 3.9.2, **Step 7** into the above tube containing streptavidin beads, mix well.
7. Incubate the tube at RT for 30 min with gentle agitation on a vortexer.
8. Place the tube on a magnetic stand for 3 min to isolate the beads. Remove and discard the supernatant.
9. Add 400 μL of 1× NTB buffer to resuspend the beads, transfer the resuspension to a new 1.5 mL low adhesion microcentrifuge tube, and agitate for 3 min.
10. Repeat **Steps 8** and **9** for a total of 2 washes.
11. Add 400 μL of nuclease-free H_2O to resuspend the beads and transfer to a new 1.5 mL low adhesion microcentrifuge tube.
12. Place the tube on a magnetic stand for 3 min to isolate the beads. Remove and discard the supernatant.
13. Add 100 μL of nuclease-free H_2O to resuspend the beads and transfer to a new 1.5 mL low adhesion microcentrifuge tube.
14. Place the tube on a magnetic stand for 3 min to isolate the beads. Remove and discard the supernatant.
15. Add 21 μL of nuclease-free H_2O to resuspend the beads. Transfer the beads to a PCR tube and place on ice.

3.11 PCR Amplification

1. Mix the following components in a PCR tube: 25 μL of NEBNext Ultra II 2× Q5 Master Mix, 4 μL of Index Primers Mix, and 21 μL of streptavidin beads.
2. Pipette to mix thoroughly.
3. Place the tube in a thermocycler and set up conditions as in Table 1.

Table 1
PCR cycling conditions

Step	Temperature	Time	Cycles
Initial denaturation	98 °C	30 s	N/A
Denaturation	98 °C	10 s	
Annealing & Extension	65 °C	1 min and 15 s	14 cycles
Final extension	65 °C	5 min	N/A
Hold	4 °C	∞	N/A

4. Quick spin the PCR tube to pellet beads. Desired DNA are in the supernatant.
5. Check the PCR product by DNA electrophoresis using a 2.2% gel. Expect to see a smear below 500 bp.
6. Transfer 45 μL of the PCR reaction to a 1.5 mL low adhesion microcentrifuge tube. Avoid taking the streptavidin beads.
7. Clean up the DNA using 45 μL of well-resuspended AMPure XP beads for each reaction as described previously in Subheading 3.8. Elute the DNA from the beads into 33 μL of 0.1× TE. Then, transfer 30 μL to a clean 1.5 mL tube.
8. Determine DNA concentration. Qubit dsDNA HS assay kit is recommended for accurate measurement.
9. Submit the DNA samples to a sequencing facility for paired-end sequencing.

4 Notes

1. Use formaldehyde within 6 months after opening to minimize oxidation.
2. Collect the cells in your desired growth phase. For most studies, exponential growth might be desired.
3. If cells are grown at a lower temperature such as 30 °C, increase crosslinking time to 45 min.
4. It is important to remove all of the supernatant. Repeat centrifugation if necessary.
5. Because of the presence of SDS, bubbles easily form during pipetting. Try to minimize it by pipetting slowly.
6. It is important to cool the reaction down before setting up the biotin labeling reaction.
7. During the incubation at 16 °C, invert the tube multiple times. We recommend a ligation time over 15 h.
8. Incubate the sample at 65 °C for more than 17 h.
9. Use cut tips to avoid pulling the organic phase.
10. Perform the sonication at 4 °C.
11. Beads can be mixed either by pipetting up and down 10 times or by vortexing for a few seconds.
12. Be careful not to disturb the beads which contain the desired DNA.
13. Over-drying the beads may result in low recovery of DNA.
14. Cool down the DNA on ice before adding the enzyme mix and the buffer in the following step.

15. As DNA input is low, the manufacturer recommends that the NEBNext adaptor for Illumina be diluted 1:5 before using.
16. Mix adaptors with the End Prep reaction mixture before adding ligation enhancer and ligation mix. This action will avoid self-ligation of adaptors.
17. If you have more than one reaction, process all the beads in the same tube.
18. If you have more than one reaction, resuspend beads with 2× NTB buffer accordingly and then distribute 25 μL of resuspension to individual 1.5 mL low adhesion microcentrifuge tubes.

Acknowledgements

We thank the Wang laboratory for the group effort to constantly optimize this protocol. This work was supported by the National Institutes of Health Grants R01GM141242, R01GM143182, and R01AI172822 (X.W.). This research is a contribution of the GEMS Biology Integration Institute, funded by the National Science Foundation DBI Biology Integration Institutes Program, Award #2022049 (X.W.).

References

1. Beliveau BJ, Boettiger AN, Avendano MS, Jungmann R, McCole RB, Joyce EF, Kim-Kiselak C, Bantignies F, Fonseka CY, Erceg J, Hannan MA, Hoang HG, Colognori D, Lee JT, Shih WM, Yin P, Zhuang X, Wu CT (2015) Single-molecule super-resolution imaging of chromosomes and in situ haplotype visualization using Oligopaint FISH probes. Nat Commun 6:7147. https://doi.org/10.1038/ncomms8147
2. Cremer T, Cremer M (2010) Chromosome territories. Csh Perspect Biol 2(3):a003889. https://doi.org/10.1101/cshperspect.a003889
3. Wang X, Montero Llopis P, Rudner DZ (2013) Organization and segregation of bacterial chromosomes. Nat Rev Genet 14(3): 191–203. https://doi.org/10.1038/nrg3375
4. McCord RP, Kaplan N, Giorgetti L (2020) Chromosome conformation capture and beyond: toward an integrative view of chromosome structure and function. Mol Cell 77(4): 688–708. https://doi.org/10.1016/j.molcel.2019.12.021
5. Dekker J, Rippe K, Dekker M, Kleckner N (2002) Capturing chromosome conformation. Science 295(5558):1306–1311. https://doi.org/10.1126/science.1067799
6. Lieberman-Aiden E, van Berkum NL, Williams L, Imakaev M, Ragoczy T, Telling A, Amit I, Lajoie BR, Sabo PJ, Dorschner MO, Sandstrom R, Bernstein B, Bender MA, Groudine M, Gnirke A, Stamatoyannopoulos J, Mirny LA, Lander ES, Dekker J (2009) Comprehensive mapping of long-range interactions reveals folding principles of the human genome. Science 326(5950):289–293. https://doi.org/10.1126/science.1181369
7. Dixon JR, Selvaraj S, Yue F, Kim A, Li Y, Shen Y, Hu M, Liu JS, Ren B (2012) Topological domains in mammalian genomes identified by analysis of chromatin interactions. Nature 485(7398):376–380. https://doi.org/10.1038/nature11082
8. Le TB, Imakaev MV, Mirny LA, Laub MT (2013) High-resolution mapping of the spatial organization of a bacterial chromosome. Science 342(6159):731–734. https://doi.org/10.1126/science.1242059
9. Marbouty M, Le Gall A, Cattoni DI, Cournac A, Koh A, Fiche JB, Mozziconacci J,

Murray H, Koszul R, Nollmann M (2015) Condensin- and replication-mediated bacterial chromosome folding and origin condensation revealed by Hi-C and super-resolution imaging. Mol Cell 59(4):588–602. https://doi.org/10.1016/j.molcel.2015.07.020

10. Rao SSP, Huntley MH, Durand NC, Stamenova EK, Bochkov ID, Robinson JT, Sanborn AL, Machol I, Omer AD, Lander ES, Aiden EL (2014) A 3D map of the human genome at kilobase resolution reveals principles of chromatin looping. Cell 159(7):1665–1680. https://doi.org/10.1016/j.cell.2014.11.021

11. Wang X, Brandao HB, Le TB, Laub MT, Rudner DZ (2017) Bacillus subtilis SMC complexes juxtapose chromosome arms as they travel from origin to terminus. Science 355(6324):524–527. https://doi.org/10.1126/science.aai8982

12. Bohm K, Giacomelli G, Schmidt A, Imhof A, Koszul R, Marbouty M, Bramkamp M (2020) Chromosome organization by a conserved condensin-ParB system in the actinobacterium Corynebacterium glutamicum. Nat Commun 11(1):1485. https://doi.org/10.1038/s41467-020-15238-4

13. Lioy VS, Junier I, Lagage V, Vallet I, Boccard F (2020) Distinct activities of bacterial condensins for chromosome management in pseudomonas aeruginosa. Cell Rep 33(5):108344. https://doi.org/10.1016/j.celrep.2020.108344

14. Hoencamp C, Dudchenko O, Elbatsh AMO, Brahmachari S, Raaijmakers JA, van Schaik T, Cacciatore AS, Contessoto VG, van Heesbeen RGHP, van den Broek B, Mhaskar AN, Teunissen H, St Hilaire BG, Weisz D, Omer AD, Pham M, Colaric Z, Yang ZZ, Rao SSP, Mitra N, Lui C, Yao WJ, Khan R, Moroz LL, Kohn A, St Leger J, Mena A, Holcroft K, Gambetta MC, Lim FBA, Farley E, Stein N, Haddad A, Chauss D, Mutlu AS, Wang MC, Young ND, Hildebrandt E, Cheng HH, Knight CJ, Burnham TLU, Hovel KA, Beel AJ, Mattei PJ, Kornberg RD, Warren WC, Cary G, Gómez-Skarmeta JL, Hinman V, Lindblad-Toh K, Di Palma F, Maeshima K, Multani AS, Pathak S, Nel-Themaat L, Behringer RR, Kaur P, Medema RH, van Steensel B, de Wit E, Onuchic JN, Di Pierro M, Aiden EL, Rowland BD (2021) 3D genomics across the tree of life reveals condensin II as a determinant of architecture type. Science 372(6545):984. https://doi.org/10.1126/science.abe2218

15. Ren ZQ, Takacs CN, Brandao HB, Jacobs-Wagner C, Wang XD (2023) Organization and replicon interactions within the highly segmented genome of Borrelia burgdorferi. PLoS Genet 19(7):e1010857. https://doi.org/10.1371/journal.pgen.1010857

16. Ren Z, Liao Q, Karaboja X, Barton IS, Schantz EG, Mejia-Santana A, Fuqua C, Wang X (2022) Conformation and dynamic interactions of the multipartite genome in agrobacterium tumefaciens. Proc Natl Acad Sci USA 119(6). https://doi.org/10.1073/pnas.2115854119

17. Val ME, Marbouty M, de Lemos Martins F, Kennedy SP, Kemble H, Bland MJ, Possoz C, Koszul R, Skovgaard O, Mazel D (2016) A checkpoint control orchestrates the replication of the two chromosomes of vibrio cholerae. Sci Adv 2(4):e1501914. https://doi.org/10.1126/sciadv.1501914

18. Montero Llopis P, Stephansky R, Wang X (2022) High-throughput imaging of Bacillus subtilis. Methods Mol Biol 2476:277–292. https://doi.org/10.1007/978-1-0716-2221-6_19

Chapter 14

An Optimized Adaptation of DamID for NGS Applications

Karen L. Reddy and Xianrong Wong

Abstract

Recent studies have implicated higher-order genome organization in the regulation of genes and cellular state. Lamina-Associated Domains (LADs) are regions of heterochromatin associated with the nuclear envelope and the nuclear lamina, a protein network involved in both nuclear organization and genome structure. LADs are developmentally regulated, and their dysregulation is associated with several diseases and pathological states, including cancer and premature aging. In addition to LADs, other nuclear protein compartments appear to scaffold or support unique chromatin environments to affect gene expression. These revelations carry profound implications for our comprehension of developmental processes and the pathogenesis of various diseases, especially given the numerous disorders already directly associated with, for example, mutations in lamin and INM proteins. This spatial compartmentalization of chromatin subtypes to unique protein compartments has led to the adoption of proximity-labeling methods, such as DamID (DNA Adenine Methyltransferase Identification), to identify these unique chromatin compartments.

Key words DamID sequencing, Lamina-Associated Domains (LADs), Genome Organization, Epigenetics, Chromatin

1 Introduction

The nuclear periphery constitutes a distinct nuclear compartment characterized by an intricate lamina framework, comprising lamin A/C and B-type lamins. This structural framework assumes pivotal roles as an anchoring and interaction hub for inner nuclear membrane (INM) proteins at the interface of the nuclear envelope (NE) and chromatin, thereby facilitating connections between chromatin, the NE, and the cytoplasm [1]. A subset of INM proteins, such as lamin B receptor (LBR), Emerin, and lamin-associated peptide 2 (Lap2), have demonstrated pronounced affinities for transcriptional repressors, chromatin modifiers, and post-translationally modified core histones [2–7]. These observations are in agreement with functional investigations firmly establishing the nuclear periphery as a region of transcriptional repression and

Valerie J. Carabetta and Olaitan Akintunde (eds.), *High Throughput Gene Screening: Methods and Protocols*,
Methods in Molecular Biology, vol. 2866, https://doi.org/10.1007/978-1-0716-4192-7_14,

regulation [3, 8–14]. Comprehensive mapping of chromatin interactions at the nuclear periphery, specifically identifying lamina-associated domains (LADs), have been made feasible by the introduction of the DNA adenine methyltransferase identification (DamID) technique [12, 15–18]. The DamID technique can be used to detect any protein–DNA interactions, much like chromatin immunoprecipitation (ChIP), but it has been used most extensively to detect interactions within protein compartments that can be difficult to immunoprecipitate, such as membrane-bound proteins or meshworks (e.g., the nuclear lamina or nucleolus [12, 16, 19–23]). In particular, DamID has brought the nuclear lamina network and other nuclear protein compartments to the forefront as a pivotal determinants in genome organization, epigenetic modulation, and gene regulation [12, 24, 25]. DamID, while most often used to identify LADs, can be used to map any protein:chromatin compartment. The protocol presented here represents an adapted version of the original DamID method [18], optimized for next-generation sequencing (NGS), including suggestions for bioinformatically identifying LADs from NGS data.

As mentioned above, DamID has emerged as a valuable tool in advancing the field of nuclear architecture and genome organization. While most of the examples and protocol below are focused on using DamID to identify lamina-proximal chromatin domains in mammalian cells, it is important to note that this approach can be used for any chromatin-interacting protein in any metazoan or plant cell type and is particularly useful for proteins that are refractory to chromatin immunoprecipitation (ChIP) or ChIP-related approaches (i.e., poor antibodies, widespread interactions in heterochromatin regions that are difficult to ChIP, or proteins that form higher-order structures, such as meshworks). To elucidate the intricate dynamics of chromatin interactions, specifically lamina-associated domains (LADs), at the nuclear periphery during differentiation, the DamID technique was used to investigate LAD signatures across an array of distinct cell types, including embryonic stem cells (ES cells), neural progenitor cells (NPCs), astrocytes (ASCs), embryonic fibroblasts (MEFs), myoblasts, as well as pro-B and T cells, among others [12, 25–28]. Collectively, these investigations underscored the presence of dynamic changes in LADs, overlaying a consistent core architectural framework consisting of the majority of LADs, which remain unchanged across cell types. Notably, genes situated within domains that establish associations with the nuclear lamina tend to undergo repression, while genes within domains that relinquish peripheral contact exhibit activation or a poised state for expression. This aligns with earlier cytological observations. The dynamic subnuclear positioning of critical developmental genes, coupled with variable LAD patterns observed across developmental lineages, offers compelling evidence for a functional characterization of LADs. This contrasts with the original, more generalized perception of LADs as predominantly "silent" regions. Like most genome-wide techniques, advancements beyond simply identifying the genomic location of LADs has centered on the integration of identified LADs with other genome-wide datasets to examine the connections between these attributes and LADs. Genome-wide datasets typically used include chromosome conformation data (such as Hi-C), transcription factor binding sites or occupancy, regions characterized by specific histone modifications (from ChIP-seq and related techniques), and DNA methylation profiles. Intersection of LAD data with such datasets has provided valuable insights into LAD characteristics and illuminated mechanisms that may underlie LAD establishment or the implications on the regulation of genes within a LAD [26, 27, 29–34].

As a label transfer technique, DamID hinges on the expression of a protein of interest fused to the Dam methyltransferase (Dam) from *Escherichia coli*. In the case of identifying nuclear lamin (NL) chromatin interactions in mammalian cells, the NL protein lamin B1 is most often used since it is ubiquitously expressed in all cell types. Within the fusion construct, the Dam component is capable of methylating adenine residues at the N6 position, resulting in the formation of $G^{me6}ATC$ tracts within genomic regions where the fusion protein interacts (Fig. 1). Given that the fusion protein is localized exclusively to the NL (in the case of lamin B1 fusions), the GATC modifications predominantly manifest within regions that have come in contact with or are in proximity to the methyltransferase at the NL, thus demarcating NL–chromatin interactions. It is important to note that the methylation modification is permanent, so even transient interactions of chromatin with the NL are modified. Subsequently, $G^{me6}ATC$-marked DNA can be selectively amplified through the utilization of the $G^{me6}ATC$-specific enzyme *Dpn*I, followed by ligation-mediated PCR, and ultimately, the GATC-specific enzyme *Dpn*II is employed to enhance the signal-to-noise ratio of the amplified library. Notably, a freely diffusible Dam ("Dam-only") is expressed in a separate set of cells in parallel to serve as a reference for normalizing random Dam/DNA interactions and addressing potential amplification biases. Thus, DNA interactions detected by DamID are initially mapped and displayed as ratios of Dam-X/Dam-only (Fig. 2).

To adapt the DamID protocol to deep sequencing technologies on the Illumina platforms (in this case the Illumina HiSeq 2000 platform and above), DNA library sizes typically must be in the range of 100 to 500 bp for efficient clustering on the flow cell and maximal sequencing read yield. Importantly, unlike ChIP-seq and its newer derivatives, such as Cut&Run [35] or Cut&Tag [36], which use sonication, micrococcal nuclease, or fragmentation, respectively, to fragment DNA, the reliance of DamID on the restriction enzyme *Dpn*I, a sequence-specific cutter, means that

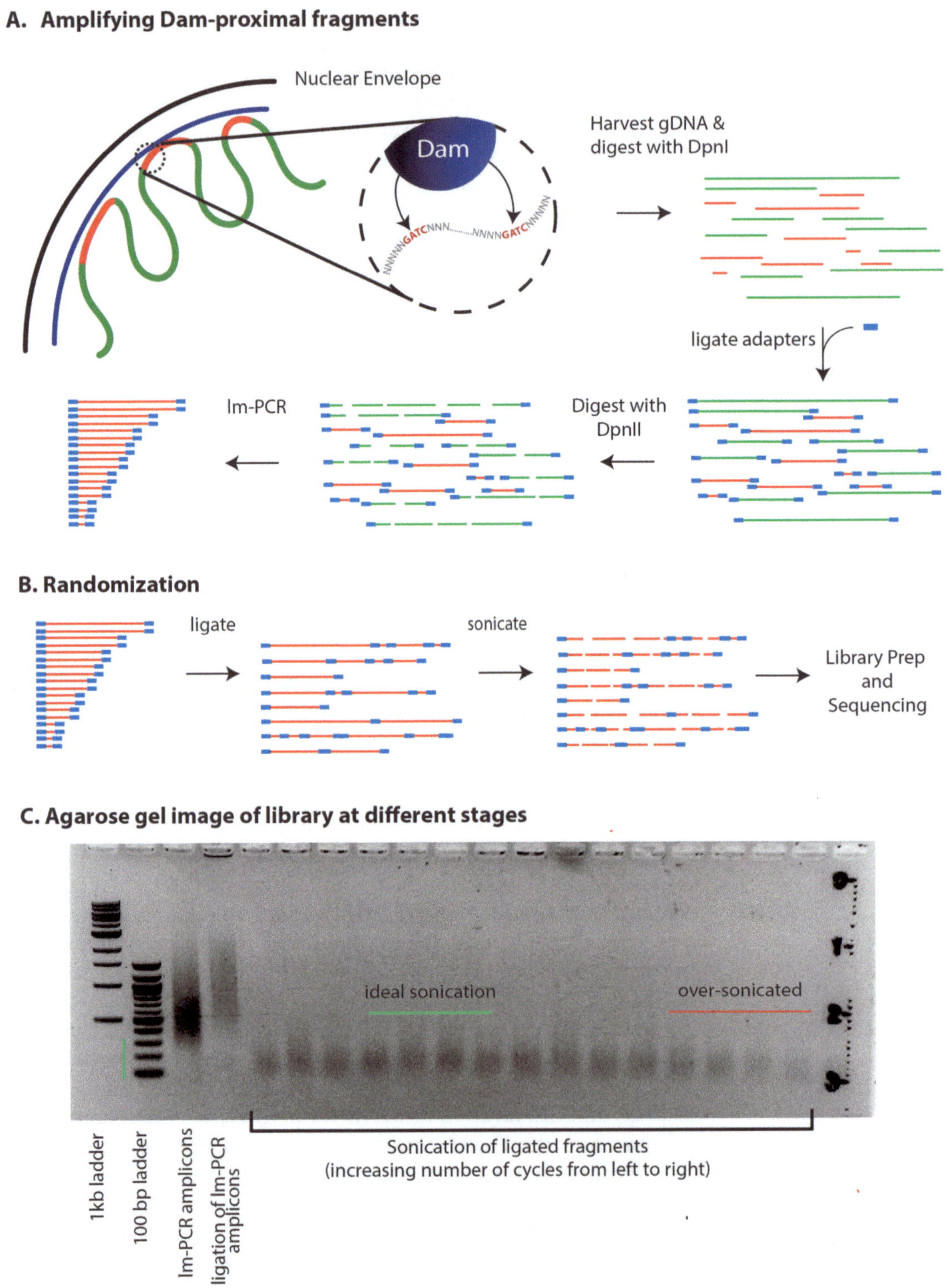

Fig. 1 Schematic of the DamID-sequencing protocol. (**a**) Expression of Dam-fusion construct and *in vivo* chromatin labeling followed by DamID protocol. $G^{me6}ATC$ marked DNA (red) are selectively amplified from the rest of the genome (green). (**b**) Randomization approach: DamID amplicons are ligated before sonicating to

each amplified size fragment derives from a specific genomic location. Thus, the "smear" of DNA after ligation-mediated PCR (lm-PCR) cannot be size-selected or directly sonicated/sheared since this would lead to significant biases in terms of genomic representation of fragments, losing both larger and smaller *Dpn*I-delimited fragments. In our distinctive adaptation of DamID-seq, we randomize our amplified material by ligating *Dpn* all adapter-flanked PCR products prior to sonication. After sonication, this yields randomized DNA fragments ranging between 100–300 bp (Fig. 1a-c). The resulting fragments are then subjected to standard DNA sequencing library preparation for 100 bp single end reads. For all DamID deep sequencing experiments, material from samples expressing freely diffusible Dam is used as a parallel control, to normalize for local DNA accessibility and sequencing bias.

Following next-generation sequencing (NGS), the analysis commonly comprises several steps, including the removal of sequencing adapter sequences resulting from the ligation and sonication of the DamID amplicons, quality trimming, and alignment to a reference genome (Fig. 2a). Ultimately, this process yields scores assigned to discrete genomic bins, which are computational divisions of the genome designed to facilitate analysis. Subsequently, the data obtained from laminB1-specific DamID (in the case of measuring lamina-proximal chromatin) is normalized against non-specific G^{me6}ATC methylation data originating from the Dam-only control. This comparative analysis is undertaken to assess the degree of LAD enrichment, and the outcomes are conveyed as $\log_2$ enrichment ratios. This binning and normalization process can be used for any DamID experiment to yield Dam-X/Dam $\log_2$ ratios. These raw genomic traces (or $\log_2$ ratios from any experimental method) can then be fed into a bioinformatic algorithm to identify LAD intervals (LADetector; Fig. 2), which facilitates the intersection of other genome-wide datasets, such as transcription factor binding sites, regions of histone modifications, DNA methylation, and other forms of genomic occupancy, with the identified LADs. Such analyses help elucidate the interplay between these genomic features and the identified LADs. We provide a full analysis pipeline that analyzes and converts DamID NGS reads to genomic DamID enrichment bins and LAD intervals based on the Circular Binary Segmentation algorithm, (https://github.com/thereddylab/pyLAD), a rapid and robust platform for

Fig. 1 (continued) 100–500 bp. (**c**) Agarose gel image of library at different stages of the protocol. The red bar indicates over-sonication. Green bars indicate optimal fragment size (100–500 bp) and ideal sonicated fragments that show loss of higher molecular weight smears (resulting in darkening of bands representing lower molecular weight). Continued sonication results in loss of sample, as indicated by the drop in intensity of the bands

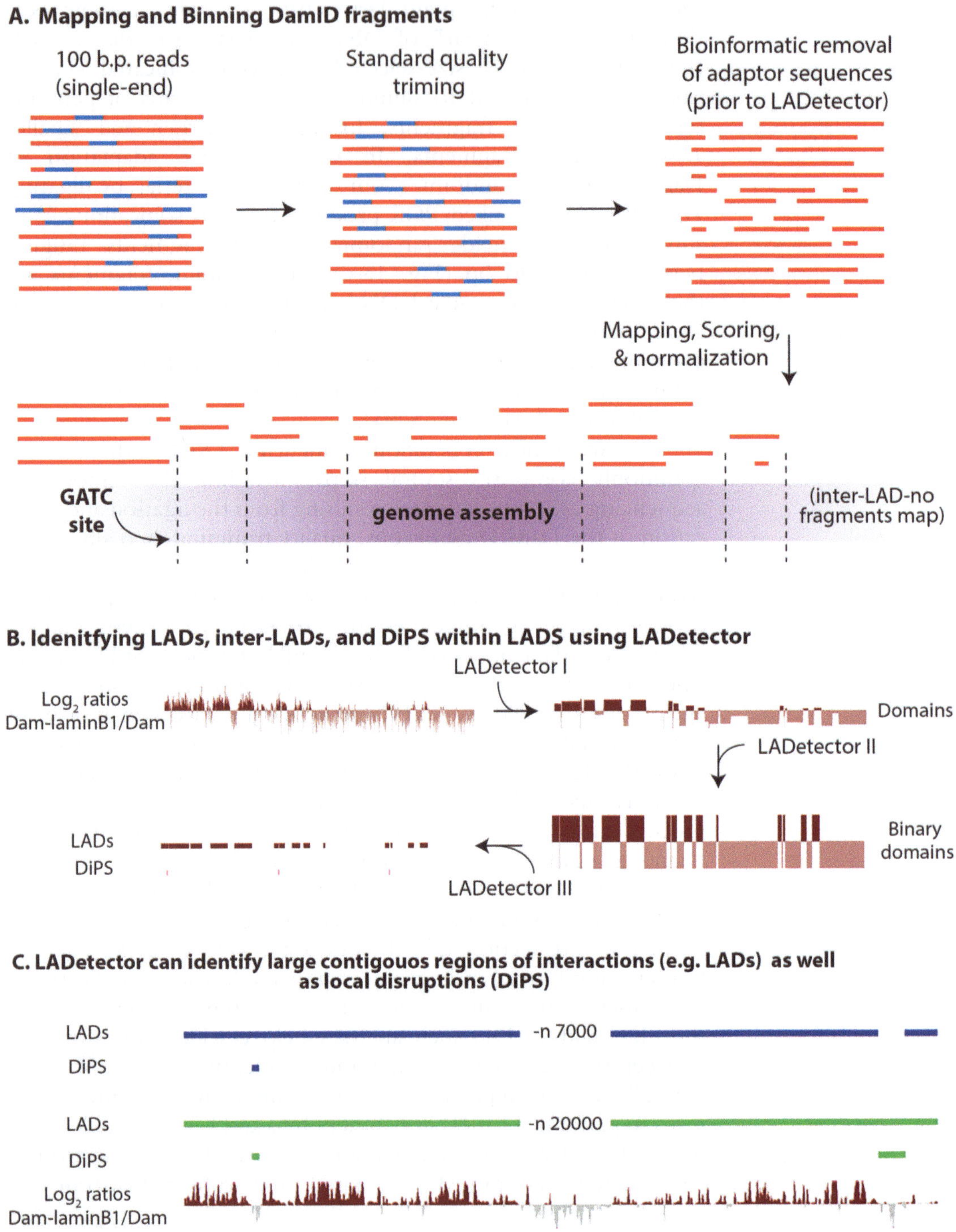

Fig. 2 (**a**) 100 bp single-end sequencing reads are quality controlled by trimming the ends of reads that fail certain criteria based on quality scores. The reads are then removed of DamID adapter sequences resulting from the randomization (ligation) procedure. Each resulting (red) fragment greater than 25 nucleotides in length is then mapped back to the genome built by Bowtie. The LADetector program can be divided into three parts: mapping, normalization, and LAD detection. (**a**) The first part, pre-normalization mapping, deals with the

identifying LADs from DamID- or ChIP-based approaches (Fig. 2b, [3, 17, 27]). This tool can also be used to detect any interactions from $\log_2$ ratios that result in broad associations which cannot be resolved by conventional and widely used peak-calling modalities. Thus, broad interactions from ChIP data can also be used as an input.

2 Materials

2.1 Virus Production

1. 1. HEK 293 T/17 cells.
2. Culture medium—DMEM high glucose (4.5 g/L) with 10% FBS, penicillin/streptomycin, 2 mM L-glutamine, and non-essential amino acids.
3. Opti-MEM.
4. Generation 2 packaging plasmids (pCMV-VSV-G and pCMV delta R8.2; Addgene).
5. DamID plasmids [pLgw V5-EcoDam ("Dam only") and pLgw V5-EcoDam mouse-lamin B1 (or "X" protein of choice, Addgene).
6. 1 M sodium butyrate.
7. Transfection reagent of choice. Transfection using FuGENE 6 transfection reagent (Promega) is described in this protocol. We have also used Lipofectamine 3000 (ThermoFisher) and electroporation approaches with success.
8. 0.45 μm syringe filters.

2.2 Transduction

1. Fetal bovine serum.
2. 10 mg/mL polybrene.

2.3 DamID (Adapted from [18])

1. Cell type of choice. Here, primary MEFs harvested from E13.5 embryos were used.

Fig. 2 (continued) removal of primer sequences and mapping of the reads to a reference genome. This is followed by counting the number of reads that have been mapped to each genomic bin, genome wide. The second part of the program normalizes the counts in each bin to the sequencing depth of both the experimental sample and the control sample. The normalized experimental scores are then further normalized to the control sample scores to yield genome wide $\log_2$ ratios. (**b**) Part three of the program uses the LADetector version 2 algorithm to extract genomic intervals for LADs and DiPs (*D*epleted *i*n *P*eripheral *S*ignal—regions, that locally lose contact with the lamina, but are otherwise in a LAD) from the $\log_2$ ratios. (**c**) An example of how LADetector identifies LADs, interLADs (non-LADs), or DiPS within LADs with different criteria [40]. Shown are examples allowing up to 7 kb (-n 7000) or 20 kb (-n 20,000) gaps in signal continuity within a LAD. Separate parameters can be set to identify gaps of a user-determined size, as DiPS. This allows for broad domain calling while also retaining the ability to identify local changes in interaction frequencies

2. Culture medium—DMEM high glucose (4.5 g/L) with 10% FBS, penicillin/streptomycin, 2 mM L-glutamine, and non-essential amino acids.
3. Phosphate Buffered Saline (PBS).
4. Genomic DNA harvest kit (QIAamp DNA Mini kit (Qiagen)).
5. 3.0 M sodium acetate, pH 5.2.
6. *Dpn*I and *Dpn*II (New England Biolabs).
7. T4 DNA ligase (5 U/μL).
8. Double-stranded adapters of annealed oligonucleotides (IDT, HPLC purified) AdRt (5′- CTAATACGACTCACTATAGGG CAGCGTGGTCGCGGCCGAGGA-3′) and AdRb (5′-TCCT CGGCCG-3′). To anneal AdRt to AdRb, add equal volumes of 100 μM of each oligo and submerge the mixture in a boiling water bath. Allow the water bath to cool slowly to room temperature.
9. AdR_PCR primer (5′-GGTCGCGGCCGAGGATC-3′, (IDT)).
10. Advantage cDNA polymerase mix (Takara) or similar.
11. 2.5 mM dNTP mix.
12. QIAquick PCR Purification Kit (Qiagen) or similar.
13. NEBNext End Repair Module (New England Biolabs).
14. Sonicator connected to a cooling system (e.g., Bioruptor UCD-200, Covaris).
15. 1.5 mL DNA LoBind microfuge tube (Eppendorf).
16. 1.5 mL TPX tubes (Diagenode).
17. 100% ethanol.
18. Glycogen (molecular biology grade, 20 mg/mL).
19. 70% ethanol.
20. Qubit fluorimeter and reagents (Qubit assay tubes and broad range mix).
21. Quant-iT PicoGreen dsDNA kit (Invitrogen).
22. NEBNext Ultra II DNA library prep kit for Illumina (NEB). Multiplexing oligos, if needed (single end index; New England Biolabs). Make sure to use the DNA prep kit and oligos compatible with your sequencer.
23. Bioanalyzer 2100 (equipment) with DNA High Sensitivity reagents (Agilent).
24. Kapa quantification complete kit for Illumina (Roche).
25. Applied Biosystems 7500 Real Time qPCR system and reagents.

3 Methods

3.1 Virus Production

3.1.1 Day -1

HEK 293 T/17 cells are seeded in 10 cm dishes. Each dish should be seeded such that the cells are at 90% confluence the next day (*see* **Note 1**). Incubate the cells in culture medium overnight at 37 °C, 5% CO_2.

3.1.2 Day 0, Transfection of HEK 293 T/17 Cells

1. Prewarm FuGENE 6 Transfection Reagent and serum-free DMEM high glucose (4.5 g/L) to room temperature.
2. For each of pLgw V5-EcoDam and pLgw V5-EcoDam mouse-LmnB1, prepare a 1.5 mL Eppendorf tube filled with 500 μL *serum-free* DMEM high glucose (4.5 g/L). Prepare the "Dam-only" control (pLgw V5-EcoDam) and Dam-X (your protein of interest, e.g., pLgw V5-EcoDam lmnB1) in separate tubes and dishes. These viruses will be used to infect target cells in parallel.
3. Add 33.8 μL of FuGENE 6 transfection (prewarmed) directly into the prepared DMEM in the tubes from **step 2** and mix by inverting the tube.
4. Incubate the FuGENE 6 Transfection Reagent/medium mixture for 5 min at room temperature.
5. Prepare the following:
   (a) For the "Dam-only" virus, mix 2.5 μg of pCMV-VSV-G, 3.75 μg pCMV delta R8.2, and 5 μg of pLgw V5-EcoDam.
   (b) For the Dam-X virus, mix 2.5 μg of pCMV-VSV-G, 3.75 μg pCMV delta R8.2, and 5 μg pLgw V5-EcoDam lamin B1.
6. Transfer the prepared DNA mixture into the FuGENE 6 Transfection Reagent/medium mixture and mix immediately.
7. Incubate the resulting mixture at room temperature for at least 15 min and up to 45 min.
8. Remove media from the seeded HEK 293 T/17 cells until 8.5 mL is left in the dishes. Add the FuGENE 6 Transfection Reagent/DNA mixture to the cells.
9. Incubate cells at 37 °C, 5% CO_2.
10. Add sodium butyrate to the transfected cells, to a final concentration of 10 mM, 3 h post-transfection for an overnight incubation at 37 °C, 5% CO_2 (*see* **Note 2**).

    Please note that from this point forward, these cell culture dishes will have live self-inactivating lentivirus particles. Please act with care and follow all lentiviral protocols and guidelines for production, use, and disposal at your institute.

3.1.3 Days 1–3, Collecting Viral Particles

1. In the morning, remove the transfection media containing sodium butyrate and wash the cells with prewarmed 1X PBS (*see* **Note 2**).
2. After washing with PBS, add 4.5 mL warm opti-MEM to each dish gently, to not dislodge the cells from the dishes. Let incubate for 12 h before first collection.
3. Days 1–3: Collect viral supernatants every 12 h and replenish the opti-MEM in each dish. Store collected supernatant at 4 °C. Remember to keep each virus separate (do not mix Dam-only and Dam-X viral supernatants).
4. Collect viral supernatants for a maximum of 3 days and pool. "Peeling" or "lifting" of cells is expected, so collect with care. Always add new warmed media to the sides of the dish very slowly. If most cells are lifting from the plate, cease collections. Viral titers are high starting at 24 h after transfection (Day 1 evening) and increase until ~72–96 h. The virus is labile at 37 °C, so collection intervals beyond 12 h do not substantively increase titers.

3.2 Transduction of Target Cells

3.2.1 Day 1

Seed primary MEFs at 30–40% confluency in two 10 cm dishes (one for Dam-only and one for Dam-X) and incubate at 37 °C, 5% CO_2 (*see* **Note 3**).

3.2.2 Day 2

1. Filter pooled viral supernatants through a 0.45 μM syringe filter. To avoid clogging the filter, supernatants can be centrifuged at low speeds to pellet any dead cells or debris. This is especially important if substantive numbers of cells were lifting off the viral production plates. The filtered supernatants can be used immediately or stored at -80 °C. Please note that freezing reduces viral transduction frequency.
2. Remove media from the dishes of primary MEFs seeded on the previous day.
3. Add 4.5 mL of viral supernatant to the MEFs. Add Dam-only supernatant to one of the dishes and Dam-X to the other.
4. Add 500 μL FBS to each of the dishes from **step 3** (this brings the media up to 10% FBS, the viral supernatants were made in serum-free media).
5. Add polybrene to a final concentration of 4 μg/mL (*see* **Note 4**).
6. Incubate the cells with viral supernatants at 37 °C, 5% CO_2 for 12–14 h.

3.2.3 Days 3–7 and beyond

1. Remove viruses and add 10 mL culture medium to each dish.
2. Culture the cells for at least another 96 h (*see* **Note 5**). Split, expand, and maintain as needed.

3.3 DamID

3.3.1 Genomic DNA Harvest

1. Harvest MEFs from each plate. We find that we recover enough DNA from a single 10 cm dish (and even smaller dishes). Remove media from cells and wash cells once with 1X PBS, remove PBS, and then add 3 mL trypsin per dish. Incubate the reaction for 5 min at 37 °C, 5% CO_2.
2. Quench the trypsin with 3 mL of culture media containing FBS. Pipet up and down to disperse the cells into the solution, taking care to rinse and capture any cells remaining attached to dish. Place this 6 mL suspension (3 mL of trypsin plus 3 mL of media for quenching) in a 15 mL capped tube.
3. Centrifuge this cell suspension at 500× *g* for 5 min and remove the resulting supernatant by aspiration. There should be an obvious cell pellet at the bottom of the tube. Avoid aspirating the pellet.
4. Snap freeze the cell pellet in liquid nitrogen (*see* **Note 6**).
5. Harvest the genomic DNA using the QIAamp DNA Mini kit, following the manufacturer's instructions. For each sample/ treatment, elute genomic DNA twice with 200 μL of (pre-warmed at 55 °C) 10 mM Tris HCl, pH 8. Pool the two elutions. This is the genomic DNA.
6. Quantify the obtained genomic DNA using spectrophotometry methods, like the nanodrop. This quantitation will inform how much 10 mM Tris HCl, pH 8 to resuspend the precipitated DNA in **step 14**.
7. Add 40 μL of 3 M sodium acetate, pH 5.2 and mix well.
8. Add 800 μL 100% ethanol to precipitate the DNA and mix well.
9. Incubate the precipitate at −80 °C for a minimum of 2 h.
10. Remove tubes from the freezer and centrifuge at ≥16,000 x *g* or maximum speed a tabletop microcentrifuge at 4 °C for 20 min.
11. Carefully remove the supernatant and wash the DNA pellets with 500 μL of 70% ethanol.
12. Centrifuge the tubes again at 16000 rpm, 4 °C for 20 min using a microcentrifuge.
13. Remove the supernatant and air dry the DNA pellet.
14. Resuspend the DNA pellet to obtain an approximate concentration of 1 μg/μL.
15. Quantify the resuspended genomic DNA again using a nanodrop or Qubit.

3.3.2 Dam Methylation Specific Amplification (See Note 7)

1. In a 10 μL reaction volume, digest 0.5–2.5 μg of genomic DNA from each treatment (Dam-only and Dam-X) in separate tubes using *Dpn*I and allow the reaction to proceed overnight at 37 °C (Table 1).

Table 1
***DpnI* digestion of methylated gDNA**

Component	Amount
Genomic DNA	0.5–2.5 μg
10× NEB cutsmart buffer	1 μL
*Dpn*I	1 μL
Nuclease-free dH_2O	Adjust to final volume of 10 μL

Table 2
Ligation of adapters to *DpnI*-digested gDNA

Component	Amount
*Dpn*I-digested DNA mix	10 μL
10× T4 ligase buffer	2 μL
T4 DNA ligase	1 μL
dsAdR (50 μM)	0.8 μL
Nuclease-free dH_2O	6.2 μL

2. Heat reactions at 80 °C for 20 min to deactivate *DpnI.*
3. Following a cooling step (a few minutes is sufficient), ligate the *Dpn*I-digested fragments (in a 20 μL reaction volume, set up as in Table 2) with double-stranded adapters (dsAdr). Incubate at 16 °C for a minimum of 4 h.
4. Incubating the reaction at 65 °C for 10 min to inactivate the ligase.
5. After the samples have been cooled for a few minutes on ice, set up a *Dpn*II digestion as in Table 3 (50 μL reaction volume), and incubate at 37 °C for 1 h.
6. Clean up the reaction from **step 5** using a PCR purification kit (*see* **Note 6**).
7. Amplify the ligated DNA pools using the primer AdR_PCR and the Advantage cDNA polymerase mix. Set up reaction as in Table 4 and the PCR conditions are listed in Table 5.
8. Clean up the PCR reaction using a PCR purification kit.
9. Quantify the concentration of the resulting enriched methylated fragments using a nanodrop or Qubit (*see* **Notes 6** and **8**).
10. Verify amplification on an agarose gel (Fig. 1C).

Table 3
***Dpn*II digestion to remove intervening sequences**

Component	Amount
Ligation mixture	20 μL
10× NEB cutsmart buffer	5 μL
NEB *Dpn*II	1 μL
Nuclease-free dH_2O	24 μL

Table 4
PCR amplification of methylated DNA fragments

Component	Amount
*Dpn*II-digested DNA	10 μL
10× cDNA PCR buffer	5 μL
Adr_PCR (50 μM)	1.25 μL
2.5 mM dNTP mix	4 μL
PCR advantage enzyme mix (50X)	1 μL
Nuclease-free dH_2O	28.75 μL

Table 5
Cycling conditions to amplify methylated gDNA

Number of cycles	Denaturation	Annealing	Extension
1	–	–	68 °C 10 min
1	94 °C 1 min	65 °C 5 min	68 °C 15 min
4	94 °C 1 min	65 °C 1 min	68 °C 10 min
14	94 °C 1 min	65 °C 1 min	68 °C 2 min

3.4 Library Randomization and Preparation of NGS (See Note 9; Fig. 1b)

1. End repair 1–5 μg of column-purified DamID material (from Subheading 3.3) using the NEBNext End Repair Module. Incubate the reaction at 20 °C for 30 min (Table 6).
2. Purify reaction using the QIAquick PCR purification kit.
3. Ligate 1 μg of the purified material with 1 μL of T4 DNA ligase in a total volume of 20 μL, at 16 °C overnight to generate a randomized library of large fragments.
4. Add 180 μL of 10 mM Tris, pH 8.0 into the ligation reaction.

Table 6
End repair of amplified fragments prior to re-ligation for randomization

Component	Amount
DamID-amplified material	1–5 μg
10x end repair buffer	10 μL
20x end repair enzyme mix	5 μL
Nuclease-free dH_2O	Up to 100 μL

5. Sonicate the diluted ligation mix to generate fragments suitable for sequencing. In this protocol, a Bioruptor® UCD-200 is used with the following settings: High power, 30 s ON, 30 s OFF for 1 h in a 1.5 mL DNA LoBind microfuge tube.
6. Transfer DNA into 1.5 mL TPX tubes and sonicate for 4 rounds of 10 min (high power, 30 s ON and 30 s OFF). Transfer the DNA to new TPX tubes after each round to prevent etching of the TPX plastic (*see* **Note 10**). The sonication procedure should yield DNA sizes ranging from 100 to 200 bp (see agarose gel in Fig. 1).
7. Precipitate the sheared DNA by adding 20 μL of 3 M sodium acetate, pH 5.2, 500 μL ethanol, and supplement with 3 μL of glycogen. Incubate the precipitate at −80 °C for at least 2 h (*see* **Note 11**).
8. Centrifuge the DNA mix at maximum speed for 10 min to pellet the sheared DNA with the carrier glycogen.
9. Wash pelleted DNA/glycogen with 70% ethanol and then centrifuge again at maximum speed for 10 min.
10. Air dry the DNA pellet.
11. Resuspend the air-dried DNA pellet in 20 μL of 10 mM Tris-HCl, pH 8.0.
12. Quantify 1 μL of the concentrated sheared DNA using the Quant-iT PicoGreen dsDNA kit.
13. Perform sequencing library preparation using the NEBNext Ultra DNA library prep kit for Illumina.
14. Determine library quality and size using a Bioanalyzer 2100 with DNA High Sensitivity reagents.
15. Lastly, quantify the libraries using the Kapa quantification complete kit for Illumina on an Applied Biosystems 7500 Real-time qPCR system (or similar).
16. Normalize the sample libraries and pool for multiplex sequencing following recommendations and procedures outlined for your sequencer and sequencing facility.

3.5 Data Analysis

See Fig. 2 and the "read me" files at https://github.com/thereddylab/LADetector and https://github.com/thereddylab/pyLAD for detailed instructions on our recommended analysis pipeline(s). Other methods for identification and analysis of LADs (and other broad domains) include Enriched Domain Detector (EDD) and Hidden Markov Modeling [17, 25, 26, 35, 37, 38]. This pipeline can be used for any analyses requiring the detection of large domains, rather than discrete localized peaks, including data from ChIP-seq or TSA-seq, another proximity labeling method [39].

4 Notes

1. Different HEK293 lines proliferate at different rates. The investigator should test a range of cell numbers to be seeded the day before transfection, to achieve 90% confluency. This is important as the cells will stop proliferating post-transfection. Additionally, too low or too high a confluency would lead to cells being easily dislodged during post-transfection PBS washes or collection of viral supernatants.
2. Sodium butyrate is a HDAC inhibitor that increases global transcription, including viral transcription, drastically increasing the titers of viral supernatants. Sodium butyrate should be removed and HEK293 cells washed prior to the first viral supernatant collection, since it is a HDAC inhibitor and so its presence would affect chromatin organization of transduced cells.
3. Primary MEFs show abnormal morphology when cultured at low confluences. Even though a 30–40% confluency results in a lower MOI, it is more important that the cells are well behaved. For immortalized cells, such as C57B/L6 3T3 fibroblasts, cells are typically seeded at a lower confluency (e.g., 10,000 cells/cm^2 the day before).
4. Polybrene increases transduction efficiency. Some cells require higher polybrene amounts for increased transduction while others, such as neurons, are sensitive to polybrene. A polybrene-only control can determine if your cells can tolerate polybrene and this should be tested if you are uncertain about the effects of using this reagent. It can be omitted, but efficiencies will be reduced. In that case, you may need to concentrate your virus.
5. Genes are usually expressed through lentiviruses only after a 48-hour period following transfection. It is advisable to extend the cell culture beyond this timeframe to provide sufficient time for the DNA adenine methyltransferase to take effect. If using

the original DamID constructs, there is no selection that can be used to select for cells that have taken up the virus. We and others have introduced selection modules to these constructs to enable selection, but this is not necessary.

6. This is a stopping point. Samples can be kept frozen at −80 °C (snap frozen cell pellets) or − 20 °C.
7. DamID amplification typically is carried out with a no *Dpn*I control and a no ligase control.
8. For post-DamID amplification, a small aliquot of the amplified material can be run on a 1% agarose gel for visual inspection (Fig. 1). For additional quality control checks prior to NGS, qPCR can be performed to check known regions of the genome that are within LADs or not within LADs, provided that this information is known.
9. In contrast to ChIP, DamID relies on the restriction enzyme *Dpn*I, ensuring that each amplified size fragment originates from specific genomic loci. Consequently, directly shearing the DamID-amplified material through sonication could introduce bias in terms of the genomic representation of fragments. Therefore, a randomization step is included in our protocol prior to sonication.
10. Transferring the DNA to new TPX tubes after each round will prevent etching of the TPX plastic. This will appear as white flakes in the sheared DNA solution.
11. Glycogen here is added as a carrier to increase the visibility of the pelleted DNA post-sonication.

Acknowledgement

This work was supported by grant GM132427 from the National Institute of General Medical Sciences awarded to KLR.

References

1. Gruenbaum Y, Margalit A, Goldman RD et al (2005) The nuclear lamina comes of age. Nat Rev Mol Cell Biol 6:21–31
2. Wong X, Cutler JA, Hoskins VE et al (2021) Mapping the micro-proteome of the nuclear lamina and lamina-associated domains. Life Sci Alliance 4
3. Zullo JM, Demarco IA, Piqué-Regi R et al (2012) DNA sequence-dependent compartmentalization and silencing of chromatin at the nuclear lamina. Cell 149:1474–1487
4. Guarda A, Bolognese F, Bonapace IM et al (2009) Interaction between the inner nuclear membrane Lamin B receptor and the heterochromatic methyl binding protein, MeCP2. Exp Cell Res 315:1895–1903
5. Haraguchi T, Holaska JM, Yamane M et al (2004) Emerin binding to Btf, a death-promoting transcriptional repressor, is disrupted by a missense mutation that causes Emery-Dreifuss muscular dystrophy. Eur J Biochem 271:1035–1045
6. Ye Q, Callebaut I, Pezhman A et al (1997) Domain-specific interactions of human HP1-type Chromodomain proteins and inner

nuclear membrane protein LBR*. J Biol Chem 272:14983–14989

7. Holaska JM, Lee KK, Kowalski AK et al (2003) Transcriptional repressor germ cell-less (GCL) and barrier to autointegration factor (BAF) compete for binding to Emerin in vitro *. J Biol Chem 278:6969–6975
8. Kosak ST, Skok JA, Medina KL et al (2002) Subnuclear compartmentalization of immunoglobulin loci during lymphocyte development. Science 296:158–162
9. Szczerbal I, Foster HA, Bridger JM (2009) The spatial repositioning of adipogenesis genes is correlated with their expression status in a porcine mesenchymal stem cell adipogenesis model system. Chromosoma 118:647–663
10. Williams RRE, Azuara V, Perry P et al (2006) Neural induction promotes large-scale chromatin reorganisation of the Mash1 locus. J Cell Sci 119:132–140
11. Meister P, Towbin BD, Pike BL et al (2010) The spatial dynamics of tissue-specific promoters during C. Elegans development. Genes Dev 24:766–782
12. Peric-Hupkes D, Meuleman W, Pagie L et al (2010) Molecular maps of the reorganization of genome-nuclear lamina interactions during differentiation. Mol Cell 38:603–613
13. Yao J, Fetter RD, Hu P et al (2011) Subnuclear segregation of genes and core promoter factors in myogenesis. Genes Dev 25:569–580
14. Reddy KL, Zullo JM, Bertolino E et al (2008) Transcriptional repression mediated by repositioning of genes to the nuclear lamina. Nature 452:243–247
15. Greil F, Moorman C, van Steensel B (2006) DamID: mapping of in vivo protein-genome interactions using tethered DNA adenine methyltransferase. Methods Enzymol 410: 342–359
16. Orian A, Abed M, Kenyagin-Karsenti D et al (2009) DamID: a methylation-based chromatin profiling approach. Methods Mol Biol 567: 155–169
17. Guelen L, Pagie L, Brasset E et al (2008) Domain organization of human chromosomes revealed by mapping of nuclear lamina interactions. Nature 453:948–951
18. Vogel MJ, Peric-Hupkes D, van Steensel B (2007) Detection of in vivo protein–DNA interactions using DamID in mammalian cells. Nat Protoc 2(6):1467–1478. https://doi.org/10.1038/nprot.2007.148
19. van Koningsbruggen S, Gierlinski M, Schofield P et al (2010) High-resolution whole-genome sequencing reveals that specific chromatin domains from most human chromosomes associate with nucleoli. Mol Biol Cell 21: 3735–3748
20. Bersaglieri C, Kresoja-Rakic J, Gupta S et al (2022) Genome-wide maps of nucleolus interactions reveal distinct layers of repressive chromatin domains. Nat Commun 13:1483
21. Vertii A, Ou J, Yu J et al (2019) Two contrasting classes of nucleolus-associated domains in mouse fibroblast heterochromatin. Genome Res 29:1235–1249
22. Bersaglieri C, Santoro R (2019) Genome organization in and around the nucleolus. Cells 8
23. Bizhanova A, Yan A, Yu J et al (2020) Distinct features of nucleolus-associated domains in mouse embryonic stem cells. Chromosoma 129:121–139
24. Vogel MJ, Peric-Hupkes D, van Steensel B (2007) Detection of in vivo protein-DNA interactions using DamID in mammalian cells. Nat Protoc 2:1467–1478
25. Chen S, Luperchio TR, Wong X et al (2018) A lamina-associated domain border governs nuclear lamina interactions, transcription, and recombination of the Tcrb locus. Cell Rep 25: 1729–1740.e6
26. Wong X, Hoskins VE, Melendez-Perez AJ et al (2021) Lamin C is required to establish genome organization after mitosis. Genome Biol 22:305
27. Harr JC, Luperchio TR, Wong X et al (2015) Directed targeting of chromatin to the nuclear lamina is mediated by chromatin state and A-type lamins. J Cell Biol 208:33–52
28. Perovanovic J, Dell'Orso S, Gnochi VF et al (2016) Laminopathies disrupt epigenomic developmental programs and cell fate. Sci Transl Med 8:335ra58
29. Wen B, Wu H, Shinkai Y et al (2009) Large histone H3 lysine 9 dimethylated chromatin blocks distinguish differentiated from embryonic stem cells. Nat Genet 41:246–250
30. Meuleman W, Peric-Hupkes D, Kind J et al (2013) Constitutive nuclear lamina-genome interactions are highly conserved and associated with a/T-rich sequence. Genome Res 23:270–280
31. Pope BD, Ryba T, Dileep V et al (2014) Topologically associating domains are stable units of replication-timing regulation. Nature 515: 402–405
32. Handoko L, Xu H, Li G et al (2011) CTCF-mediated functional chromatin interactome in pluripotent cells. Nat Genet 43:630–638
33. Dixon JR, Selvaraj S, Yue F et al (2012) Topological domains in mammalian genomes identified by analysis of chromatin interactions. Nature 485:376–380

34. Nora EP, Lajoie BR, Schulz EG et al (2012) Spatial partitioning of the regulatory landscape of the X-inactivation centre. Nature 485:381–385
35. Skene PJ, Henikoff S (2017) An efficient targeted nuclease strategy for high-resolution mapping of DNA binding sites. Elife 6:e21856
36. Fu Z, Jiang S, Sun Y et al (2024) Cut&tag: a powerful epigenetic tool for chromatin profiling. Epigenetics 19:2293411
37. Lund E, Oldenburg AR, Collas P (2014) Enriched domain detector: a program for detection of wide genomic enrichment domains robust against local variations. Nucleic Acids Res 42:e92
38. Shah PP, Keough KC, Gjoni K et al (2023) An atlas of lamina-associated chromatin across twelve human cell types reveals an intermediate chromatin subtype. Genome Biol 24:16
39. Chen Y, Zhang Y, Wang Y et al (2018) Mapping 3D genome organization relative to nuclear compartments using TSA-Seq as a cytological ruler. J Cell Biol 217(11): 4025–4048. https://doi.org/10.1083/jcb.201807108
40. Luperchio TR, Sauria MEG, Wong X, et al (2017) Chromosome conformation paints reveal the role of lamina association in Genome organization and regulation. https://www.biorxiv.org/content/biorxiv/early/2017/03/30/122226

Part V

Applications of NGS Technology

Chapter 15

How Have Massively Parallel Sequencing Technologies Furthered Our Understanding of Oncogenesis and Cancer Progression?

Diogo Ann Onuselogu, Saskia Benz, and Suparna Mitra

Abstract

Massively parallel sequencing technologies have been a boon to many fields of biological science, including oncology. Cancer is an umbrella term for many diseases featuring abnormal cellular growth due to genetic and epigenetic aberrations. Advances in sequencing technology allow for interrogation of the DNA and RNA of cancer cells and other cells in the tumor microenvironment down to a single-base resolution. However, these strides come after a rich history of ground-breaking biological assays, like the discovery of the Philadelphia chromosome in the context of leukemia. Many specific genetic and epigenetic modifications have been implicated in oncogenesis, cancer progression, and response to treatment. Sequencing technologies have also helped to associate populations of bacteria in the microbiome to cancer development and prognosis. However, all this new information, especially when procured via high-throughput methods, comes at the cost of being more computationally and staff-resource intensive. There is also more risk to the privacy of the individuals with sequenced genomes. Notwithstanding, the overall benefit of sequencing technologies can greatly outweigh the risks with careful advancements and continued focus on the goal: helping those affected by cancer via precision medicine. Cancer biology has been and will continue to be elucidated by sequencing innovations in ways unimaginable without it.

Key words Next-generation sequencing, Cancer, Oncogenesis, Technology, Modern history

1 Cancer and Massively Parallel Sequencing

1.1 Cancer Genes and Gene Products

Cancer is dysregulated cell growth initiated and supported by genetic mutations. These aberrations in cellular biological machinery work by blocking stop gates that would normally suppress or tightly regulate such proliferation. Furthermore, cancer is not a disease with a single clinical presentation, but an umbrella of phenotypes for cancers in different locations, with different biological or environmental triggers, and varying grades of severity. In the same vein, cancer treatment can be either general, like radiotherapy or anti-proliferative chemotherapy, or targeted against a specific gene or protein unique to a cancer type [1]. For this reason,

Valerie J. Carabetta and Olaitan Akintunde (eds.), *High Throughput Gene Screening: Methods and Protocols*, Methods in Molecular Biology, vol. 2866, https://doi.org/10.1007/978-1-0716-4192-7_15,

being able to determine which genes and gene products are associated with cancer diagnosis, prognosis, or treatment is key to managing and treating the disease [2]. To do so, the information encoded by biological molecules in cancer must be decoded, read, understood, analyzed, and then acted upon if possible and necessary.

1.2 Massively Parallel Sequencing (MPS)

Investigating biology at the sequence level began seven decades ago with the 1950's advent of amino acid sequencing [3]. Since then, technological capacities and scientific understanding of biological sequences have flourished in all aspects, with unsurprising benefits to the field of oncology.

A Brief History of DNA Sequencing

Massively parallel sequencing, also known as next-generation sequencing (NGS) or second-generation sequencing, was developed in the late 1990s as a high-throughput alternative to the electrophoresis-based sequencing methods of first-generation sequencing (FGS). Now, synonymous with Sanger sequencing, FGS methods for DNA sequencing were created on the foundations of RNA and amino acid sequencing breakthroughs [3, 4]. FGS was revolutionized in the late 1970s by two teams: Sanger et al. [5] and Maxam, Gilbert [6]. Both groups' methods involved radioactive labeling and separation of DNA fragments. Sanger's method quickly gained popularity (hence the eponym) due to its easily produced and easily interpreted results [4].

Though FGS led to great advances in molecular biology, MPS revolutionized the field as technology advanced faster than Moore's Law. MPS/NGS allows for an unbiased selection of a plethora of reads to be sequenced at the same time. This is done utilizing "multiplexing" protocols, like experimental setups that allow multiple samples of DNA to be sequenced at the same time without mixing them up during analysis. Such protocols allow for millions of short DNA fragments (typically 100–150 base pairs (bp)) to be sequenced fast enough that the three billion bp long human genome can now be sequenced, with key features reported in just 30 hours [7]. This is an impressive improvement from the shotgun-sequencing approached used by Craig Venter and Celera that took nine months [8], the Human Genome Project that took more than a decade [9], or even the initial rate that the pioneers of DNA sequencing operated at—one base a month [3, 10].

Commercial Contributions

Though sequencing was originally monopolized by astute academic institutions and the company Applied Biosystems, several companies played a key role in the research and development of NGS after completion of the Human Genome Project [3]. For example,

454 Life Science's NGS machine utilized PCR-amplification-in-oil emulsion techniques and pyrosequencing. The former involved random fragmentation of a genome and creating several copies of individual fragments ("amplification") in reaction droplets to boost signal detection during sequencing. Afterwards, pyrosequencing determined each newly added base by the light given off by the reaction upon complimentary binding and release of a pyrophosphate [11]. Theirs was the first commercially available platform and began the "democratization" of massively parallel sequencing [3, 4].

Soon afterwards, Solexa produced a NGS platform utilizing "bridge-amplification" of fragmented and adapter-ligated DNA bound to a flow chip, the technology of which came from the Swiss company Manteia. The amplification was solid-phase, as opposed to 454's nanodroplet-in-oil method, and utilized dense DNA library chips to hold these amplicons and produce even denser DNA polyony colonies ("polonies," or clusters of amplified fragments to be sequenced [12]). Solexa was eventually bought by the company Illumina. Already renowned for its microarray technology, Illumina is now also a household name in the NGS sphere [4].

Cancer Context: DNA Sequencing Today

Today, MPS DNA sequencing allows for whole genomic, functional genomic, exomic, and transcriptomic studies. These techniques are especially helpful when comparing healthy and disease cell or tissue states [2]. Of the above, whole genome sequencing (WGS) is the most comprehensive method, and thus the most expensive financially and computationally. The ideal cost usually associated with the current affordability and accessibility of WGS and its applications for clinical use is $1,000 USD. This lends to the term of "the thousand-dollar genome."

MPS technology has a plethora of potential utility in investigating mechanisms, therapeutic strategies, and overall improving clinical outcomes. Different versions of MPS have varying advantages and drawbacks, as explored by Benz et al. [13]. For example, Illumina sequencing harnesses high-throughput results with lowest relative cost; however, sample loading provides a challenge technically. In terms of cost, Illumina sequencing costs approximately $0.13 per million bases for MiSeq and between $0.02–0.07 per million bases for HiSeq. Comparatively, longer read nanopore sequencing such as The MinION™ costs less than $1 per million bases [13].

A 2023 analysis including the full breakdown of metagenomic applications, alongside options for sequencing techniques, full cost, and read statistic comparisons is available by Benz et al. [13]. Precision cancer medicine (PCM) has evolved using novel techniques including NGS to allow tumor analysis and more specific targeting. One considerable challenge facing this personalized treatment style is the cost of PCM; despite the reducing costs of NGS, treatments themselves remain high. However, one suggestion to improve cost-effectiveness is to tailor the use of PCM to the clinical case, for example NGS testing is recommended in cancers such as ovarian, prostate, and chololangiocarcinomas [14].

Machine sequencing per se is not the only cost contributor as research staff and clinicians are needed to collect biospecimens, prepare them, analyze readouts, and report the data. Schwarze and colleagues [15] performed a micro-costing analysis on genomic sequencing in cancer and rare disease cases. They uncovered all the costs involved in clinical-grade sequencing at the Oxford Molecular Diagnostics Centre. Noting that cancer genomes need matched-normal genomes and that rare diseases usually require genomes from each individual in a parent-proband trio, Schwarze et al. reported overall costs of £6841 and £7050 per case for a cancer patient and rare disease patient, respectively [15].

Whole genome *genotyping* studies, like genome-wide association studies (GWAS), have also been used to elucidate oncogenesis, response to therapy, and survival. GWAS mainly interrogates single nucleotide polymorphisms (SNPs), which are single-base DNA changes that are not uncommon in a population. Thus, it opens doors to investigations of SNPs' effects on transcript regulation, protein expression, and epigenetic modifications in the context of cancer [16].

Cancer Context: RNA and Amino Acid Sequencing Today

As field changing as the ability to sequence DNA has been for biology and medicine, DNA itself is a remarkably stable molecule both chemically on its own and biologically within an organism. Estimated mutation rates in humans are in the low tens to low hundreds of mutations per genome with each generation. It stands to be repeated that the genome is around three billion base pairs long [17–19]. Mutation rates vary for different populations, types of mutations, regions of the genome, and disease status (i.e., cancer) [20].

Beyond genomics studies, transcriptomics and proteomics deal with much more dynamic biological profiles (Fig. 1). Transcriptomics involves the investigation of gene expression by way of RNA-based analyses. This includes RNA splice variant analysis and gene fusion identification. Gene fusions are particularly important in oncological contexts because they serve as a source of "driver" or causal oncogenic variants when they result in protein overexpression or hyperactivity [21].

Within an organism, transcriptomes vary between tissues, cell types, and even neighboring cells (Fig. 1, [22, 23]). RNA-sequencing (RNA-seq) has also become massively

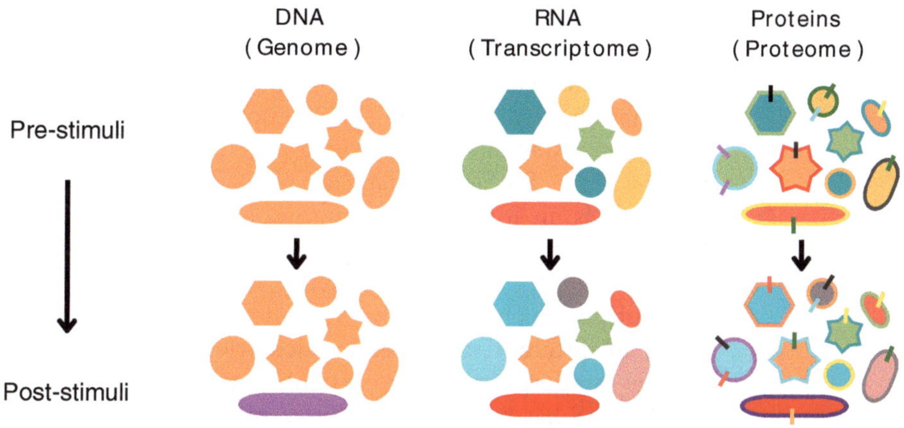

Fig. 1 Symbolic representation of the dynamic nature of the transcriptome and proteome relative to the genome in response to stimuli, like infection and UV light. All cells depicted are from the same organism but may be of different types and subtypes. Generally, genomes are quite stable, but genetic mutations that are not repaired and persist are possible, and such mutations may be pathogenic. This is the genetic basis of cancer. The RNA transcripts and protein products of genes are naturally varied but may also respond differently after stimulus exposure. For the sake of simplicity, only cell membrane proteins are depicted through the colored outlines. Various types of protein post-translational modifications are represented by differently colored dashed lines

parallelized with the onset of NGS. This is done by purifying nucleic acids, isolating RNA from DNA, and converting the RNA to complementary DNA (cDNA) for NGS sequencing. It has been used to differentiate gene expression profiles of cancerous cells from non-cancerous cells [24]. The ability to perform single cell RNA-seq (scRNA-seq) in a high-throughput manner has allowed for high-resolution transcriptomic analyses of various organs and the tumor microenvironment (TME), exemplified by the Human Cell Atlas [25] and The Cancer Genome Atlas (TCGA), respectively [26, 27]. Even further, scRNA-seq also facilitates trajectory analysis, which is computationally estimated developmental lineage paths between cell types, and spatial transcriptomics analysis, like lower-throughput RNA analyses of several single cells of a slide-mounted sample [23]. Both technologies can be used to inform investigations of oncogenesis and cancer progression.

Proteomics lags behind genomics and transcriptomics in the context of massive parallelization [28]. Regarding terminology, it should be noted that MPS or NGS are generally limited to nucleotide sequencing. However, the high-throughput characteristic of MPS technologies obligates the inclusion of ideally high-throughput sequencing in proteomics for the purposes of this literature review. That said, the proteome is more diverse than the transcriptome given the potential for post-translational modifications, including degradation-associated ubiquitination, protein phosphorylation, and glycosylation, among others. Furthermore, proteins cannot be amplified by a PCR equivalent, so parallel sequencing of multiplexed proteins would need precision like that of single molecule sequencing for DNA [4, 29]. Although liquid chromatography and mass spectrometry (LC-MS) was once the proteomics standard, it was unable to capture rare species well [28, 29]. Surface-Enhanced Laser Desorption-Ionization (SELDI) is an ionization technique allowing for high-sensitivity and high-throughput protein arrays when paired with mass spectrometry (MS). Ionized proteins with specific weights travel through an ion channel with different masses. These varying masses result in different speeds and trajectories that correspond with a unique "fingerprint" for a particular protein molecule. Together, SELDI-MS has been used to differentiate cancer proteomes from healthy proteomes, albeit with controversial results and the inability to robustly identify biomarkers [30–32].

Recent efforts bode well for the future of truly massively parallel proteomics. For example, Swaminathan et al. [33] introduced sparse, multiplex amino acid sequencing labeling only cysteine and lysine residues. Increasing sparse labeling markers from two to four could theoretically increase identified proteome coverage from 19.7% to 98.2% [34]. On another front, Matreyek and colleagues [35] have developed the generalizable methodology VAMP-seq, which determines protein Variant Abundance by Massively Parallel Sequencing. As proof-of-concept, the gene *Phosphatase and Tensin homolog* (*PTEN*), which is implicated in oncogenesis, was found to have 1138 variants associated with low protein abundance. Notably, transcriptomic mRNA expression data serves as a varyingly accurate proxy for protein expression data, due to the stability/instability of mRNA and protein molecules, as well as translation efficiency [36]. However, VAMP-seq technology obviates the need for such a proxy in a massively parallel manner. Given the mutation-based pathology and vast deregulation of protein levels associated with cancer [37, 38], developments in VAMP-seq technology have clear implications for oncology.

1.2.1 Types of Mutations and Variants in Cancer

The Genetic Mutational Landscape

As mentioned earlier, cancer is a disease driven by genetic mutations. It is this genetic instability and its effect on oncogenes, tumor suppressor genes, and DNA maintenance genes that has been acknowledged as the impetus for the development of the hallmarks of cancer in neoplastic tissues [39]. Though oncogenesis is primarily associated with somatic mutations, certain germline variants are associated with increased risk during screening or found upon genomic analysis at diagnosis (e.g., BRCA1/2 for breast cancer, [40, 41]). Overall, genetic mutations in cancers can be observed at varying resolutions, like point mutations, gene translocations, inversions, copy number variation, chromosomal rearrangement, and abnormal ploidy (Fig. 2, [2]).

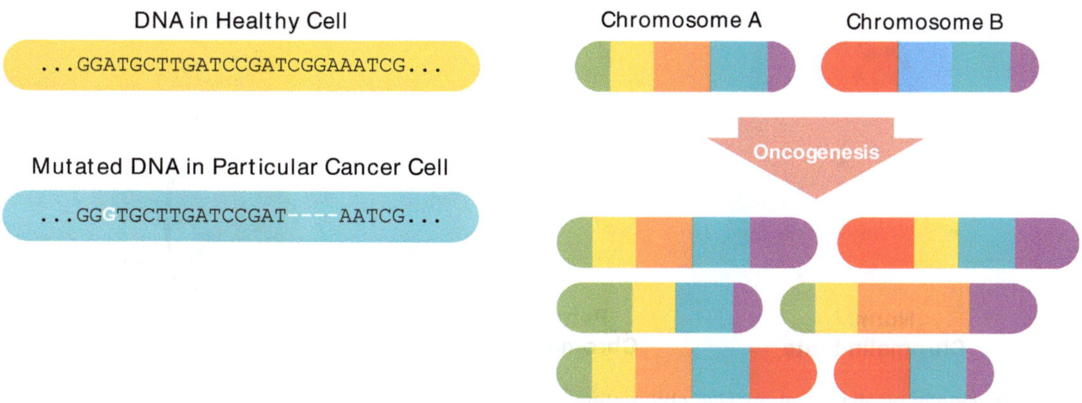

Fig. 2 Depictions of potential genetic mutations encouraging tumorigenesis including point mutations above and gene translocations, copy number variation, inversions, chromosomal re-arrangement, and abnormal ploidy below

Table 1
Breakdown by type of candidate mutations in breast cancers of 103 women from Vietnam and Mexico. SNP = single nucleotide polymorphism. Adapted from Supplementary Table 5 in ref. [42]

Type of mutation (subtypes)	Number detected	Proportion
Missense	3153	63.25%
Silent/synonymous	1157	23.21%
Nonsense	242	4.85%
Deletions (frame shift, in-frame)	194 (140, 54)	3.89% (2.81%, 1.08%)
Insertions (frame shift, in-frame)	110 (96, 14)	2.21% (1.93%, 0.28%)
Splice site (SNP, deletion, insertion)	97 (84, 9, 4)	1.95% (1.69%, 0.18%, 0.08%)
Fusion genes (out of frame, in frame)	21 (20, 1)	0.42% (0.40%, 0.02%)
Read-through	11	0.22%
Total	4985	100.00%

Banerji et al. examined the normal-matched cancer genomes of 103 females with breast cancer through WGS, whole exome sequencing (WES), or both [42]. Afterwards, they published a defining paper on the mutational landscape of breast cancer revealing nearly 5000 candidate mutations. Unsurprisingly, the mutations were predominantly non-synonymous substitutions, which meant they lead to changes in translated amino acid sequences. The least common mutation type was read-through (Table 1). Banerji and colleagues also deduced the number of allele copies for each variant and the average ploidy of the breast cancer genomes from each patient through their WGS/WES data [41]. Such a feat

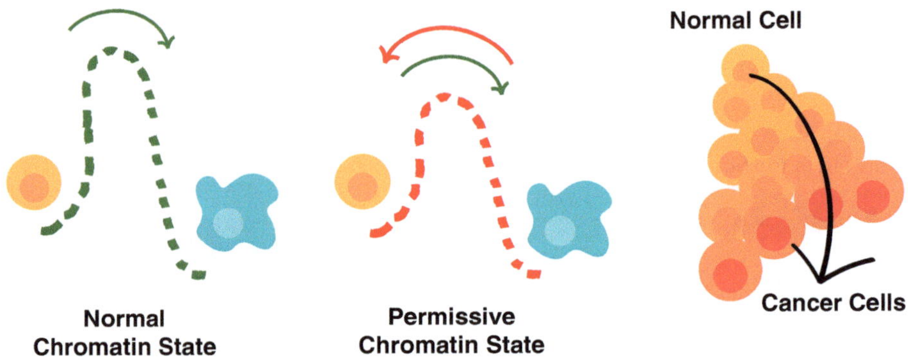

Fig. 3 Epigenetic plasticity, permissive chromatin, and driver variants. Left: Cell state stability given normal chromatin. The peak represents a hill in the symbolic Waddington landscape of cellular differentiation. Given normal chromatin homeostasis, cell states cannot easily be converted. Middle: Permissive chromatin effectively facilitates cell state conversion, represented by a lowered peak, via alternative gene regulation pathways indicated in red. Right: Out of many epigenetic changes, only a few will drive tumorigenesis (the circles of red shade) and acquisition of the canonical hallmarks of cancer. Adapted from graphical abstract in ref. [44]

at this extent would have been painstaking for first-generation sequencing paired with microarray technology to accomplish.

Non-mutational Changes in Cancer

As alluded to in the "Cancer Context" sections, cancer cells differ from healthy cells beyond genetics. Transcriptome, proteome, and even epigenome changes have been observed in cancer cells and between individual cancer cells of the same neoplasm [43–45]. As transcriptomic and proteomic differences have been discussed in the context of sequencing, this section shall briefly consider epigenomic changes.

MPS-facilitated epigenomic sequencing includes the following techniques and their successors: chromatin immunoprecipitation sequencing (ChIP-seq), assays for transposase-accessible chromatin using sequencing (ATAC-seq), and bisulphite sequencing (BS-seq). ChIP-seq can be used to study histone modifications after an antibody-binding and isolation of specific marks on the histone proteins. ATAC-seq studies chromatin accessibility via removing all non-histone-protected regions of DNA prior to sequencing and observing the coverage patterns of the remaining DNA. BS-seq interrogates DNA methylation at cytosines across the whole genome level, allowing for "methylome" assays.

Epigenomic studies can also be conducted at the single cell level. Akin to scRNA-seq, single-cell epigenomics allows for cell-by-cell analysis and investigations of oncogenesis [46, 47]. The reason is that a constant genome, yet diverse epigenome, determines the developmental fates of healthy and cancer cells alike. Hence, epigenetic modifications affect cell state instability and tumorigenesis (Fig. 3). For example, changes in DNA methylation and chromatin accessibility states have been associated with cancer. DNA methylation and densely packed chromatin ("heterochromatin") are associated with gene silencing through repression of transcriptional activity. In cancer, a global hypomethylation signature arises that has been linked to "permissive" chromatin, i.e., chromatin more readily facilitating of cell state conversion by way of alternative gene regulation pathways [43, 45].

2 Understanding Cancer Better

2.1 Elucidating Cancer Biology Prior to MPS Technologies

Pre-MPS technologies contributed extensively to our fundamental understanding of cancer and cancer genetics. Late nineteenth century and early twentieth century research observed abnormal karyotypes in cancer cells [1]. Throughout the twentieth century, even prior to FGS, many oncological discoveries were made, like connecting cancer to carcinogenic compounds and connecting such compounds to activating liver enzymes. After realizing DNA's role as a biological instruction manual, activated carcinogens were found to bind to DNA, forming adducts and potentially damaging it [48].

Another key finding is the identification of the recurring Philadelphia chromosome in chronic myeloid leukemia (CML). It is created by a translocation between chromosomes 9 and 22 [49]. However, Rowley noted that cytogenetic methods without additional technologies, such as fluorescent labels, were not always sufficient to make karyotype-informed observations on karyo-abnormal cancer cells [49]. Furthermore, she mentioned that though the formation of the Philadelphia chromosome did not necessarily result in loss of genetic material, technology available at the time was not able to deduce the pathological mechanism linking the de novo chromosome to CML.

With the advent of FGS, discoveries linking DNA abnormalities to cancer were of more refined resolution [2]. In the case of the Philadelphia chromosome, 15 years after its discovery, FGS was used to elucidate the molecular pathology of the translocation that had high prevalence in certain chronic leukemias. Kurzrock et al. sequenced the new fusion gene of the de novo chromosome, identifying its 5' *BCR* head and 3' *ABL* tail as well as various isotypes of the *BCR/ABL* gene. The ABL protein is a receptor tyrosine kinase that usually needs to undergo activation prior to signal transduction. Though the function of the BCR gene product is uncertain, when fused to the ABL tyrosine kinase, it constitutively activates the signal transducing kinase contributing to oncogenic cell survival (Fig. 4, [50]).

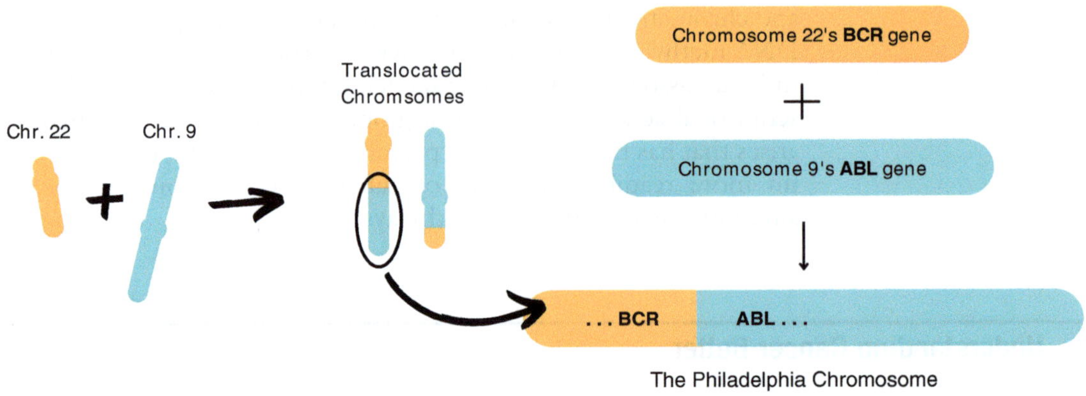

Fig. 4 Abstract representations of the development of the Philadelphia Chromosome [50]

2.1.1 Elucidating Cancer Biology with MPS Technologies

FGS facilitated a plethora of scientific discoveries and is certainly still relevant today. However, in the context of cancer, first-generation methods have certain weaknesses that MPS/NGS technologies have successfully overcome. For example, rare somatic variants are hard to detect using FGS methods. This is especially relevant to tumors that cannot be easily resected or once resected, do not provide much tumor tissue for pathological analysis. Solid tissue or liquid biopsies from patients with such cancers may not easily detect cancer-associated gene or gene product variants. Furthermore, cancer genomes are quite heterogenous, both within and between cancer types. However, the high sequencing depth of NGS compared to FGS technologies allows for the identification of reads with rare alleles and deals with heterogenous samples during read alignment. Such "over-sampling" also allows for accurately sequenced reads less hindered by experimental noise [1, 2]. Furthermore, NGS is magnitudes faster than FGS and allows for a massively parallel investigation of the cancer genome or transcriptome at much higher resolution for many more cells and their biological material.

Not only can multiple genomes or transcriptomes be analyzed in a single experimental run, but biological data from multiple experiments can be conducted at the same time due to multiplexing and barcode technology. Such parallelism has facilitated large-scale efforts to further elucidate cancer at the highest resolution. Two such efforts are the United States' TCGA (as previously mentioned) and the United Kingdom's Cancer Genome Project (CGP). TCGA has been a key leader in oncology research. As of February 2021, TCGA has published 72 journal articles; analyzed 20,000 cancer matched-normal samples; and accrued 2.5 petabytes of genomic, transcriptomic, epigenetics, and proteomic data since its beginnings in 2006 (Fig. 5). Many of their published articles concern landscape or comprehensive analyses of particular cancers or cell populations—of which they have analyzed 33 cancer types

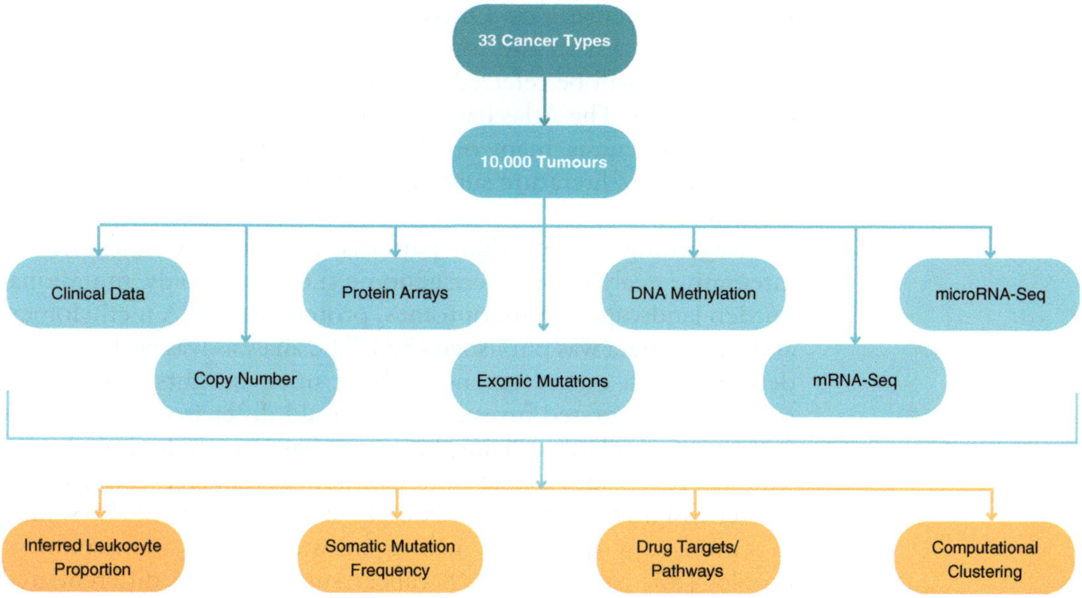

Fig. 5 Flowchart representation of the comprehensive cancer database by the TCGA. TCGA/Pan-Cancer Atlas [53]. Integrated analysis across the datasets revealed information about leukocyte proportion, mutation frequency, and pharmacological networks across the cancer and tumor types. Integrated analysis also allowed for high-level clustering of the tumors

[51]. As part of TCGA's research, the Pan-Cancer Atlas was formed, elucidating cell-of-origin influences on eventual cancer type, oncogenic processes, and the roles of various signaling pathways [52].

For the CGP, it is more focused on genome-wide oncology research and the role somatic mutations play in cancer. Through investigating acquired mutations, they aim to understand the genes, mutational development, and clonal evolution of cancers. The project is older than the TCGA, having published its first paper in 1992. Since then, the project has amassed over 430 publications, 403 of which have been published in 2007 or afterwards. The year 2007 is the first year the CGP published genomics research in addition to its then traditional and FGS-based genetics research [54]. As part of the CGP, the Catalogue of Somatic Mutations in Cancer (COSMIC) was created. COSMIC has grown over the years, now with anomalies in 748 genes described (as of September 2024) and the genes' pathological contributions aligned to the ten hallmarks of cancer [39, 55]. In 2010, COSMIC began integrating whole genome resequencing data into its catalogue which has totaled to 35,480 whole genomes and exomes by 2018 [55, 56]. Clearly, neither TCGA or the CGP would not be where they are today without their multi-omics, MPS-facilitated, and ground-breaking oncological big data research.

Examples: Focusing on Oncogenesis

Given the breath of the topic, exemplary insights from the Pan Cancer Atlas will be referred to in this exploration of oncogenesis through MPS. The Atlas has published more than 20 papers elucidating oncogenesis, many of which do so using MPS-technologies or data. Below, literature summarizing two of their key focuses are discussed.

Regarding cell-of-origin patterns, an overview of the Atlas' integrative analyses of cancer genome (i.e., aneuploidy, mutational burden landscape), transcriptomes, proteomes, DNA methylomes, and clinical data was published [53]. The analysis was performed to determine if a cancer taxonomy built upon insights from multi-omics and integrated investigations would differ from the current taxonomy standard utilizing histology-based pathological classifications. To do so, Hoadley and colleagues employed clustering of cluster assignments algorithms that observed grouping by similar mutation, copy-number, signaling pathway, and microenvironment profiles [53]. As a result, 28 cancer subtypes were differentiated outside of classical taxonomical categorizations. Though many cell-of-origin subtypes re-clustered after the integrated analysis, significant crossover was noted, especially at the organ system level. As an example, 13% of breast cancers formed a separate cluster from most breast cancer cases due to differential transcriptomic and methylomic profiles [53]. Both the regrouping and the identification of crossovers between taxonomical groups provide a corroborating yet innovative, high-level, pan-cancer, perspective on the heterogeneity of cell-of-origin cancer subtype molecular signatures in a potentially clinically actionable way.

Another Pan Cancer Atlas key paper is an overview on oncogenic processes by Ding and colleagues [21]. Through their similarly multi-omics and MPS-utilizing study, they noted that germline variants influence the somatic mutation profile in cancers in a pathway-dependent manner notwithstanding the fact that oncogenic mutations are occasionally present in healthy people [21]. Furthermore, cancerous driver mutations also influence the TME. For example, tumors with driver mutations in the *BRAF* gene were associated with higher TME cytotoxic T lymphocyte proportions than tumors with *NRAS* gene driver mutations. Such discoveries are of significance to precision medicine and cancer immunotherapies as higher cytotoxic T lymphocyte proportions are also associated with better prognosis [21].

Examples: Researching Cancer Progression

NGS technologies have also helped to elucidate the biology of cancer progression, aiding determinations of diagnosis, treatment, and prognosis [57]. Multiple prognostic and biomarker studies have emerged since the advent of MPS technologies, a few non-TCGA examples of which are discussed below.

Recently, an NGS mutation panel of 22 cancer driver genes was used to stratify clinical outcome by allele frequency in patients with advanced or metastatic cancers. Instead of taking biopsies from the 68 patients' solid tumors, cell free DNA (cfDNA) was collected for liquid biopsy. High concentrations of serum cfDNA were associated with a poorer survival, with a Hazard Ratio (4.81, p-value = 0.372). Furthermore, patients showing signs of increased mutations in their cfDNA (i.e., a higher mutational burden) had significantly poorer prognoses (p-value = 0.026, [58]).

In the realm of transcriptomics, circular RNA (circRNA) has been proposed as a breast cancer biomarker. Though not experimentally tested, a meta-analysis of more than 240 journal articles revealed different circRNA concentration levels in breast cancer subtypes and chemo-resistant cells [59]. The research by Jahani and colleagues suggested that certain circRNA states facilitated breast cancer proliferation via RNA regulatory pathways. However, studies characterizing the function and role of circRNAs in cancer are still needed.

Another transcriptomic example involved scRNA-seq and an investigation of metabolic pathway changes in human models of breast cancer [60]. The study, by Davis and colleagues, used xenograft mouse models and revealed two main points. First, micro-metastases from the xenografted models possessed distinct differential expression and metabolomics profiles that Davis et al. linked to poor prognostic outcome (Fig. 6). Second, differentially expressed genes were subjected to gene set enrichment analysis to reveal mitochondrial oxidative phosphorylation as the most upregulated pathway, potentially discordant with the accepted Warburg effect associated with tumors [61]. The fact that inhibiting this pathway pharmacologically reduced lung metastases severity [60] is of clinical significance and should be more extensively researched with current understandings of the Warburg effect in mind.

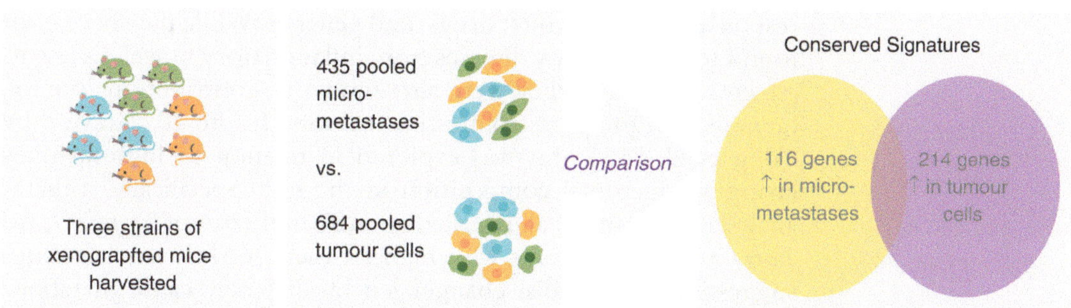

Fig. 6 Micro-metastatic and tumor cells from three different human patient-derived-xenograft models underwent supervised analysis comparing markers of differential gene expression. With significance $P < 0.05$, 330 genes were identified as differentially expressed across the models: 116 genes were relatively upregulated in the micro-metastases from the lung and lymph node and 214 genes were upregulated in the cells of eventual primary tumor. Adapted from Fig. 3a in *ref.* [60]

Highlight: Microbiome and Cancer

In addition to studying tumors and host genomes, MPS/NGS has given us an unprecedented ability to study those microorganisms that exist within living organisms. Literature now suggests that these microbes may contribute to certain cancers and may also influence response to therapeutics. The relationship between microbes and cancer is complex. While most are not directly causal, microbes have been found to modulate immune function and promote carcinogenesis [62]. For example, Kadosh et al. highlighted the role of gallic acid, produced by microbes, in advancing p53 mutations towards carcinogenesis [63]. The study investigated the Trp53 gene in mouse models, which encodes p53. They found that gallic acid, produced by gut microbiota, resulted in malignant phenotypic changes of a previously tumor-suppressive p53 mutation. Their proposed mechanism was through the metabolite hyperactivating WNT signaling and increasing the connection between TCF4 and chromatin, in relation to intestinal cancer [63]. This therefore illustrates how the microbiome can alter the overall functionality of a cancer mutation. Studies have also shown some role of bacteriophages and fungi within gastrointestinal cancer, as well as the presence of intratumoral bacteria, with varying activity in cancers [62].

Zhang et al. [64] studied how microbial changes may occur during carcinogenesis leading to gastric cancer and discovered increased abundance of certain bacterial species in patients. Importantly, they found that the structure of gastric microbes was altered along cancer progression from superficial gastritis up to gastric cancer. For example, higher *Helicobacter pylori* colonization was found in the cardia of patients with gastric cancer. While the study had a low patient cohort ($n = 47$), further study of microbial biomarkers could improve targeted treatments and patient prognosis.

Microorganisms found in the human gut change dynamically in response to lifestyle, diet, drugs, and genetics. While they have been found to impact many diseases from inflammatory bowel disease to pancreatic disease, they were also shown to contribute to tumorigenesis, including gastrointestinal tumors. In the 2021 review by Tong et al. [65], this was explored in relation to high-fat diets changing microbial composition in the gut. Specifically, a meta-analysis by Bisanz et al. revealed an increased ratio of *Firmicutes* to *Bacteroidetes* in patients with high-fat diets [66]. Overall findings suggest such microbial changes led by diet may cause metabolic reprogramming and alter known carcinogenic molecules.

As mentioned, changes to microbial composition and their modulation of immunity have been shown to promote disease. However, literature also indicates their protective potential for immunotherapy and novel cancer treatment [62]. Overall, this research demonstrates the extensive relevance of the microbiome within cancer. However, substantial challenges exist in advancing such observations towards effective clinical cancer interventions. Part of overcoming this gap involves NGS, alongside analysis of gene expression, proteins, and overall microbial function in carcinogenesis. NGS techniques can be therefore used not only for genomics, but a vast array of meta-omics techniques, providing novel tools to explore how microbes interact and behave. Metagenomics is the evaluation of microbial communities through DNA sequencing. While metagenomics analyzes the genetic content of microbial communities, metatranscriptomics reveals their metabolic activity in a given place and time, and how such activity changes [67]. Metatranscriptomics can therefore be used to investigate changes in gene expression within cancers, particularly to evaluate relationships between microbes and their human hosts. Metaproteomics and metabolomics use mass spectrometry for the analysis of thousands of proteins and peptides, or metabolites, respectively. Lipidomics is a more recent technique emerging from metabolomics, specifically analyzing lipid metabolites. Each field has applications in cancer research as briefly discussed below [68].

A multi-omics analysis approach was used by Chang et al. [69] to investigate the lung microbiomes of lung cancer patients. Nine bacterial species were found in association with reduced survival rates and metatranscriptomic changes in gene expression, correlated with tumor immune cell enrichment. In addition, Banavar et al. [70] used metatranscriptomic technology to build upon microbial 16S analysis from literature on oral cancer. Numerous functional pathways and microbial taxa were found in association with oral cancer. For example, high abundance of *Prevotella* and *Fusobacterium* was discovered in saliva from oral cancer patients. This study proposed its diagnostic classifier as the leading technique available, and as a potential early diagnostic method for earlier oral cancer diagnosis. Overall, this suggests that the association between oral cancer and microbes is based less around microbial composition than overall microbial community function. In this study [70], the data had increased granularity, thus creating a metatranscriptomic signature as a model to potentially diagnose oral cancer earlier, improving clinical outcomes. This paved the way towards usage of metatranscriptomics for the study of microbial populations in health and disease.

Regarding metagenomics and metabolomics, a 2019 study by Yachida et al. sampled the human gut microbiome of 616 patients, in relation to colorectal cancer (CRC, [71]). Highlights of the metagenomic shifts revealed elevated abundance (p-value <0.005) of *Actinomyces odontolyticus* and *Atopobium parvulum* species in intramucosal carcinomas and/or polypoid adenomas. Moreover, an increased abundance (p-value <0.005) of *Fusobacterium nucelatum* species was found in intramucosal colorectal carcinomas and advancing clinical stages. In terms of metabolomics, increased levels of both phenylalanine and branched-chain amino acids ($P < 0.005$) were also found within intramucosal carcinoma. Deoxycholate was likewise found to be augmented in intramucosal carcinomas and/or multiple polypoid adenomas. Such findings show the value in MPS/NGS in identifying both metabolomic and metagenomic biomarkers which may shift in the transition towards CRC.

Another metabolomic study, involved the short-chain fatty acid (SCFA) metabolites produced by gut microbiota, butyrate, acetate, and propionate. Using this technique, Nomura et al. [72] showed an association with progression-free survival in patients with solid tumors following anti-PD-1 immune checkpoint inhibitor treatment. While this study had a small sample size ($n = 52$), it nonetheless illustrates the role of metabolomics within cancer with emphasis on improving therapeutics. This study's findings, in relation to microbial SCFAs, suggest further benefit in investigating how this impacts host immunity, and whether additional targets exist to improve anti-PD1 cancer treatment.

Additionally, Jiang et al. [73] developed a metaproteomic technique alongside their electrophoretic mass spectrometry to investigate the oral microbiome in lung cancer. With this analysis, findings showed an increased abundance of both *Fusobacterium nucleatum* and *Prevotella histicola* within the oral cavity. Findings also showed both dysregulated maintenance of cell redox homeostasis and inhibition of the immune system within disease. Such microbes, among others, were found in association with lung cancer, with metaproteomics being used to reveal their altered functions, such as reproduction and energy metabolism. In particular, this study [73] emphasized the use of this method as a strategy to improve the use of metaproteomics for microbial analysis, particularly in relation to cancer, but with wider implications to other pathologies as well.

Currently used for discovery of novel therapeutics, lipidomics is utilized for drug development alongside overcoming resistance to drug treatments. Lipidomics has been used to find lipid-based biomarkers for cancer. An example of such is non-small cell lung cancer where lipid biomarkers can be used in a specific and sensitive manner. Relating to microbes, lipid rafts as membrane microdomains can be identified, which are key to microbial life cycles, enabling microorganisms to induce inflammation and colonize [74]. Therefore, lipidomics could be used to analyze which microbes are present during carcinogenesis for targeted therapies and prevention.

With the above in mind, microbiome–host interactions within cancer are also an exciting and growing field of study, regarding therapeutic treatments, restoration of dysbiosis, and improving outcomes to cancer drugs. For example, evidence shows that one of the most fatal and prevalent cancers, CRC, can be targeted by modulating the gut microbiota for prevention and improved prognosis. Fecal microbiota transplant is one such technique, studied in relation to CRC [75]. Metagenomic analysis by Yu et al. [76] highlighted several species, such as *Fusobacterium nucleatum and Parvimonas micra*, that are involved in colonic carcinogenesis through stimulating β-catenin signaling and the infiltration of host immune cells [76]. Therefore, fecal transplantation of microbiota from healthy individuals may increase levels of health-promoting microorganisms and restore homeostasis. While the literature is still emerging for CRC, benefits have already been documented in treating *Clostridioides difficile* infections by microbiota replacement [77].

Probiotics have also been part of the treatment for CRC, through modifying the gut microbiome in patients. Probiotics have been found to inhibit harmful enzymes, degrade mutagens, promote more beneficial microbes, and maintain homeostasis in the colon. Species such as *Lachnospiraceae species* and *Bifidobacterium animalis* have been identified as protective against colorectal cancer and even inducing tumor cell death. Moreover, prebiotics, dietary interventions, postbiotics, and microbial metabolites as nanotechnology are each novel microbial-related methods to target CRC [75, 78]. Further research is required to build upon these novel techniques for targeting microbes within cancer or treating cancers with microorganisms to ensure clinical applications are effective. However, this is a promising area of work in which MPS-fueled meta-omics plays a crucial role to facilitate its progression.

2.2 MPS Limitations in Oncology

As brilliant as MPS technologies have been for the elucidation of cancer biology, they do not come without their drawbacks. Limitations of NGS use in oncology can be due to the nature of the technology or of cancer genetics itself. One simple limitation is that NGS technologies do not work well on formalin-fixed paraffin-embedded tissues that are several years old, as they are prone to DNA substitution artifacts during sequencing [79, 80]. Second, given the high-throughput nature of its experimentation and data, MPS pipelines are much more computationally intensive than first-generation sequencing. Handling, analyzing, storing, and securing the volume of data produced by NGS all contribute to cost in ways largely unneeded by FGS. However, it should be noted that the consumables of an experiment tend to be the most expensive contributor to cost, not the price of data management and analysis [15, 81].

NGS experiments, especially those that do not target candidate regions but provide genome- or exome-wide coverage, may also lead to the discovery of incidental, yet potentially actionable findings. "Discoveries" of variants not related to an individual's known or screened pathologies can pose a challenge to scientists and clinicians [82]. Whether or not to divulge such information to individuals or their relatives becomes an issue of privacy and confidentiality. Furthermore, limitations also arise as an issue of cancer genetics. Since cancerous mutations are usually somatic rather than germline, their low allelic frequencies may be hard for default NGS workflows to differentiate them from experimental artifacts [83]. This may be further exacerbated by heterogeneity and mutation profile of tumor cells [84]. Yet, it must be acknowledged that known driver mutations may also be present in normal tissues, potentially complicating cancer and matched-normal tissue comparisons [79]. Certain structural variants in cancer genomes may be harder to identify accurately as well. Changes in highly repetitive regions like microsatellites or large indels are examples [83, 85, 86]. However, constantly improving experimentation and workflow pipelines have increased NGS accuracy in this regard [86]. Additionally, longer reads from third-generation or single-molecule sequencing show high potential as a solution to this limitation [82].

3 Conclusions

Certainly, the contribution of massively parallel and next-generation sequencing should not be taken for granted. The past half century has seen an explosion of knowledge created by the ability to sequence multitudinous biological macromolecules at the same time, accurately, and quickly. Unsurprisingly, the field of oncological research has not been ignored by such technological advances. Though proteomic methods lag behind other "omics" technologies in terms of massively parallel interrogation, this has not precluded the co-development of proteomics- and NGS-informed oncological research. Multi-omics data has elucidated oncogenesis and cancer progression in ways, perhaps imaginable yet simply impossible without MPS technologies. The resultant large biological and clinical datasets permit the asking and answering of questions of large significance and impact. Indeed, such elucidations have great implications for precision cancer medicine or "personalized" cancer care.

References

1. Stratton MR, Campbell PJ, Futreal PA (2009) The cancer genome. Nature 458(7239): 719–724. https://doi.org/10.1038/nature07943
2. Meyerson M, Gabriel S, Getz G (2010) Advances in understanding cancer genomes through second-generation sequencing. Nat Rev Genet 11(10):685–696. https://doi.org/10.1038/nrg2841
3. Shendure J, Balasubramanian S, Church GM, Gilbert W, Rogers J, Schloss JA et al (2017) DNA sequencing at 40: past, present and future. Nature 550(7676):345–353. https://doi.org/10.1038/nature24286
4. Heather JM, Chain B (2016) The sequence of sequencers: the history of sequencing DNA. Genomics 107(1):1–8. https://doi.org/10.1016/j.ygeno.2015.11.003

5. Sanger F, Nicklen S, Coulson AR (1977) DNA sequencing with chain-terminating inhibitors. Proc Natl Acad Sci USA 74(12):5463–5467. https://doi.org/10.1073/pnas.74.12.5463
6. Maxam AM, Gilbert W (1977) A new method for sequencing DNA. Proc Natl Acad Sci USA 74(2):560–564. https://doi.org/10.1073/pnas.74.2.560
7. Illumina: Human Whole-Genome Sequencing: Unbiased, accurate coverage of the human genome. https://emea.illumina.com/techniques/sequencing/dna-sequencing/whole-genome-sequencing/human.html (2020). Accessed 5 May 2021
8. Venter JC, Adams MD, Myers EW, Li PW, Mural RJ, Sutton GG et al (2001) The sequence of the human genome. Science 291(5507):1304–1351. https://doi.org/10.1126/science.1058040
9. Lander ES, Linton LM, Birren B, Nusbaum C, Zody MC, Baldwin J et al (2001) Initial sequencing and analysis of the human genome. Nature 409(6822):860–921. https://doi.org/10.1038/35057062
10. Gilbert W, Maxam A (1973) The nucleotide sequence of the lac operator. Proc Natl Acad Sci USA 70(12):3581–3584. https://doi.org/10.1073/pnas.70.12.3581
11. Margulies M, Egholm M, Altman WE, Attiya S, Bader JS, Bemben LA et al (2005) Genome sequencing in microfabricated high-density picolitre reactors. Nature 437(7057):376–380. https://doi.org/10.1038/nature03959
12. Fedurco M, Romieu A, Williams S, Lawrence I, Turcatti G (2006) BTA, a novel reagent for DNA attachment on glass and efficient generation of solid-phase amplified DNA colonies. Nucleic Acids Res 34(3):e22. https://doi.org/10.1093/nar/gnj023
13. Benz S, Mitra S (2023) From genomics to metagenomics in the era of recent sequencing technologies. In: Metagenomic data analysis. Springer, pp 1–20
14. Christofyllakis K, Bittenbring JT, Thurner L, Ahlgrimm M, Stilgenbauer S, Bewarder M et al (2022) Cost-effectiveness of precision cancer medicine-current challenges in the use of next generation sequencing for comprehensive tumour genomic profiling and the role of clinical utility frameworks. Mol Clin Oncol 16(1):1–4
15. Schwarze K, Buchanan J, Fermont JM, Dreau H, Tilley MW, Taylor JM et al (2020) The complete costs of genome sequencing: a microcosting study in cancer and rare diseases from a single center in the United Kingdom. Genet Med 22(1):85–94. https://doi.org/10.1038/s41436-019-0618-7
16. Bosse Y, Amos CI (2018) A decade of GWAS results in lung cancer. Cancer Epidemiol Biomarkers Prev 27(4):363–379. https://doi.org/10.1158/1055-9965.EPI-16-0794
17. Nachman MW, Crowell SL (2000) Estimate of the mutation rate per nucleotide in humans. Genetics 156(1):297–304
18. Roach JC, Glusman G, Smit AF, Huff CD, Hubley R, Shannon PT et al (2010) Analysis of genetic inheritance in a family quartet by whole-genome sequencing. Science 328(5978):636–639. https://doi.org/10.1126/science.1186802
19. Xue Y, Wang Q, Long Q, Ng BL, Swerdlow H, Burton J et al (2009) Human Y chromosome base-substitution mutation rate measured by direct sequencing in a deep-rooting pedigree. Curr Biol 19(17):1453–1457. https://doi.org/10.1016/j.cub.2009.07.032
20. Harris K, Pritchard JK (2017) Rapid evolution of the human mutation spectrum. Elife 6:e24284. https://doi.org/10.7554/eLife.24284
21. Ding L, Bailey MH, Porta-Pardo E, Thorsson V, Colaprico A, Bertrand D et al (2018) Perspective on oncogenic processes at the end of the beginning of cancer genomics. Cell 173(2):305–20 e10. https://doi.org/10.1016/j.cell.2018.03.033
22. Mortazavi A, Williams BA, McCue K, Schaeffer L, Wold B (2008) Mapping and quantifying mammalian transcriptomes by RNA-Seq. Nat Methods 5(7):621–628. https://doi.org/10.1038/nmeth.1226
23. Kalisky T, Oriel S, Bar-Lev TH, Ben-Haim N, Trink A, Wineberg Y et al (2018) A brief review of single-cell transcriptomic technologies. Brief Funct Genomics 17(1):64–76. https://doi.org/10.1093/bfgp/elx019
24. Maher CA, Kumar-Sinha C, Cao X, Kalyana-Sundaram S, Han B, Jing X et al (2009) Transcriptome sequencing to detect gene fusions in cancer. Nature 458(7234):97–101. https://doi.org/10.1038/nature07638
25. Regev A, Teichmann SA, Lander ES, Amit I, Benoist C, Birney E et al (2017) The human cell atlas. Elife 6:6. https://doi.org/10.7554/eLife.27041
26. Weinstein JN, Collisson EA, Mills GB, Shaw KR, Ozenberger BA, Ellrott K et al (2013) The cancer genome atlas pan-cancer analysis project. Nat Genet 45(10):1113–1120. https://doi.org/10.1038/ng.2764
27. Xu M, Li Y, Li W, Zhao Q, Zhang Q, Le K et al (2020) Immune and stroma related genes in

breast cancer: a comprehensive analysis of tumor microenvironment based on the cancer genome atlas (TCGA) database. Front Med (Lausanne) 7:64. https://doi.org/10.3389/fmed.2020.00064

28. Collins BC, Aebersold R (2018) Proteomics goes parallel. Nat Biotechnol 36(11): 1051–1053. https://doi.org/10.1038/nbt.4288
29. Tang L (2018) Next-generation peptide sequencing. Nat Methods 15(12):997. https://doi.org/10.1038/s41592-018-0240-7
30. Schlichtemeier SM, Nahm CB, Xue A, Gill AJ, Smith RC, Hugh TJ (2019) SELDI-TOF MS analysis of hepatocellular carcinoma in an Australian cohort. J Surg Res 238:127–136. https://doi.org/10.1016/j.jss.2019.01.008
31. Yigitbasi T, Calibasi-Kocal G, Buyukuslu N, Atahan MK, Kupeli H, Yigit S et al (2018) An efficient biomarker panel for diagnosis of breast cancer using surface-enhanced laser desorption ionization time-of-flight mass spectrometry. Biomed Rep 8(3):269–274. https://doi.org/10.3892/br.2018.1042
32. Tan HT, Lee YH, Chung MC (2012) Cancer proteomics. Mass Spectrom Rev 31(5): 583–605. https://doi.org/10.1002/mas.20356
33. Swaminathan J, Boulgakov AA, Hernandez ET, Bardo AM, Bachman JL, Marotta J et al (2018) Highly parallel single-molecule identification of proteins in zeptomole-scale mixtures. Nat Biotechnol 36:1076. https://doi.org/10.1038/nbt.4278
34. Swaminathan J, Boulgakov AA, Marcotte EM (2015) A theoretical justification for single molecule peptide sequencing. PLoS Comput Biol 11(2):e1004080. https://doi.org/10.1371/journal.pcbi.1004080
35. Matreyek KA, Starita LM, Stephany JJ, Martin B, Chiasson MA, Gray VE et al (2018) Multiplex assessment of protein variant abundance by massively parallel sequencing. Nat Genet 50(6):874–882. https://doi.org/10.1038/s41588-018-0122-z
36. Maier T, Guell M, Serrano L (2009) Correlation of mRNA and protein in complex biological samples. FEBS Lett 583(24): 3966–3973. https://doi.org/10.1016/j.febslet.2009.10.036
37. Vogelstein B, Papadopoulos N, Velculescu VE, Zhou S, Diaz LA Jr, Kinzler KW (2013) Cancer genome landscapes. Science 339(6127): 1546–1558. https://doi.org/10.1126/science.1235122
38. van de Vijver MJ, Peterse JL, Mooi WJ, Wisman P, Lomans J, Dalesio O et al (1988) Neu-protein overexpression in breast cancer. Association with comedo-type ductal carcinoma in situ and limited prognostic value in stage II breast cancer. N Engl J Med 319(19): 1239–1245. https://doi.org/10.1056/NEJM198811103191902
39. Hanahan D, Weinberg RA (2011) Hallmarks of cancer: the next generation. Cell 144(5): 646–674. https://doi.org/10.1016/j.cell.2011.02.013
40. King MC, Marks JH, Mandell JB, New York Breast Cancer Study Group (2003) Breast and ovarian cancer risks due to inherited mutations in BRCA1 and BRCA2. Science 302(5645): 643–646. https://doi.org/10.1126/science.1088759
41. Qing T, Mohsen H, Marczyk M, Ye Y, O'Meara T, Zhao H et al (2020) Germline variant burden in cancer genes correlates with age at diagnosis and somatic mutation burden. Nat Commun 11(1):2438. https://doi.org/10.1038/s41467-020-16293-7
42. Banerji S, Cibulskis K, Rangel-Escareno C, Brown KK, Carter SL, Frederick AM et al (2012) Sequence analysis of mutations and translocations across breast cancer subtypes. Nature 486(7403):405–409. https://doi.org/10.1038/nature11154
43. Black JRM, McGranahan N (2021) Genetic and non-genetic clonal diversity in cancer evolution. Nat Rev Cancer 21:379. https://doi.org/10.1038/s41568-021-00336-2
44. Flavahan WA, Gaskell E, Bernstein BE (2017) Epigenetic plasticity and the hallmarks of cancer. Science 357(6348). https://doi.org/10.1126/science.aal2380
45. Bateman NW, Conrads TP (2018) Recent advances and opportunities in proteomic analyses of tumour heterogeneity. J Pathol 244(5): 628–637. https://doi.org/10.1002/path.5036
46. Clark SJ, Lee HJ, Smallwood SA, Kelsey G, Reik W (2016) Single-cell epigenomics: powerful new methods for understanding gene regulation and cell identity. Genome Biol 17:72. https://doi.org/10.1186/s13059-016-0944-x
47. Ku CS, Naidoo N, Wu M, Soong R (2011) Studying the epigenome using next generation sequencing. J Med Genet 48(11):721–730. https://doi.org/10.1136/jmedgenet-2011-100242
48. Loeb LA, Harris CC (2008) Advances in chemical carcinogenesis: a historical review and

prospective. Cancer Res 68(17):6863–6872. https://doi.org/10.1158/0008-5472.CAN-08-2852

49. Rowley JD (1973) Letter: a new consistent chromosomal abnormality in chronic myelogenous leukaemia identified by quinacrine fluorescence and Giemsa staining. Nature 243(5405):290–293. https://doi.org/10.1038/243290a0
50. Kurzrock R, Gutterman JU, Talpaz M (1988) The molecular genetics of Philadelphia chromosome-positive leukemias. N Engl J Med 319(15):990–998. https://doi.org/10.1056/NEJM198810133191506
51. National Cancer Institute: The Cancer Genome Atlas Program (2021). https://www.cancer.gov/about-nci/organization/ccg/research/structural-genomics/tcga. Accessed 10 May 2021
52. Kruger R (2018) Charting a course to a cure. Cell 173(2):277. https://doi.org/10.1016/j.cell.2018.03.048
53. Hoadley KA, Yau C, Hinoue T, Wolf DM, Lazar AJ, Drill E et al (2018) Cell-of-origin patterns dominate the molecular classification of 10,000 tumors from 33 types of cancer. Cell 173(2):291–304. e6
54. The Cancer Genome Project: The Cancer Genome Project (2021). https://www.sanger.ac.uk/group/cancer-genome-project/. Accessed 10 May 2021
55. Tate JG, Bamford S, Jubb HC, Sondka Z, Beare DM, Bindal N et al (2019) COSMIC: the catalogue of somatic mutations in cancer. Nucleic Acids Res 47(D1):D941–D9D7. https://doi.org/10.1093/nar/gky1015
56. Forbes SA, Bindal N, Bamford S, Cole C, Kok CY, Beare D et al (2011) COSMIC: mining complete cancer genomes in the catalogue of somatic mutations in cancer. Nucleic Acids Res 39(Database issue):D945–D950. https://doi.org/10.1093/nar/gkq929
57. Gagan J, Van Allen EM (2015) Next-generation sequencing to guide cancer therapy. Genome Med 7(1):80. https://doi.org/10.1186/s13073-015-0203-x
58. Palmieri M, Baldassarri M, Fava F, Fabbiani A, Gelli E, Tita R et al (2020) Two-point-NGS analysis of cancer genes in cell-free DNA of metastatic cancer patients. Cancer Med 9(6):2052–2061. https://doi.org/10.1002/cam4.2782
59. Jahani S, Nazeri E, Majidzadeh AK, Jahani M, Esmaeili R (2020) Circular RNA; a new biomarker for breast cancer: a systematic review. J Cell Physiol 235(7–8):5501–5510. https://doi.org/10.1002/jcp.29558
60. Davis RT, Blake K, Ma D, Gabra MBI, Hernandez GA, Phung AT et al (2020) Transcriptional diversity and bioenergetic shift in human breast cancer metastasis revealed by single-cell RNA sequencing. Nat Cell Biol 22(3):310–320. https://doi.org/10.1038/s41556-020-0477-0
61. Liberti MV, Locasale JW (2016) The Warburg effect: how does it benefit cancer cells? Trends Biochem Sci 41(3):211–218. https://doi.org/10.1016/j.tibs.2015.12.001
62. Sepich-Poore GD, Zitvogel L, Straussman R, Hasty J, Wargo JA, Knight R (2021) The microbiome and human cancer. Science 371(6536):eabc4552
63. Kadosh E, Snir-Alkalay I, Venkatachalam A, May S, Lasry A, Elyada E et al (2020) The gut microbiome switches mutant p53 from tumour-suppressive to oncogenic. Nature 586(7827):133–138
64. Zhang X, Li C, Cao W, Zhang Z (2021) Alterations of gastric microbiota in gastric cancer and precancerous stages. Front Cell Infect Microbiol 11:559148
65. Tong Y, Gao H, Qi Q, Liu X, Li J, Gao J et al (2021) High fat diet, gut microbiome and gastrointestinal cancer. Theranostics 11(12):5889
66. Bisanz JE, Upadhyay V, Turnbaugh JA, Ly K, Turnbaugh PJ (2019) Meta-analysis reveals reproducible gut microbiome alterations in response to a high-fat diet. Cell Host Microbe 26(2):265–72. e4
67. Mitra S, Rupek P, Richter DC, Urich T, Gilbert JA, Meyer F et al (2011) Functional analysis of metagenomes and metatranscriptomes using SEED and KEGG. BMC Bioinf 12(1):1–8
68. Cullin N, Antunes CA, Straussman R, Stein-Thoeringer CK, Elinav E (2021) Microbiome and cancer. Cancer Cell 39(10):1317–1341
69. Chang Y-S, Hsu M-H, Tu S-J, Yen J-C, Lee Y-T, Fang H-Y et al (2021) Metatranscriptomic analysis of human lung metagenomes from patients with lung cancer. Genes 12(9):1458
70. Banavar G, Ogundijo O, Toma R, Rajagopal S, Lim YK, Tang K et al (2021) The salivary metatranscriptome as an accurate diagnostic indicator of oral cancer. NPJ Genom Med 6(1):105
71. Yachida S, Mizutani S, Shiroma H, Shiba S, Nakajima T, Sakamoto T et al (2019) Metagenomic and metabolomic analyses reveal distinct stage-specific phenotypes of the gut microbiota in colorectal cancer. Nat Med 25(6):968–976
72. Nomura M, Nagatomo R, Doi K, Shimizu J, Baba K, Saito T et al (2020) Association of short-chain fatty acids in the gut microbiome with clinical response to treatment with

nivolumab or pembrolizumab in patients with solid cancer tumors. JAMA Netw Open 3(4): e202895-e

73. Jiang X, Zhang Y, Wang H, Wang Z, Hu S, Cao C et al (2022) In-depth metaproteomics analysis of oral microbiome for lung cancer. Research 2022:9781578

74. Yan F, Zhao H, Zeng Y (2018) Lipidomics: a promising cancer biomarker. Clin Transl Med 7:1–3

75. Saeed M, Shoaib A, Kandimalla R, Javed S, Almatroudi A, Gupta R et al (2022) Microbe-based therapies for colorectal cancer: advantages and limitations. In: Seminars in cancer biology. Elsevier, pp 652–665

76. Yu J, Feng Q, Wong SH, Zhang D, Liang QY, Qin Y et al (2017) Metagenomic analysis of faecal microbiome as a tool towards targeted non-invasive biomarkers for colorectal cancer. Gut 66(1):70–78

77. Choi HH, Cho Y-S (2016) Fecal microbiota transplantation: current applications, effectiveness, and future perspectives. Clinical Endos 49(3):257–265

78. Gao Z-Y, Cui Z, Yan Y-Q, Ning L-J, Wang Z-H, Hong J (2021) Microbe-based management for colorectal cancer. Chin Med J 134(24):2922–2930

79. Colomer R, Mondejar R, Romero-Laorden N, Alfranca A, Sanchez-Madrid F, Quintela-Fandino M (2020) When should we order a next generation sequencing test in a patient with cancer? EClinicalMedicine 25:100487. https://doi.org/10.1016/j.eclinm.2020.100487

80. Salk JJ, Schmitt MW, Loeb LA (2018) Enhancing the accuracy of next-generation sequencing for detecting rare and subclonal mutations. Nat Rev Genet 19(5):269–285. https://doi.org/10.1038/nrg.2017.117

81. Gordon LG, White NM, Elliott TM, Nones K, Beckhouse AG, Rodriguez-Acevedo AJ et al (2020) Estimating the costs of genomic sequencing in cancer control. BMC Health Serv Res 20(1):492. https://doi.org/10.1186/s12913-020-05318-y

82. Xue Y, Ankala A, Wilcox WR, Hegde MR (2015) Solving the molecular diagnostic testing conundrum for Mendelian disorders in the era of next-generation sequencing: single-gene, gene panel, or exome/genome sequencing. Genet Med 17(6):444–451. https://doi.org/10.1038/gim.2014.122

83. Di Resta C, Ferrari M (2018) Next generation sequencing: from research area to clinical practice. EJIFCC 29(3):215–220

84. Basho RK, AKJAJoHO E (2015) Clinical applications and limitations of next-generation sequencing. 11(3)

85. Zavodna M, Bagshaw A, Brauning R, Gemmell NJ (2014) The accuracy, feasibility and challenges of sequencing short tandem repeats using next-generation sequencing platforms. PLoS One 9(12):e113862. https://doi.org/10.1371/journal.pone.0113862

86. Daber R, Sukhadia S, Morrissette JJ (2013) Understanding the limitations of next generation sequencing informatics, an approach to clinical pipeline validation using artificial data sets. Cancer Genet 206(12):441–448. https://doi.org/10.1016/j.cancergen.2013.11.005

Chapter 16

The Use of Next-Generation Sequencing in Personalized Medicine

Liya Popova and Valerie J. Carabetta

Abstract

The revolutionary progress in development of next-generation sequencing (NGS) technologies has made it possible to deliver accurate genomic information in a timely manner. Over the past several years, NGS has transformed biomedical and clinical research and found its application in the field of personalized medicine. Here we discuss the rise of personalized medicine and the history of NGS. We discuss current applications and uses of NGS in medicine, including infectious diseases, oncology, genomic medicine, and dermatology. We provide a brief discussion of selected studies where NGS was used to respond to wide variety of questions in biomedical research and clinical medicine. Finally, we discuss the challenges of implementing NGS into routine clinical use.

Key words High-throughput sequencing, Massively parallel sequencing, Next-generation sequencing, Ultra-high-throughput sequencing, Second generation sequencing, Deep sequencing, Personalized medicine

1 Introduction

1.1 The Rise of Genomic Medicine

The variability of individual genomes is responsible for different phenotypic traits, but it also can be more complex in determining predisposition to certain diseases [1]. According to the National Human Genome Research Institute, "genomic medicine" is a fast-growing field that involves the use of an individual's genomic information as part of their care, such as for diagnostic or therapeutic decision-making. From a genetics perspective, the majority of diseases have genetic components and can be divided into three categories: (1) Chromosomal, with deletion or addition of entire chromosomes, (2) single-gene (Mendelian) diseases, inherited as familial autosomal or X-linked traits, and (3) complex diseases, which are the results of interactions between certain genetic traits and environmental factors [2]. Since the completion of the Human Genome Project, there has been shift toward pharmacogenomics as

Valerie J. Carabetta and Olaitan Akintunde (eds.), *High Throughput Gene Screening: Methods and Protocols*, Methods in Molecular Biology, vol. 2866, https://doi.org/10.1007/978-1-0716-4192-7_16,

a key part of personalized medicine to differentiate responders and non-responders to certain treatments. Integration of genomic medicine sets foot into new stage of clinical medicine implementing genomic-based diagnosis and personalized assessment and treatment [3]. Nowadays, more and more physicians use NGS testing, which allows for prediction of disease predisposition, prognosis, and susceptibility to treatment. Advances in sequencing technology allowed for marked decreases in costs of genetic testing, from the three billion dollars cost of the Human Genome Project in 2004. Today, whole exome and genome sequencing are performed within a clinically useful timeframe and acceptable costs, making the transition from clinical research trials to availability as standard clinical tests outside of global genomic study initiatives [4]. One important application of sequencing technology is in the cancer genomics field and has led to the identification of previously unknown pathogenic mutations. For example, mutations in *BMPR1A* and *SMAD4* have now been linked to colon cancer, and *BRCA1* and *BRCA2* mutations were shown to be associated with early breast and ovarian cancer. These cancers can be prevented or managed by timely detection of mutations and identification of common genetic variants. This allows for the stratification into high and low risk categories for cancer development and the implementation of appropriate clinical interventions, such as early screening for colon cancer, mastectomy, and oophorectomy [5].

Fast development of gene therapy and genome editing platforms during the past decade have given hope for the treatment of some cancers and Mendelian disorders. For example, chimeric antigen receptor (CAR)-T cell immunotherapy, which is based on genetic engineering of T-cells, has demonstrated efficacy in treating hematologic malignancies and solid tumors, such as B-cell leukemia and lymphoma [6]. Another example is the two most common monogenic disorders, sickle cell disease and β-thalassemia, which could potentially be cured by autologous hematopoietic stem cell transplantation after genetic modification and increasing the expression of non-sickling β-globin or anti-sickling β-like globins. This would lead to restoration of functional, nonpathogenic hemoglobin-containing red blood cells [7]. The recent approval of cystic fibrosis transmembrane conductance regulator (*CFTR*) mutation correctors, such as combinations of ivacaftor, lumacaftor, elexacaftor, and tezacaftor, by the Food and Drug Administration (FDA) for the treatment of cystic fibrosis in children with homozygous F508del mutations is an exciting example of gene therapy transitioning from an idea to extensive research to full clinical application. This class of medicines works stabilizing the mutated CFTR, thereby increasing production of specific ion channels and receptors, improving the lung function by decreasing the build-up of thick mucus. The F508del mutation is the most common and this treatment can potentially help 70% of patients suffering from cystic fibrosis [8].

Undoubtedly, the development of genomic medicine will continue to impact clinical medicine as the technology advances, setting high hopes for future therapies and possible cures. It is already clear that to reach its full potential, we need to expand the investment in basic science and foundational research. NGS and genome editing have already led to some major breakthroughs, but there remains a lot of work to be done in the identification of every gene and their variants function, and understanding their role in development of disease [5].

1.2 A Brief History of NGS

The development of sequencing technology began in the late 1970s, with the introduction of two techniques. Maxam-Gilbert sequencing, also known as the partial chemical degradation method, involves chemical cleavage of DNA molecules that are resolved by electrophoresis and bands are visualized using radioactivity [9]. The other, more popular technique is Sanger sequencing. This approach is based on DNA polymerase inserting chain-terminating dideoxynucleotides (ddNTPs) into the growing chain. Incorporation of a ddNTP stops the elongation reaction, resulting in the formation of various length extension products, which can be detected by electrophoresis [10, 11]. However, significant technological advances occurred rapidly since then and revolutionized Sanger sequencing. The Sanger method was further improved by integration of polymerase chain reaction (PCR) and the use of fluorophores and capillary gel electrophoresis. These techniques were used for the first sequence of the human genome, completed in 2003 as part of the human genome project [12].

The era of NGS, also called second-generation sequencing, began in 2005 with a pyrosequencing method that allowed the achievement of a 100-fold increase in throughput (Fig. 1). Pyrosequencing is a technique where a DNA fragment is attached to a bead in an emulsion and luminescence is used to detect

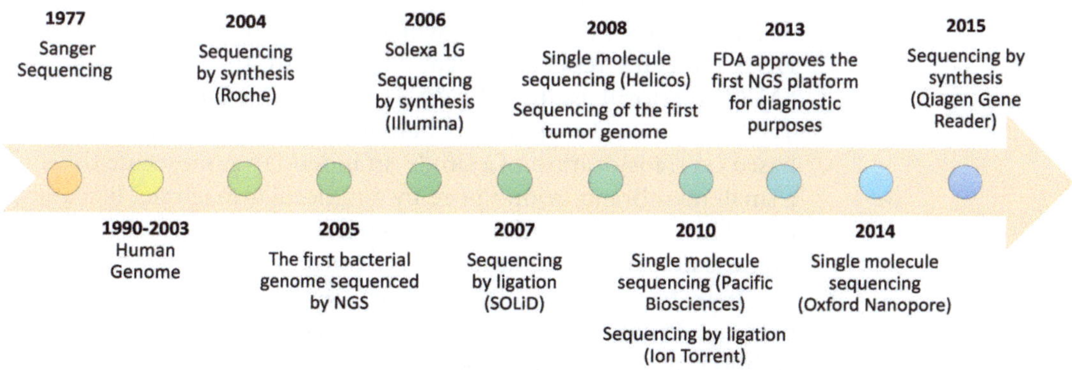

Fig. 1 Timeline of commercial NGS platforms. A brief history of the instruments introduced in the market between 2004 and 2015 and developmental milestones. NGS; next-generation sequencing

pyrophosphate (PP_i) release as a nucleotide is incorporated [13]. This sequencing by synthesis (SBS) platform was commercialized by 454 Life Sciences (Roche), but production was discontinued only 10 years later [12, 14]. A variation in pyrosequencing was created by Ion Torrent Systems (Life Technologies), which produced a Personal Genome Machine (PGM) sequencer that detects protons released during the reactions (pH changes) instead of fluorescence. This technique allowed for a significant reduction in costs and increased speed of sequencing [15]. In 2006, Solexa (Illumina) produced the Genome Analyzer, a new platform based on cyclic-reversible termination of the polymerization reaction by use of fluorescently labeled 3′-O-azidomethyl-dNTPs. The reaction is "cyclic-reversible" because during the reaction, it temporarily terminates, allowing the removal of unincorporated bases, so that the added nucleotides yield a fluorescent signal [15]. One year later, the sequencing by ligation (SBL) platform was launched by Applied Biosystems (Life Technologies), known as the SOLiD (Small Oligonucleotide Ligation and Detection) system. As the name implies, the ligase-mediated technique determines sequence based on ligation, not nucleotide addition [12, 16]. These technologies paved the way for further innovations and improvements that could ultimately lead to personalized medicine in clinical practice.

The next significant breakthrough in NGS was single-molecule real-time (SMRT) sequencing, also known as third-generation sequencing, from Pacific Biosciences [17]. Although single-molecule sequencing introduces a possible difficulty in detecting a signal from the reaction, SMRT can achieve high precision, with a medial accuracy of 99.3% after multiple reads [18]. The downside of single-pass sequence reads is the relatively high error rate (11%–15%), which can be reduced by a sufficient number of sequencing passes to achieve >99% accuracy, or by combining different continuous long reads from the same location [18–20]. The principle behind the SMRT technique is based on the usage of a strand-displacing DNA polymerase, which allows sequencing of the same DNA molecule repeatedly [15, 21, 22]. Another commonly used SMRT platform is the MinION, released by Oxford Nanopore Technologies. The approach behind the nanopore technology is based on translocation of a single-stranded DNA molecule through a small membrane-bound pore by a molecular machine, like a helicase. During translocation, the electrical potential across the membrane is detected. The MinION is a portable sequencer that allows the user to generate high-accuracy sequencing data in a short period of time [15, 23, 24]. These advances in DNA sequencing technology led to an exponential growth in performance and increase in cost-efficiency, which enabled the revolution in genomic medicine and has made the possibility of personalized medicine achievable.

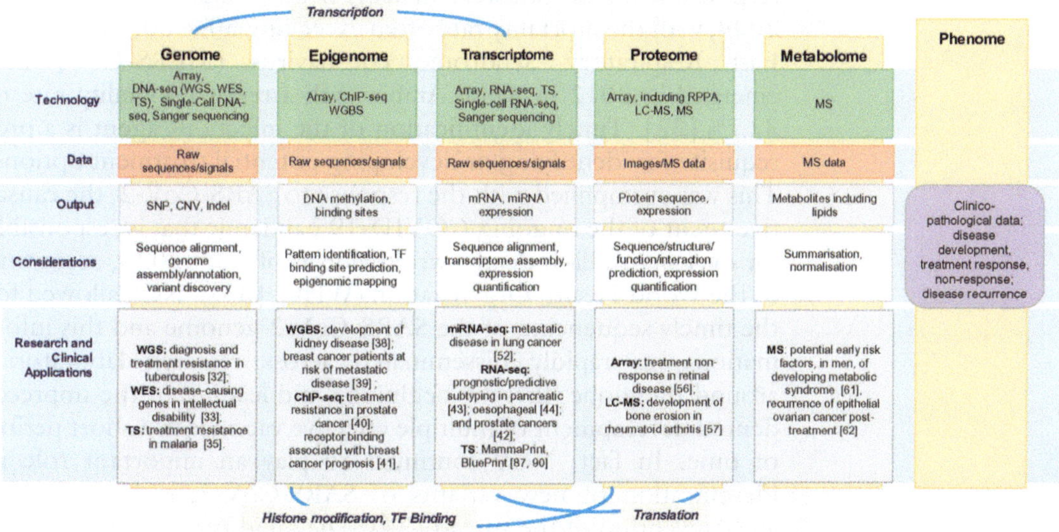

Fig. 2 Schematic representation of five "omics" levels. The genome, epigenome, transcriptome, proteome, and metabolome are represented, with technologies, types of data, data outputs, and important applications summarized. These five levels together comprise the patient characteristics or phenome. Important connections between the different levels are also represented, including transcription, histone modification, and transcription factor (TF)-binding, and translation. Reproduced from [129]. Copyright, The Authors, 2018. Published by Oxford University Press. Distributed under the terms of the Creative Commons Attribution License (http://creativecommons.org/licenses/by/4.0/)

In recent years, exponential advances in sequencing technologies, accessibility, and decreasing costs have led to the beginning of large-scale projects in personalized medicine and translational biomedical research of clinical genomics (Fig. 2). As a tool to detect different biomarkers, NGS has possible applications in most areas of personalized medicine, from infectious diseases to dermatology. Here we will discuss the exciting aspects of the multidisciplinary applications of NGS technology in personalized medicine.

2 The Use of NGS for the Management of Infectious Diseases

2.1 NGS as a Diagnostic Tool

The failure of current diagnostic tests to detect the etiology of infectious processes is not uncommon [25–29]. Timely diagnosis and treatment of a pathogen can be crucial for survival and prevention of transmission. Emerging pathogens represent a large threat to public health in the era of globalization and rapid migration between countries and continents. The application of NGS allows for accurate detection of both existing and novel pathogens in a timely manner. For example, NGS technologies were used to characterize the genomes and for the detection of Bas-Congo rhabdovirus, MERS-CoV, and SARS-CoV-2. Bas-Congo rhabdovirus was responsible for the outbreak of acute hemorrhagic fever in Africa in 2009, with the mortality rate often exceeding 50% [30, 31]. Middle East Respiratory Syndrome coronavirus (MERS-CoV) first emerged in 2012 in Saudi Arabia with a reported fatality rate of 32.7% [32]. Timely identification of the infectious agent is a prerequisite for identifying or developing potential treatment options. This was exemplified with the response to SARS-CoV-2, the causative agent of the ongoing COVID-19 pandemic that is responsible for over 6.9 million deaths worldwide as of April 2023, according to the World Health Organization (WHO, [33]). NGS allowed for the timely sequencing of the SARS-CoV-2 genome and this information was rapidly disseminated across research laboratories around the globe, thereby beginning and leading to the unprecedented development of multiple effective vaccines in a short period of time. In fact, NGS continues to play an important role in identification of new variants of SARS-CoV-2, early detection, and prevention of the spread [34]. The rapid turnaround and cost effectiveness of NGS has made it possible to rapidly respond to global emergencies.

2.2 NGS Contributions to Epidemiology of Infectious Agents

NGS is also beneficial towards the management of infectious diseases on a smaller scale. Practical application of high-resolution sequencing for clinical epidemiology has allowed for deep analysis of mutants, transmission pathways, and evolution of bacterial populations. For example, NGS was used to study outbreaks of *Staphylococcus aureus* in hospital settings and its spread to the community [35]. In one study of an outbreak of methicillin-resistant *S. aureus* (MRSA) in the neonatal intensive care unit, whole-genome sequencing of 14 isolates and subsequent reconstruction of a phylogenic tree revealed that outbreak-related isolates were resistant to gentamicin and mupirocin. This was unusual for MRSA strains across the United Kingdom and differed from the typical antibiogram pattern of resistance to cefoxitin, erythromycin, and ciprofloxacin. Köser et al. performed whole genome sequencing to create an evolutionary tree of outbreak and non-outbreak strains, which revealed clusters based on antibiotic resistance and toxins [36]. This information could have been used to inform treatment decisions and contain the outbreak faster.

In 2010, in Haiti, whole-genome sequencing was used to detect the source of an epidemic of *Vibrio cholerae* O1, given that there was no documented evidence of any previous outbreaks on the island. Katz et al. used NGS to discover that the strains were derived from a common ancestor, and a later epidemiological analysis showed that the outbreak originated from Nepal, where United Nations peacekeepers were working in unsanitary conditions [37]. Katz et al. further demonstrated that only de novo mutations were responsible for genetic polymorphisms that led to drug resistance in these clinical isolates, without a major contribution from horizontal gene transfer [38]. However, this is not a universal observation, as NGS analysis of *V. cholerae* strains originating from the Queensland waterway in Australia demonstrated that the waterway serves as a reservoir for virulence genes that can be transferred rapidly between highly-divergent strains via mobile genetic elements [39]. These studies demonstrate two powerful uses of NGS, in tracing the origins of outbreaks and in the understanding of how virulence genes spread among populations.

There was an outbreak of *Escherichia coli* O104:H4 that spread throughout Germany on contaminated bean sprouts in 2011. One of the clinical features of the infection was colonic ischemia, which is unusual for this type of infection, in addition to bloody diarrhea, with the complication of hemolytic-uremic syndrome. NGS allowed for the quick identification of the outbreak strain, which was enteroaggregative *E. coli* that had acquired Shiga-toxin 2 and antibiotic resistance genes that were responsible for the unusual clinical presentation [40]. A final important application of NGS towards our understanding of epidemiology of antibiotic-resistant species was in agriculture. NGS was used to demonstrate that the use of antibiotics in farm animals resulted in the occurrence of antibiotic-resistant strains. Furthermore, it was found that animal and poultry husbandry can be a potential reservoir for antibiotic-resistant human pathogens [41, 42], which was one of the first demonstrations of this phenomenon. Clearly, NGS has made valuable contributions to our understanding of epidemiology and its use will only continue to grow moving forward.

2.3 NGS and the Evolution of Drug Resistance

NGS is powerful for the study of selection targets or the dynamics of genome evolution that occur in a population over time. Applications for the study of evolution might include genome-wide association studies (GWAS) or transposon insertion sequencing (Tn-seq, [43, 44]). Whole-genome sequencing of *S. aureus* clinical isolates has aided in our understanding of the mutations that accumulate to drive the evolution of drug resistance in vivo [45]. In one study of cystic fibrosis patients, GWAS analysis revealed a high frequency of mutations in the alternate sigma factor SigB, isolated from *S. aureus* strains from chronically infected patient airways [46]. There seems to be an in vivo selective pressure to attenuate SigB activity, which is interesting because SigB regulates the general stress response and expression of virulence factors, like capsule formation and alfa-toxin production [47, 48]. SigB is also important for intracellular peristence, as *sigB* mutants poorly formed persisters, which are slow growing, drug-tolerant cells that exist in a population of genetically identical cells. Therefore, while examining the evolution of drug resistance, the NGS data also potentially opened the door for the design of new antibacterial therapeutics to control refractory infections [48]. In *Vibrio cholera,* the occurrence of strains resistant to commonly used antibiotics remains a large problem. Using NGS to study clinical isolates of *V. cholerae* in Laos, Iwanaga et al. determined the mechanism of antibiotic resistance. They showed that the collection of *V. cholerae* strains acquired the SXT constin element, an antibiotic resistance cassette encoding resistance to chloramphenicol, tetracycline, streptomycin, and sulfamethoxazole [49, 50]. With this information, it may be valuable to screen isolates by PCR for the presence of the SXT constin element, which could inform treatment strategy.

One controversial question about the acquisition of drug-resistance genes by *Mycobacterium tuberculosis* was answered using NGS. For a long time, it was believed that drug resistance can only emerge in patients with active tuberculosis (TB) disease, while bacteria in the latent state have little capacity for mutation since they are dormant or slowly growing. People with latent TB are infected with *M. tuberculosis*, but they do not have clinical symptoms and therefore cannot spread the disease to others. Someone with active TB disease has clinical symptoms and can spread *M. tuberculosis* to others. Because of this, latent infections are typically treated with one first-line antibiotic, isoniazid. The use of high-density, whole-genome sequencing allowed for the assessment of the capability of *M. tuberculosis* to acquire mutations and evolve over the course of infection. It was found that *M. tuberculosis* acquires mutations during latency despite the markedly low replication rate. Surprisingly, mutation rate and the number of chromosomal mutations observed during latency was like that observed in *M. tuberculosis* in the active state. This finding suggests that monotherapy for latent tuberculosis is a risk factor for emergence of resistance and there is a need to reconsider how latent TB infections are treated [51]. The use of NGS can improve and directly influence physician treatment choices.

2.4 Metagenomics

Another major challenge in management of infectious diseases is the detection of etiological pathogens from complex, polymicrobial patient specimens. In vivo, bacteria live in multispecies communities, and identifying the clinically relevant species is difficult. This is because standard laboratory growth techniques are not always sufficient for the cultivation of all the microbial species from a specimen. Clinically significant organisms might be slow-growing or get out-competed by other strains when co-cultured under artificial laboratory conditions [52]. The use of a metagenomic analysis allows for the analysis of these communities without prior culturing, which also improves turnaround time for results. It has been shown that deep sequencing of 16S rRNA, with appropriate downstream bioinformatic analysis yields high accuracy data and allows for the detection of bacterial species in heterogeneous clinical specimens that cannot be identified using conventional clinical laboratory methods [53]. This approach for the recognition of novel, disease-causing species has the potential to become a diagnostic tool for use in clinical practice. In addition, this type of analysis could be used for epidemiologic surveys and improvement of sequence databases. The largest advantage of metagenomic analysis is the rapid turnaround, as it removes the 16–24 h required for culturing, and it is relatively inexpensive, making it ideal for routine clinical use.

Metagenomics have also been used for the characterization of the human microbiome. The normal human microbiome is diverse and represents symbiotic relationships between host and microbes, like bacteria and fungi, that form a complex community. Disruption of normal microbiota has been linked to many diseases, including depression, Alzheimer's disease, autoimmune diseases, and coagulation disorders [54]. Metagenomic analysis has been widely applied to determine the composition and diversity of the gut, vaginal, oral, skin, and placental microbiome in health and disease (reviewed in [55–57]). Other valuable information gained from this type of study is the determination of essential regulators or factors that are necessary for interspecies or host interactions, and the distribution of antibiotic resistance genes or virulence genes among populations [55]. As soon as the essential regulatory genes and factors are identified, it will open the door for potential biomarker screening or therapeutic development. For example, these could lead to the identification of novel, pathogen-specific bacteriophages that could be used for the treatment of antibiotic-resistant bacterial pathogens [58, 59].

Periodontitis is a polymicrobial biofilm-induced disease that leads to dysbiosis of the oral microbiome, then to inflammation and bone loss [60, 61]. In order to determine the underlying mechanisms of dysbiosis, Yost et al. characterized gene expression patterns using a meta-transcriptomic analysis of the entire oral microbiome during the progression of periodontitis [62]. In samples with periodontal disease progression, analysis demonstrated changes in gene expression in major periodontal pathogens, including the upregulation of genes associated with transport, proteolysis, and the protein kinase C-activating G-protein coupled receptor signaling pathway. Comparison of follow-up non-progressive periodontal disease and baseline samples did not identify any differentially expressed genes, implying that clinically stable sites do not have changes in microbial gene expression. Additionally, specific pathogens associated with periodontitis showed individual patterns of genes expression. For example, flagellar biosynthesis genes were upregulated in *Treponema denticola*, and biotin synthesis genes and capsular polysaccharide biosynthesis were upregulated in *Porphyromonas gingivalis* [62]. This is one example of how metagenomics in combination with other NGS technologies could become a powerful tool to understand complex communication and relationships among polymicrobial communities.

2.5 NGS in Vaccine Design

One of the most important applications of NGS in clinical research might be new vaccine design. Most vaccines that helped to combat the severe infectious diseases of twentieth century were developed based on Pasteur's approach from 1885. This involved the isolation of the pathogenic microorganism, followed by inactivation, and finally injection (reviewed in [63]). Many vaccines that are still in use today were developed using this formula, including the killed-inactivated polio vaccine, and live attenuated vaccines against measles, mumps, and rubella (MMR), varicella zoster, and rotavirus. In subsequent years, new technologies became available, which led to considerable progress in vaccinology. One breakthrough was the development of recombinant vaccines, where surface antigens are expressed via plasmids in bacterial or yeast hosts for insertion into virus-like particles [64]. This avoided the possibility of reversion of the attenuated strains to wildtype and subsequent infection. The Hepatitis B (HBV), Human papilloma virus (HPV), and pertussis vaccine were all created using this technology.

The next step in the evolution of vaccine development is reverse vaccinology, or genome-based antigen discovery. This approach is largely based on data from NGS analysis of the pathogen of interest's genome, where the sequencing of genomes from a large number of strains, serotypes, or related species allows for the identification of broadly distributed, potential antigens for vaccine development. One benefit of NGS was the discovery of new antigen determinants in pathogens that can avoid host immune responses, such as *Streptococcus pneumoniae*, *Streptococcus agalactiae*, *S. aureus*, and *Chlamydia pneumoniae* [65, 66]. While promising antigens have been identified, these have not been developed into effective vaccines for practical use. Even though reverse vaccinology may be useful for some pathogens, its potential is limited for pathogens with highly variable surface antigens, like human immunodeficiency virus (HIV) or respiratory syncytial virus (RSV), for which structural vaccinology is a more promising approach [63]. Currently, the evolution of vaccinology has led to synthetic biology, that uses in vitro synthesis of DNA and RNA for vaccine development. One of the most well-known examples of synthetic biology is the mRNA vaccines developed against SARS-CoV-2. NGS was used to analyze the genome and transcriptome of the virus. The coding sequence of a specific spike protein was selected as a vaccine candidate. The target gene was inserted into plasmid, transcribed into an mRNA molecule, and packaged into liposomes to protect the mRNA [67, 68]. The success of these vaccines provides strong evidence that NGS can be used for future vaccine development.

3 NGS in Genomic Medicine

3.1 The Use of NGS to Study Single Gene Disorders

One of the most exciting clinical applications of NGS is for genomic medicine. Although it is a relatively new discipline, genomic medicine is already transforming therapeutic approaches and the management of patients with genetic diseases, especially single gene, Mendelian disorders, and certain cancers. Rapid testing of all genes related to a disease is particularly useful in patients with rare diseases, such as hereditary ataxia, amyotrophic lateral sclerosis [69], an unusual clinical course, or an atypical response to treatment [70]. In addition, for cases where whole exome sequencing (WES) is not successful or accurate, RNA-seq is being explored as an alternative diagnostic strategy for simple Mendelian disorders [71]. Since 2012, a global initiative brought together major institutions to support the opening of Centers for Mendelian Genomics, where the application of NGS technologies has been used for gene discovery [72]. For example, WES was used to identify a mutation in the dihydroorotate dehydrogenase (*DHODH)* gene, which is responsible for postaxial acrofacial dysostosis, also known as Miller syndrome [73]. Interestingly, this approach led to the discovery of a coincidental mutation in dynein axonemal heavy chain 5 (*DNAH5*) gene, which is associated with primary ciliary dyskinesia, in two patients with Miller syndrome that provided the underlying cause of their recurrent respiratory tract infections. The identification of gene mutations in poorly understood diseases will allow for improved diagnostics and improved care [74].

One of the most successful applications of the use of a DNA-based sequencing approach in genomic medicine was for inherited platelet diseases. The first worldwide studies in this field were initiated by BRIDGE (Bleeding and Platelet Disorder consortium) in 2011, where NGS was used for inherited platelet diseases to create a gene platform based on whole exome and whole genome sequencing of patients with the disease [75]. Since then, NGS has been extensively used to establish the diagnosis, discover new alleles resulting in inherited platelet disorders and for genetic counseling [76]. For example, the use of WES identified new variants of diaphanous–related formin 1 gene (*DIAPH1*) that lead to sensorineural hearing loss-thrombocytopenia syndrome [77]. In another study, Gorski et al. used WES to uncover novel genetic variants that were responsible for the development of neutralizing antibodies against factor VIII in patients with severe hemophilia A. Factor VIII is commonly used as a replacement therapy in such patients. 17 variants were identified that were associated with a high risk of development of neutralizing antibodies and 11 variants were identified as protective. From this work, they identified a possible susceptibility locus for antibody development against factor VIII, in the *LCT* locus, located on chromosome 2q21 [78]. This study demonstrated that NGS is effective for the identification of rare variants in the population and can be used to identify new susceptibility loci for genetic diseases.

3.2 The Use of NGS to Study Complex Traits

Understanding non-Mendelian, multigene diseases can also benefit from the application of NGS, with recent advances being made in cognitive, neurological, cardiovascular, and psychiatric disorders. To detect genetic variations to diagnose a specific disorder or group of disorders for congenital heart diseases, targeted gene panels have been developed. For example, 13 genes associated with common congenital heart disease are available as a NGS congenital heart disease panel for prenatal testing (Table 1) [79, 80]. The majority of congenital heart disease occurs sporadically, but certain environmental factors and genetic disorders have been associated with increased risk [81]. For example, *GATA4* mutations are associated with familial ventricular septal defects [82], while mutation in transcription factor *NKX2–5* was found in families with atrial septal defects and conduction delays [83]. In DiGeorge syndrome, a chromosomal microdeletion of 22q11 results in the loss of T-box transcription factor *TBX1*, which is responsible for the associated aortic arch defects and is associated with majority of cases [84]. Although NGS should be a powerful tool for the detection and diagnosis of inherited cardiac disease, obstacles arose when trying to apply NGS to detect disease alleles in sporadic cases. This is because large-scale screening on the population level revealed a high quantity of potentially pathogenic variations in healthy individuals [85]. Thus, large population linkage studies are required to determine the contribution that these alleles play in the development of cardiac disease. Others propose that target gene panels remain the most practical for use in a clinical setting, but WES has great value in the research environment for novel gene discovery and the identification of rare variants [86, 87]. WES and RNA-seq were recently applied to examine rare copy number variants (CNVs), which are associated with congenital heart defects. From this study of approximately 2400 families, association of CNVs with specific sub-types of congenital heart disease were identified and five novel CNVs were identified [88]. Clearly, NGS analysis will continue to be important for our foundational knowledge of congenital heart disease, even if its use is limited for routine clinical diagnostics.

3.3 Applications of NGS for Neonates and Pre-Neonatal Diagnostics

Another exciting possible clinical applications of NGS is in neonatal and pediatric medicine to establish diagnosis of genetic diseases in a timely manner. This is especially important where a delayed or phenotype-based clinical decision may lead to a substantial increase in morbidity or mortality [89, 90]. One children's hospital study in neonatal and pediatric intensive care units (ICUs), analyzed the diagnostic rate of infants younger than 4 months old with

Table 1
Commonly used genes in prenatal testing panels of congenital heart disorders

CHD Genes	Disorder	Inheritance pattern
CHD7[b]	CHARGE syndrome[b] Isolated CHD52	AD
ELN[a]	Supravalvular aortic stenosis	AD
GATA4[a]	Atrial septal defect 2 Atrioventricular septal defect 4 Tetralogy of Fallot Ventricular septal defect 1	AD
GATA6[b]	Atrial septal defect 9 Atrioventricular septal defect 5 Pancreatic agenesis and congenital heart defects[b] Persistent truncus arteriosus Tetralogy of Fallot	AD
GDF1[b]	Right atrial isomerism[b] Congenital heart defects, multiple types, 6	AR AD
JAG1[b]	Alagille syndrome[b] Tetralogy of Fallot	AD
NKX2–5[a]	Atrial septal defect 7 Conotruncal heart malformations, variable Hypoplastic left heart syndrome 2 Tetralogy of Fallot Ventricular septal defect 3	AD
NKX2-6[a]	Conotruncal heart malformations Persistent truncus arteriosus	AR
NOTCH1[b]	Aortic valve disease 1 Adams-Oliver syndrome 5[b]	AD
NR2F2[a]	Congenital heart defects, multiple types, 4	AD
TBX1[b]	DiGeorge syndrome[b] Velocardiofacial syndrome[b] Conotruncal anomaly face syndrome[b] Tetralogy of Fallot	AD
TBX5[b]	Holt-Oram syndrome[b] Atrioventricular septal defect, atrial septal defect, ventricular septal defect	AD
ZIC3[b]	Congenital heart defects, nonsyndromic, 1, X-linked Heterotaxy, visceral, 1, X-linked[b] VACTERL association, X-linked[b]	XLR

AD autosomal dominant; *AR* autosomal recessive; *XLR* X-linked recessive; *NS-CHD* non-syndromic congenital heart disease. Adapted from Ref. [80].

[a]Gene associated with NS-CHD only

[b]Gene associated with NS-CHD and extracardiac presentation

suspected genetic disorders and congenital anomalies. Rapid whole-genome sequencing, which could be completed in 50 h, was compared to standard genetic testing techniques, like comparative genomic hybridization arrays and fluorescence in situ hybridization (FISH). 20 out of the 35 critically ill infants were diagnosed with a genetic disease by rapid whole-genome sequencing and 3 out of 32 by standard genetic testing. 18 of those 20 diagnoses established by whole-genome sequencing were not identified with standard genetic testing, and nine of them were diseases that were not considered on the original differential diagnosis list. Furthermore, 13 cases led to a modification of the acute clinical management, including palliative care, medications, and genetic counselling [91]. This clearly highlights the need for the rapid, accurate diagnosis of genetic diseases, as it can lead to improved management and patient outcomes. In another single hospital study, WES was performed for 278 infants in the first 100 days of their life, who were hospitalized in the neonatal ICU (NICU), with suspected genetic diseases based on clinical indications. 102 infants met the criteria for molecular diagnosis and approximately 39 of them had an atypical or unrecognized clinical presentation. Establishment of the molecular diagnoses altered subsequent medical care, leading to medication changes, transplant consideration, and initiated new subspecialist care, which was not contemplated before the NGS analysis [92]. The use of NGS for the diagnosis and management of genetic diseases among neonates will likely become widespread in the near future.

Another revolutionary application of NGS is in its use for prenatal diagnostics. The discovery of circulating cell-free fetal DNA in the plasma of pregnant women made it possible to detect paternally inherited alleles by selective targeting PCR analysis. However, this methodology has not been accurate for detecting fetal genetic aberrations or maternally inherited genetic disorders [93]. The implementation of NGS allowed for screening for multiple mutations in a non-invasive way. This includes the difficult-to-detect fetal autosomal recessive, monogenic disorders, and aneuploidies. With the increasing affordability and a rapid turnaround time, the possibility of instituting non-invasive prenatal screening using NGS technologies as a common screening test in the routine prenatal clinical practice is becoming more realistic [94]. In fact, NGS is being applied to diagnose fetuses with congenital heart defects [95]. However, technical challenges likely remain, such as the difficulty in distinguishing fetal and maternal genomes. In addition, there are ethical considerations and the establishment of clear guidelines for testing and the implementation of genetic counseling before and after the testing are required.

4 NGS in Oncology

4.1 NGS for Cancer Diagnostics

NGS technologies can promote advancements in the field of cancer research, including uncovering fundamental knowledge about the complex mechanisms of pathogenesis and fostering new therapeutics. NGS can detect genetic abnormalities at the single gene level, changes in gene expression and facilitate the molecular description of tumors. Together, this information could lead to revision of the current approaches in establishing diagnosis, treatment, and prognosis of neoplastic diseases [96, 97]. Over the past few decades with the application of newer NGS technologies, there have been many important breakthroughs. One such breakthrough was the revision of breast cancer diagnosis and management based on genetic subtypes. Specific genetic abnormalities have been associated with different outcomes, and whole genome sequencing and transcriptomic analyses prompted the reevaluation of the risk of relapse and response to treatments [98]. This led to the discovery of breast cancer molecular subtypes that benefit from adjuvant chemotherapy. For example, estrogen receptor-a (ERa) is a ligand-dependent transcription factor that regulates cell growth genes and plays an important role in breast cancer progression. More than 70% of breast cancer cells express ERa which is encoded by the gene *ESR1* [99]. Recently, chromatin immunoprecipitation followed by NGS (ChIP-seq) was used by Ross-Innes et al. to determine ER-binding sites in tumors from patients with different clinical outcomes This study revealed that in patients with Tamoxifen-resistant (a drug used to treat hormone receptor-positive breast cancer) breast cancers, there was a unique pattern of ER-binding sites compared to drug-susceptible cancer cells [100]. ChIP-seq analysis of clinical samples showed that tamoxifen resistance is related to ER binding affinity, with differential ER binding genome profiles associated with distinct clinical outcomes. For patients with good clinical outcomes, there were ~ 500 ER binding events in the genome and in patients with poor outcomes and metastases, there were 1192 events, representing the acquisition of additional, tumor-specific ER-binding regions. This is one of the seminal studies that detected the association between ER-binding in ER+ breast cancers and clinical outcomes [100]. In addition, this analysis led to the identification of a new resistance mechanism in ER+ breast cancer cells, which could become a screening tool to determine if tamoxifen will be an effective chemotherapy.

One of the few solid cancers that is difficult to analyze by NGS is melanoma, due to a high number of mutations from ultraviolet light. Hodis et al. performed whole genome sequencing and identified novel candidate genes (*SNX31, PPP6C, TACC1, STK19, RAC1, ARID2*) by comparing baseline mutation level to mutation in genes of interest [101]. WES analysis of 147 tumors revealed that

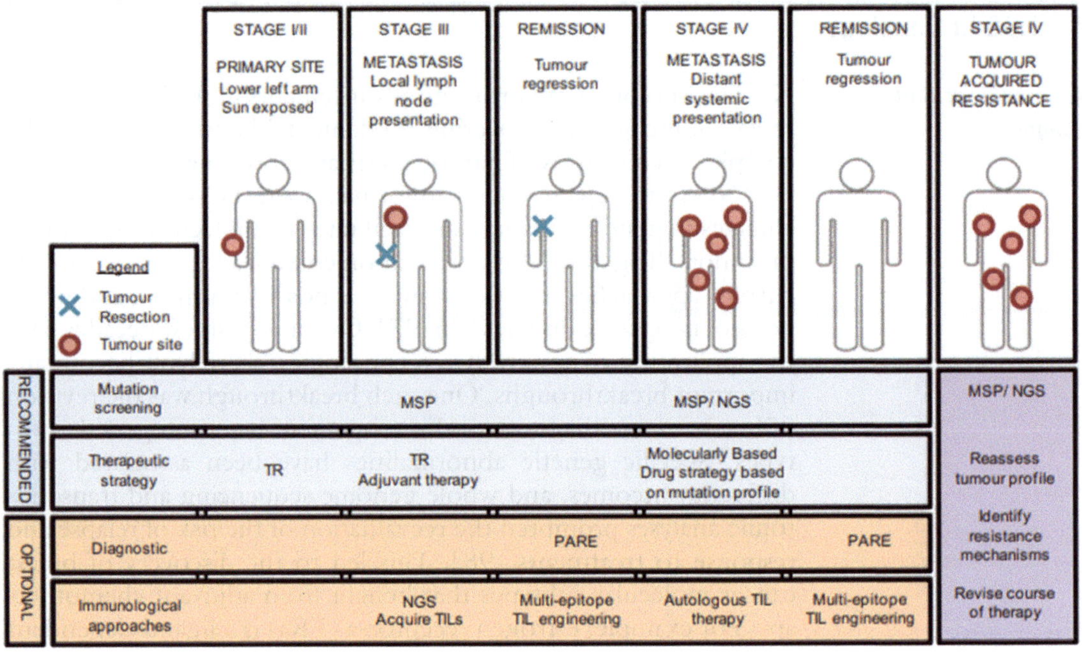

Fig. 3 Mutational profiling in personalized therapy. Presentation of a patient with melanoma at different tumor stages. Included are recommended screening and therapeutic managements, which vary depending on tumor progression. NGS has become an important part of advanced melanoma care. MSP, melanoma-specific mutation panel; NGS, next-generation sequencing; TR, tumor resection; PARE, personalized analysis with rearranged ends; TIL, tumor infiltrating lymphocyte. Reproduced with permission from Ref. [103]. Copyright 2012, Elsevier

12% had mutations in *PPP6C* with additional mutations in *BRAF* or *NRAS*. Mutation in *BRAF* is known to be involved in the pathogenesis of metastatic melanoma. This discovery and approval of the BRAF kinase inhibitor Vemurafenib started a new era in personalized therapeutics of melanoma. Application of NGS to identify treatable mutations such as *BRAF* in each patient is a promising approach in personalized treatment that has already been implemented at oncological centers (Fig. 3, [102, 103]).

4.2 NGS to Study the Underlying Genetics of Cancer and Metastasis

Androgen hormones, whose effect are mediated through androgen receptors (AR), play a significant role in the progression of prostate cancer. Like the ER receptors in breast cancer, mutations of the ARs in early prostate cancer are infrequent but are commonly found in advanced androgen-independent tumors. The abnormal activation of ARs leads to series of intracellular events that initiate transcriptional activity and promote uncontrolled growth. ChIP-seq was used to study AR binding across the genomes of prostate cancer cell lines. Analysis revealed that NKX3–1 is a novel transcription factor that cooperates with ARs to regulate transcription of genes that drive prostate cancer progression [104, 105]. Thus, the use of

NGS technology can uncover proteins that work together to regulate transcription of growth-promoting genes, which aids in our fundamental understanding and possibly will lead to novel therapeutics.

Despite the advances in surgical treatment and chemotherapeutics, lung cancer is still one of the deadliest cancers, with high lethality due to metastatic disease [106]. Often, the disease is diagnosed in later stages when it becomes symptomatic and already resulted in extrapulmonary metastases. One focus of lung cancer research is genome analysis and gene expression, with the goal of gaining a better understanding of the progression to metastatic disease. Recent studies have shown that microRNAs (miRNAs) have been linked to aberrant gene expression in cancer cells of different origins [107]. miRNAs are short, 22 nucleotide molecules that play important regulatory roles in various fundamental biological processes, like cellular proliferation, apoptosis, and differentiation [108]. So far, three miRNAs (has-miR-183-3p, has--miR-183-5p, has-miR-574-5p) have been linked to lung adenocarcinoma metastasis. NGS and qRT-PCR was used to confirm differential expression of these miRNAs in cancer samples with and without metastases. Expression profiling analysis showed decreased expression of has-miR-183-3p and has-miR-183-5p in metastatic samples of lung adenocarcinoma. Conversely, has-miR-574-5p was increased in metastatic samples. miRNA-seq was used to identify numerous additional miRNAs linked with tumor survival and progression [109].

The expression of other types of small RNAs have also been linked to the presence of distant metastases in lung adenocarcinoma cells. Daugaard et al. performed small RNA sequencing to profile the transcriptomes of lung adenocarcinoma cells [110]. PIWI-interacting RNAs (piRNAs) are 26–31 nucleotides in length and are not well conserved. They are known to be involved in transposon silencing, genome rearrangement, epigenetics, stem-cell maintenance, and tumorigenesis [111, 112]. piRNAs and miRNAs both play a significant role in the development of distant lung cancer metastases. For piRNAs, there was a significant association between the decreased expression of piR-57,125 and distant metastases. For miRNAs, decreased expression of miR-30a-3p and increased expression of miR-210-3p was linked with distant metastases [110]. Recently, it was shown that tumor cells release miRNAs into blood, where they stay stable [113]. Thus, genome-wide expression analysis of circulating miRNAs and possibly piRNAs from serum can be used as biomarkers for metastatic progression and overall prognosis.

4.3 Challenges in Implicating NGS Technology in Clinical Practice

The potential of NGS technology in a clinical setting is vast. However, the disputes about the use of NGS in diagnosing, monitoring, and prognosis of different cancers are ongoing. The discovery of new cancerous biomarkers because of deep genomic screens provides support for the implementation of NGS into clinical practice, although it would still need extensive validation and approval [114]. The diagnosis, treatment, and management of leukemias or lymphomas is one example where clinical practice would benefit from NGS implementation. When these cancers develop, there usually is a single T or B cell clone that uncontrollably proliferates, which has a unique T- or B-cell receptor sequence. Since this malignant clone can compose more than a half of the entire B- or T-cell population in cancer patients, the clonality of receptors is used for diagnosis of blood cancers and tracking residual disease following treatment [115]. NGS analysis of T- and B-cell receptor genes from blood has been proposed for the assessment of treatment response and risk stratification in patients with acute lymphoblastic leukemia, with minimal residual disease. The use of NGS instead of allele-specific, quantitative real-time PCR would be more informative and accurate in the characterization of the lymphoid population to monitor disease evolution and detect relapse as early as possible [116].

For the diagnosis of non-Hodgkin's lymphoma, a controversial approach was proposed to detect free DNA in plasma as a source of cancer-specific DNA instead of white blood cells. The challenge with diagnosing lymphomas is the absence of known point mutations and difficulty in detecting circulating tumor cells. This proposal is based on the fact that all lymphomas have rearrangements of their immunoglobulin genes. The idea is to use NGS to identify heavy chain immunoglobulin (*IgH*) gene rearrangement by capturing and sequencing the *IgH* genomic regions. With knowledge of the rearranged sequences, circulating fragments DNA can be detected in plasma, thereby working as a tumor progression marker. This approach has the potential for clinical application to diagnose and monitor disease progression, but still requires large-scale clinical trials [117].

In a study of acute myeloid leukemia patients, a NGS panel was tested for accuracy compared to routine, clinically approved methods, such as hybridization probes and fragment analysis by Sanger sequencing of positive samples. A 19-gene NGS panel targeting specific acute myeloid leukemia genes was used for a cohort of 162 patients. Comparison with conventional techniques showed 100% concordance, and NGS was able to detect more clinically relevant mutations. This NGS panel was highly sensitive and specific, with cutoff points of 3% and 5% for point mutations and insertions/deletions, respectively. This panel was proposed for use in diagnostic and prognostic classification [118].

Although NGS appears to be a breakthrough technique in establishing the molecular diagnosis of malignancies, it is not ideal. Whole-genome sequencing provides large amount of data with single nucleotide resolution, but the costs might still be too high for routine clinical implementation in all practices [119]. For example, the cost of NGS in clinical settings in the United States between 2016 and 2019 were on average between $438 and $3,700 for tumor mutation burden tests and from $1,722 to $2,249 for hereditary cancers [120]. WES is a more cost-effective alternative to whole-genome sequencing, but it also can be expensive and there is a still a need to analyze large amounts of data [119]. The discovery of new biomarkers based on large sequencing data requires the implementation of large gene screening panels, which would be challenging to introduce into routine clinical practice because of the high-volume of tests and slower turnaround time. This would require additional resources, costs, and expertise to interpret high complexity data. Currently only smaller, targeted gene panels are considered practical for tumor classification. For the development of future NGS clinical panels, the role of mutations in oncogenesis, detection of clinically important polymorphisms, translocations or gene fusions, turnaround time, and costs of laboratory testing must all be considered [121, 122].

5 NGS Applications in Dermatology

The use of NGS in investigative dermatology started with the application of NGS to determine the underlying immune response associated with specific insults. Gaide et al. used a technique called immunosequencing, which combines multiplexed PCR with NGS to monitor V(D)J recombination of the T-cell receptor gene [123]. Using immunosequencing, they examined the T-cell response after skin vaccination and were able to clarify the clonal origin. It was discovered that resident memory T cells that reside in the peripheral tissue and the central memory T-cells from the lymph nodes have the same T-cell progenitor, which creates memory T cells with different effector properties, but the same antigen specificity in different compartments. Peripheral skin T cells display a rapid hypersensitivity response, while central lymph node T cells are responsible for a delayed response. These data provide the underlying mechanism for allergic contact dermatitis and explain the propensity for recurrence and the refractory properties of this disease [123, 124].

NGS of the T-cell receptor has also been used for both diagnostic and prognostic purposes for cutaneous T-cell lymphoma (CTCL, [125]). Detection of a malignant clone is crucial in making the diagnosis of CTCL and the widely used diagnostic test of T-cell receptor g PCR only detects specific subsets. It was recently demonstrated that NGS is more specific and sensitive in the detection of malignant T-cells and can accurately distinguish it from benign inflammatory diseases. Another application that NGS was useful for was in treatment response monitoring and identification of disease recurrence. As NGS can be customized for personal medicine, it could be used to monitor individual responses to therapy and detect prognostic or diagnostic biomarkers at any stage of the disease.

6 Conclusions

NGS is a powerful tool that accelerates new discoveries and gives broader opportunities for research in the fields of genomics and personalized medicine. The discussion about the potential application of NGS in the field of personalized medicine and day-to-day clinical practice has been ongoing. The great excitement arises from the possibility to improve diagnostic and prognostic approaches in certain areas of medicine, like genetic disease and oncology. Ultimately, these improvements could lead to earlier diagnosis, earlier detection of relapse, and more favorable clinical outcomes. Perhaps the most successful clinical application of NGS is in oncology, in which the adaptation of diagnostic and therapeutic strategies is based on gene signatures and mutational profiles of individuals [96, 97, 113]. Despite the technological advances, most disease treatments are based on "standard of care" approaches, which might not work for every patient. In addition, patients that are unresponsive to standard treatments are met in clinical practice quite often, especially in the fields of cancer or rare diseases. Unfortunately, in such clinical cases, rounds of ineffective treatment can lead to delays and unfavorable outcomes before the appropriate treatment has been identified. Implementation of NGS, together with clinical findings and laboratory tests, could help to diagnose individual variations and predict certain predispositions and responsiveness to distinct treatment schemes [126]. For example, the development of cancer targeted therapies became possible after the identification of specific, causative DNA mutations. The first whole-genome sequencing of a cancer patient was performed in 2008. The patient had acute myelogenous leukemia, and sequencing revealed ten genetic mutations that were relevant to the disease, opening the door to genomics as a fundamental weapon in the battle of cancer [127]. Since then, whole-genome sequencing has been expanded to other cancers and led to worldwide cancer sequencing projects. It is not just oncology that has benefited from the implementation of NGS into clinical practice. Establishing a diagnosis of a rare genetic disease has been a challenge for decades, leaving doctors with only a few diagnostic approaches and an absence of tools to find disease variants. NGS overcomes these obstacles, offering greater sensitivity and increasing the likelihood accurate diagnosis of rare diseases, including the detection of different genetic variants. For example, different variations in genetic mutations in neonatal diabetes mellitus will respond differently to available treatments, and NGS allows for their accurate detection [128]. The use of genomic information has pushed the evolution of gene therapy and expanded feasibility of personalized medicine.

Although NGS is an excellent tool and can be extremely beneficial in clinical settings, it is still facing multiple obstacles on the way to routine practice. These obstacles include the clinical laboratory environment, secure storage of large amounts of generated data, and the need for validation of testing platforms on broad populations. Additional future challenges include improving the turnaround time for the delivery of laboratory results, and the need to educate clinicians to analyze the data effectively and accurately. However, the clinical application of NGS should be prioritized because of the therapeutic potential. Knowledge of the genetic causes of a medical condition through NGS can lead to improved therapeutic options, even if it is not a cure. Its adaptation and adjustment into clinical practice requires broader expansion, faster processing of large amounts of data, and a standardized, affordable educational platform for appropriate personnel. Unquestionably, NGS is a powerful tool and its broad implementation into clinical practice will be of benefit to both physicians and patients.

Acknowledgement

This work was supported by grant GM138303, awarded to V.J.C.

References

1. Weischenfeldt J, Symmons O, Spitz F, Korbel JO (2013) Phenotypic impact of genomic structural variation: insights from and for human disease. Nat Rev Genet 14(2): 125–138. https://doi.org/10.1038/nrg3373
2. Lazaridis KN, Petersen GM (2005) Genomics, genetic epidemiology, and genomic medicine. Clin Gastroenterol Hepatol 3(4): 320–328. https://doi.org/10.1016/s1542-3565(05)00085-6
3. Hasanzad M, Sarhangi N, Naghavi A, Ghavimehr E, Khatami F, Ehsani Chimeh S, Larijani B, Aghaei Meybodi HR (2022) Genomic medicine on the frontier of precision medicine. J Diabetes Metab Disord 21(1):853–861. https://doi.org/10.1007/s40200-021-00880-6
4. Horton RH, Lucassen AM (2019) Recent developments in genetic/genomic medicine. Clin Sci (Lond) 133(5):697–708. https://doi.org/10.1042/cs20180436
5. Shendure J, Findlay GM, Snyder MW (2019) Genomic medicine-progress, pitfalls, and promise. Cell 177(1):45–57. https://doi.org/10.1016/j.cell.2019.02.003
6. Jogalekar MP, Rajendran RL, Khan F, Dmello C, Gangadaran P, Ahn BC (2022) CAR T-cell-based gene therapy for cancers: new perspectives, challenges, and clinical developments. Front Immunol 13:925985. https://doi.org/10.3389/fimmu.2022.925985
7. Leonard A, Tisdale JF, Bonner M (2022) Gene therapy for hemoglobinopathies: Beta-

thalassemia, sickle cell disease. Hematol Oncol Clin North Am 36(4):769–795. https://doi.org/10.1016/j.hoc.2022.03.008
8. Condren ME, Bradshaw MD (2013) Ivacaftor: a novel gene-based therapeutic approach for cystic fibrosis. J Pediatr Pharmacol Ther 18(1):8–13. https://doi.org/10.5863/1551-6776-18.1.8
9. Maxam AM, Gilbert W (1977) A new method for sequencing DNA. Proc Natl Acad Sci USA 74(2):560–564. https://doi.org/10.1073/pnas.74.2.560
10. Crossley BM, Bai J, Glaser A, Maes R, Porter E, Killian ML, Clement T, Toohey-Kurth K (2020) Guidelines for Sanger sequencing and molecular assay monitoring. J Vet Diagn Invest 32(6):767–775. https://doi.org/10.1177/1040638720905833
11. Sanger F, Nicklen S, Coulson AR (1977) DNA sequencing with chain-terminating inhibitors. Proc Natl Acad Sci 74(12):5463–5467. https://doi.org/10.1073/pnas.74.12.5463
12. Park S-J, Saito-Adachi M, Komiyama Y, Nakai K (2016) Advances, practice, and clinical perspectives in high-throughput sequencing. Oral Dis 22(5):353–364. https://doi.org/10.1111/odi.12403
13. Ronaghi M, Karamohamed S, Pettersson B, Uhlén M, Nyrén P (1996) Real-time DNA sequencing using detection of pyrophosphate release. Anal Biochem 242(1):84–89. https://doi.org/10.1006/abio.1996.0432
14. Loman NJ, Misra RV, Dallman TJ, Constantinidou C, Gharbia SE, Wain J, Pallen MJ (2012) Performance comparison of benchtop high-throughput sequencing platforms. Nat Biotechnol 30(5):434–439. https://doi.org/10.1038/nbt.2198
15. Reuter Jason A, Spacek DV, Snyder Michael P (2015) High-throughput sequencing technologies. Mol Cell 58(4):586–597. https://doi.org/10.1016/j.molcel.2015.05.004
16. Mardis ER (2017) DNA sequencing technologies: 2006–2016. Nat Protoc 12(2):213–218. https://doi.org/10.1038/nprot.2016.182
17. Rhoads A, Au KF (2015) PacBio sequencing and its applications. Genomics Proteomics Bioinformatics 13(5):278–289. https://doi.org/10.1016/j.gpb.2015.08.002
18. Roberts RJ, Carneiro MO, Schatz MC (2013) The advantages of SMRT sequencing. Genome Biol 14(7):405. https://doi.org/10.1186/gb-2013-14-7-405
19. Hu R, Sun G, Sun X (2016) LSCplus: a fast solution for improving long read accuracy by short read alignment. BMC Bioinformatics 17(1):451. https://doi.org/10.1186/s12859-016-1316-y
20. Eid J, Fehr A, Gray J, Luong K, Lyle J, Otto G, Peluso P, Rank D, Baybayan P, Bettman B, Bibillo A, Bjornson K, Chaudhuri B, Christians F, Cicero R, Clark S, Dalal R, Dewinter A, Dixon J, Foquet M, Gaertner A, Hardenbol P, Heiner C, Hester K, Holden D, Kearns G, Kong X, Kuse R, Lacroix Y, Lin S, Lundquist P, Ma C, Marks P, Maxham M, Murphy D, Park I, Pham T, Phillips M, Roy J, Sebra R, Shen G, Sorenson J, Tomaney A, Travers K, Trulson M, Vieceli J, Wegener J, Wu D, Yang A, Zaccarin D, Zhao P, Zhong F, Korlach J, Turner S (2009) Real-time DNA sequencing from single polymerase molecules. Science 323(5910):133–138. https://doi.org/10.1126/science.1162986
21. Kchouk M, Gibrat J-F, Elloumi M (2017) Generations of sequencing technologies: from first to next generation. Biol Med 9:1–8
22. Levene MJ, Korlach J, Turner SW, Foquet M, Craighead HG, Webb WW (2003) Zero-mode waveguides for single-molecule analysis at high concentrations. Science 299(5607):682–686. https://doi.org/10.1126/science.1079700
23. Haque F, Li J, Wu H-C, Liang X-J, Guo P (2013) Solid-state and biological nanopore for real-time sensing of single chemical and sequencing of DNA. Nano Today 8(1):56–74. https://doi.org/10.1016/j.nantod.2012.12.008
24. Wang Y, Yang Q, Wang Z (2015) The evolution of nanopore sequencing. Front Genet 5. https://doi.org/10.3389/fgene.2014.00449
25. Fox T, Geppert J, Dinnes J, Scandrett K, Bigio J, Sulis G, Hettiarachchi D, Mathangasinghe Y, Weeratunga P, Wickramasinghe D, Bergman H, Buckley BS, Probyn K, Sguassero Y, Davenport C, Cunningham J, Dittrich S, Emperador D, Hooft L, Leeflang MM, McInnes MD, Spijker R, Struyf T, Van den Bruel A, Verbakel JY, Takwoingi Y, Taylor-Phillips S, Deeks JJ (2022) Antibody tests for identification of current and past infection with SARS-CoV-2. Cochrane Database Syst Rev 11(11):Cd013652. https://doi.org/10.1002/14651858.CD013652.pub2

26. Kaplon J, Théry L, Bidalot M, Grangier N, Frappier J, Aho Glélé LS, de Rougemont A, Ambert-Balay K (2020) Diagnostic accuracy of four commercial triplex immunochromatographic tests for rapid detection of rotavirus, adenovirus, and norovirus in human stool samples. J Clin Microbiol 59(1). https://doi.org/10.1128/jcm.01749-20
27. Bonura F, Mascarella C, Filizzolo C, Bonura C, Ferraro D, Di Bernardo F, Collura A, Martella V, Giammanco GM, De Grazia S (2021) Evaluation of the diagnostic performances of two commercially available assays for the detection of enteric adenovirus antigens. Diagn Microbiol Infect Dis 101(3): 115459. https://doi.org/10.1016/j.diagmicrobio.2021.115459
28. Merckx J, Wali R, Schiller I, Caya C, Gore GC, Chartrand C, Dendukuri N, Papenburg J (2017) Diagnostic accuracy of novel and traditional rapid tests for influenza infection compared with reverse transcriptase polymerase chain reaction: a systematic review and meta-analysis. Ann Intern Med 167(6): 394–409. https://doi.org/10.7326/m17-0848
29. Cristovam E, Almeida D, Caldeira D, Ferreira JJ, Marques T (2017) Accuracy of diagnostic tests for Legionnaires' disease: a systematic review. J Med Microbiol 66(4):485–489. https://doi.org/10.1099/jmm.0.000454
30. Chiu C, Fair J, Leroy EM (2013) Bas-Congo virus: another deadly virus? Future Microbiol 8(2):139–141. https://doi.org/10.2217/fmb.12.145
31. Grard G, Fair JN, Lee D, Slikas E, Steffen I, Muyembe JJ, Sittler T, Veeraraghavan N, Ruby JG, Wang C, Makuwa M, Mulembakani P, Tesh RB, Mazet J, Rimoin AW, Taylor T, Schneider BS, Simmons G, Delwart E, Wolfe ND, Chiu CY, Leroy EM (2012) A novel rhabdovirus associated with acute hemorrhagic fever in central Africa. PLoS Pathog 8(9):e1002924. https://doi.org/10.1371/journal.ppat.1002924
32. Zhang AR, Shi WQ, Liu K, Li XL, Liu MJ, Zhang WH, Zhao GP, Chen JJ, Zhang XA, Miao D, Ma W, Liu W, Yang Y, Fang LQ (2021) Epidemiology and evolution of Middle East respiratory syndrome coronavirus, 2012–2020. Infect Dis Poverty 10(1):66. https://doi.org/10.1186/s40249-021-00853-0
33. Dashboard WCC. https://data.who.int/dashboards/covid19/cases. Accessed April 25 2023
34. Grimaldi A, Panariello F, Annunziata P, Giuliano T, Daniele M, Pierri B, Colantuono C, Salvi M, Bouché V, Manfredi A, Cuomo MC, Di Concilio D, Tiberio C, Fiorenza M, Portella G, Cimmino I, Sorrentino A, Fusco G, Granata MR, Cerino P, Limone A, Atripaldi L, Ballabio A, Cacchiarelli D (2022) Improved SARS-CoV-2 sequencing surveillance allows the identification of new variants and signatures in infected patients. Genome Med 14(1):90. https://doi.org/10.1186/s13073-022-01098-8
35. Eyre DW, Golubchik T, Gordon NC, Bowden R, Piazza P, Batty EM, Ip CLC, Wilson DJ, Didelot X, O'Connor L, Lay R, Buck D, Kearns AM, Shaw A, Paul J, Wilcox MH, Donnelly PJ, Peto TEA, Walker AS, Crook DW (2012) A pilot study of rapid benchtop sequencing of *Staphylococcus aureus* and *Clostridium difficile* for outbreak detection and surveillance. BMJ Open 2(3): e001124. https://doi.org/10.1136/bmjopen-2012-001124
36. Köser CU, Holden MTG, Ellington MJ, Cartwright EJP, Brown NM, Ogilvy-Stuart AL, Hsu LY, Chewapreecha C, Croucher NJ, Harris SR, Sanders M, Enright MC, Dougan G, Bentley SD, Parkhill J, Fraser LJ, Betley JR, Schulz-Trieglaff OB, Smith GP, Peacock SJ (2012) Rapid whole-genome sequencing for investigation of a neonatal MRSA outbreak. N Engl J Med 366(24): 2267–2275. https://doi.org/10.1056/NEJMoa1109910
37. Frerichs RR, Keim PS, Barrais R, Piarroux R (2012) Nepalese origin of cholera epidemic in Haiti. Clin Microbiol Infect 18(6):E158–E163. https://doi.org/10.1111/j.1469-0691.2012.03841.x
38. Katz LS, Petkau A, Beaulaurier J, Tyler S, Antonova ES, Turnsek MA, Guo Y, Wang S, Paxinos EE, Orata F, Gladney LM, Stroika S, Folster JP, Rowe L, Freeman MM, Knox N, Frace M, Boncy J, Graham M, Hammer BK, Boucher Y, Bashir A, Hanage WP, Domselaar GV, Tarr CL (2013) Evolutionary dynamics of *Vibrio cholerae* O1 following a single-source introduction to Haiti. mBio 4(4): e00398–e00313. https://doi.org/10.1128/mBio.00398-13
39. Bhandari M, Rathnayake IU, Huygens F, Jennison AV (2023) Clinical and environmental *vibrio cholerae* non-O1, non-O139 strains from Australia have similar virulence and antimicrobial resistance gene profiles. Microbiol Spectrum 11(1):e02631–e02622. https://doi.org/10.1128/spectrum.02631-22
40. Rohde H, Qin J, Cui Y, Li D, Loman NJ, Hentschke M, Chen W, Pu F, Peng Y, Li J,

Xi F, Li S, Li Y, Zhang Z, Yang X, Zhao M, Wang P, Guan Y, Cen Z, Zhao X, Christner M, Kobbe R, Loos S, Oh J, Yang L, Danchin A, Gao GF, Song Y, Li Y, Yang H, Wang J, Xu J, Pallen MJ, Wang J, Aepfelbacher M, Yang R (2011) Open-source genomic analysis of Shiga-toxin-producing *E. coli* O104:H4. N Engl J Med 365(8): 718–724. https://doi.org/10.1056/NEJMoa1107643

41. Lowder BV, Guinane CM, Ben Zakour NL, Weinert LA, Conway-Morris A, Cartwright RA, Simpson AJ, Rambaut A, Nübel U, Fitzgerald JR (2009) Recent human-to-poultry host jump, adaptation, and pandemic spread of *Staphylococcus aureus*. Proc Natl Acad Sci 106(46):19545–19550. https://doi.org/10.1073/pnas.0909285106
42. Price LB, Stegger M, Hasman H, Aziz M, Larsen J, Andersen PS, Pearson T, Waters AE, Foster JT, Schupp J, Gillece J, Driebe E, Liu CM, Springer B, Zdovc I, Battisti A, Franco A, Żmudzki J, Schwarz S, Butaye P, Jouy E, Pomba C, Porrero MC, Ruimy R, Smith TC, Robinson DA, Weese JS, Arriola CS, Yu F, Laurent F, Keim P, Skov R, Aarestrup FM (2012) *Staphylococcus aureus* CC398: host adaptation and emergence of methicillin resistance in livestock. mBio 3(1): e00305–e00311. https://doi.org/10.1128/mBio.00305-11
43. Brockhurst MA, Colegrave N, Rozen DE (2011) Next-generation sequencing as a tool to study microbial evolution. Mol Ecol 20(5): 972–980. https://doi.org/10.1111/j.1365-294X.2010.04835.x
44. Kobras CM, Fenton AK, Sheppard SK (2021) Next-generation microbiology: from comparative genomics to gene function. Genome Biol 22(1):123. https://doi.org/10.1186/s13059-021-02344-9
45. Mwangi MM, Wu SW, Zhou Y, Sieradzki K, de Lencastre H, Richardson P, Bruce D, Rubin E, Myers E, Siggia ED, Tomasz A (2007) Tracking the *in vivo* evolution of multidrug resistance in *Staphylococcus aureus* by whole-genome sequencing. Proc Natl Acad Sci 104(22):9451–9456. https://doi.org/10.1073/pnas.0609839104
46. McAdam PR, Holmes A, Templeton KE, Fitzgerald JR (2011) Adaptive evolution of *Staphylococcus aureus* during chronic endobronchial infection of a cystic fibrosis patient. PLoS One 6(9):e24301. https://doi.org/10.1371/journal.pone.0024301
47. Meier S, Goerke C, Wolz C, Seidl K, Homerova D, Schulthess B, Kormanec J, Berger-Bächi B, Bischoff M (2007) Sigma B and the sigma B-dependent *arlRS* and *yabJ-spoVG* loci affect capsule formation in *Staphylococcus aureus*. Infect Immun 75(9):4562–4571. https://doi.org/10.1128/iai.00392-07
48. Moisan H, Brouillette E, Jacob CL, Langlois-Bégin P, Michaud S, Malouin F (2006) Transcription of virulence factors in *Staphylococcus aureus* small-colony variants isolated from cystic fibrosis patients is influenced by SigB. J Bacteriol 188(1):64–76. https://doi.org/10.1128/jb.188.1.64-76.2006
49. Iwanaga M, Toma C, Miyazato T, Insisiengmay S, Nakasone N, Ehara M (2004) Antibiotic resistance conferred by a class I integron and SXT constin in *Vibrio cholerae* O1 strains isolated in Laos. Antimicrob Agents Chemother 48(7):2364–2369. https://doi.org/10.1128/aac.48.7.2364-2369.2004
50. Mutreja A, Kim DW, Thomson NR, Connor TR, Lee JH, Kariuki S, Croucher NJ, Choi SY, Harris SR, Lebens M, Niyogi SK, Kim EJ, Ramamurthy T, Chun J, Wood JLN, Clemens JD, Czerkinsky C, Nair GB, Holmgren J, Parkhill J, Dougan G (2011) Evidence for several waves of global transmission in the seventh cholera pandemic. Nature 477(7365):462–465. https://doi.org/10.1038/nature10392
51. Ford CB, Lin PL, Chase MR, Shah RR, Iartchouk O, Galagan J, Mohaideen N, Ioerger TR, Sacchettini JC, Lipsitch M, Flynn JL, Fortune SM (2011) Use of whole genome sequencing to estimate the mutation rate of *Mycobacterium tuberculosis* during latent infection. Nat Genet 43(5):482–486. https://doi.org/10.1038/ng.811
52. Schlaberg R, Simmon KE, Fisher MA (2012) A systematic approach for discovering novel, clinically relevant bacteria. Emerg Infect Dis 18(3):422–430. https://doi.org/10.3201/eid1803.111481
53. Salipante SJ, Sengupta DJ, Rosenthal C, Costa G, Spangler J, Sims EH, Jacobs MA, Miller SI, Hoogestraat DR, Cookson BT, McCoy C, Matsen FA, Shendure J, Lee CC, Harkins TT, Hoffman NG (2013) Rapid 16S rRNA next-generation sequencing of polymicrobial clinical samples for diagnosis of complex bacterial infections. PLoS One 8(5): e65226. https://doi.org/10.1371/journal.pone.0065226
54. Altveş S, Yildiz HK, Vural HC (2020) Interaction of the microbiota with the human body in health and diseases. Biosci Microbiota Food Health 39(2):23–32. https://doi.org/10.12938/bmfh.19-023

55. Belizário JE, Napolitano M (2015) Human microbiomes and their roles in dysbiosis, common diseases, and novel therapeutic approaches. Front Microbiol 6. https://doi.org/10.3389/fmicb.2015.01050
56. Gilbert JA, Blaser MJ, Caporaso JG, Jansson JK, Lynch SV, Knight R (2018) Current understanding of the human microbiome. Nat Med 24(4):392–400. https://doi.org/10.1038/nm.4517
57. Fan Y, Pedersen O (2021) Gut microbiota in human metabolic health and disease. Nat Rev Microbiol 19(1):55–71. https://doi.org/10.1038/s41579-020-0433-9
58. Bai Z, Zhang Y-z, Miyano S, Yamaguchi R, Fujimoto K, Uematsu S, Imoto S (2022) Identification of bacteriophage genome sequences with representation learning. Bioinformatics 38(18):4264–4270. https://doi.org/10.1093/bioinformatics/btac509
59. Ács N, Gambino M, Brøndsted L (2020) Bacteriophage enumeration and detection methods. Front Microbiol 11. https://doi.org/10.3389/fmicb.2020.594868
60. Hajishengallis G (2014) The inflammophilic character of the periodontitis-associated microbiota. Mol Oral Microbiol 29(6): 248–257. https://doi.org/10.1111/omi.12065
61. Duran-Pinedo AE, Chen T, Teles R, Starr JR, Wang X, Krishnan K, Frias-Lopez J (2014) Community-wide transcriptome of the oral microbiome in subjects with and without periodontitis. ISME J 8(8):1659–1672. https://doi.org/10.1038/ismej.2014.23
62. Yost S, Duran-Pinedo AE, Teles R, Krishnan K, Frias-Lopez J (2015) Functional signatures of oral dysbiosis during periodontitis progression revealed by microbial metatranscriptome analysis. Genome Med 7(1): 27. https://doi.org/10.1186/s13073-015-0153-3
63. De Gregorio E, Rappuoli R (2014) From empiricism to rational design: a personal perspective of the evolution of vaccine development. Nat Rev Immunol 14(7):505–514. https://doi.org/10.1038/nri3694
64. Nascimento IP, Leite LC (2012) Recombinant vaccines and the development of new vaccine strategies. Braz J Med Biol Res 45(12):1102–1111. https://doi.org/10.1590/s0100-879x2012007500142
65. Luciani F (2016) High-throughput sequencing and vaccine design. Rev Sci Tech 35(1): 53–65. https://doi.org/10.20506/rst.35.1.2417
66. Sette A, Rappuoli R (2010) Reverse vaccinology: developing vaccines in the era of genomics. Immunity 33(4):530–541. https://doi.org/10.1016/j.immuni.2010.09.017
67. Chirico F, Teixeira da Silva JA, Tsigaris P, Sharun K (2022) Safety & effectiveness of COVID-19 vaccines: a narrative review. Indian J Med Res 155(1):91–104. https://doi.org/10.4103/ijmr.IJMR_474_21
68. Fiolet T, Kherabi Y, MacDonald CJ, Ghosn J, Peiffer-Smadja N (2022) Comparing COVID-19 vaccines for their characteristics, efficacy and effectiveness against SARS-CoV-2 and variants of concern: a narrative review. Clin Microbiol Infect 28(2):202–221. https://doi.org/10.1016/j.cmi.2021.10.005
69. Ibañez K, Polke J, Hagelstrom RT, Dolzhenko E, Pasko D, Thomas ERA, Daugherty LC, Kasperaviciute D, Smith KR, Deans ZC, Hill S, Fowler T, Scott RH, Hardy J, Chinnery PF, Houlden H, Rendon A, Caulfield MJ, Eberle MA, Taft RJ, Tucci A (2022) Whole genome sequencing for the diagnosis of neurological repeat expansion disorders in the UK: a retrospective diagnostic accuracy and prospective clinical validation study. Lancet Neurol 21(3): 234–245. https://doi.org/10.1016/s1474-4422(21)00462-2
70. Miller NA, Farrow EG, Gibson M, Willig LK, Twist G, Yoo B, Marrs T, Corder S, Krivohlavek L, Walter A, Petrikin JE, Saunders CJ, Thiffault I, Soden SE, Smith LD, Dinwiddie DL, Herd S, Cakici JA, Catreux S, Ruehle M, Kingsmore SF (2015) A 26-hour system of highly sensitive whole genome sequencing for emergency management of genetic diseases. Genome Med 7: 100. https://doi.org/10.1186/s13073-015-0221-8
71. Yépez VA, Gusic M, Kopajtich R, Mertes C, Smith NH, Alston CL, Ban R, Beblo S, Berutti R, Blessing H, Ciara E, Distelmaier F, Freisinger P, Häberle J, Hayflick SJ, Hempel M, Itkis YS, Kishita Y, Klopstock T, Krylova TD, Lamperti C, Lenz D, Makowski C, Mosegaard S, Müller MF, Muñoz-Pujol G, Nadel A, Ohtake A, Okazaki Y, Procopio E, Schwarzmayr T, Smet J, Staufner C, Stenton SL, Strom TM, Terrile C, Tort F, Van Coster R, Vanlander A, Wagner M, Xu M, Fang F, Ghezzi D, Mayr JA, Piekutowska-Abramczuk D, Ribes A, Rötig A, Taylor RW, Wortmann SB, Murayama K, Meitinger T, Gagneur J, Prokisch H (2022) Clinical implementation of

RNA sequencing for Mendelian disease diagnostics. Genome Med 14(1):38. https://doi.org/10.1186/s13073-022-01019-9

72. Bamshad MJ, Shendure JA, Valle D, Hamosh A, Lupski JR, Gibbs RA, Boerwinkle E, Lifton RP, Gerstein M, Gunel M, Mane S, Nickerson DA (2012) The centers for Mendelian genomics: a new large-scale initiative to identify the genes underlying rare Mendelian conditions. Am J Med Genet A 158a(7):1523–1525. https://doi.org/10.1002/ajmg.a.35470
73. Biesecker LG (2010) Exome sequencing makes medical genomics a reality. Nat Genet 42(1):13–14. https://doi.org/10.1038/ng0110-13
74. Ng SB, Buckingham KJ, Lee C, Bigham AW, Tabor HK, Dent KM, Huff CD, Shannon PT, Jabs EW, Nickerson DA, Shendure J, Bamshad MJ (2010) Exome sequencing identifies the cause of a mendelian disorder. Nat Genet 42(1):30–35. https://doi.org/10.1038/ng.499
75. Nurden AT, Nurden P (2018) High-throughput sequencing for rapid diagnosis of inherited platelet disorders: a case for a European consensus. Haematologica 103(1):6–8. https://doi.org/10.3324/haematol.2017.182295
76. Heremans J, Freson K (2018) High-throughput sequencing for diagnosing platelet disorders: lessons learned from exploring the causes of bleeding disorders. Int J Lab Hematol 40(Suppl 1):89–96. https://doi.org/10.1111/ijlh.12812
77. Stritt S, Nurden P, Turro E, Greene D, Jansen SB, Westbury SK, Petersen R, Astle WJ, Marlin S, Bariana TK, Kostadima M, Lentaigne C, Maiwald S, Papadia S, Kelly AM, Stephens JC, Penkett CJ, Ashford S, Tuna S, Austin S, Bakchoul T, Collins P, Favier R, Lambert MP, Mathias M, Millar CM, Mapeta R, Perry DJ, Schulman S, Simeoni I, Thys C, Gomez K, Erber WN, Stirrups K, Rendon A, Bradley JR, van Geet C, Raymond FL, Laffan MA, Nurden AT, Nieswandt B, Richardson S, Freson K, Ouwehand WH, Mumford AD (2016) A gain-of-function variant in DIAPH1 causes dominant macrothrombocytopenia and hearing loss. Blood 127(23):2903–2914. https://doi.org/10.1182/blood-2015-10-675629
78. Gorski MM, Blighe K, Lotta LA, Pappalardo E, Garagiola I, Mancini I, Mancuso ME, Fasulo MR, Santagostino E, Peyvandi F (2016) Whole-exome sequencing to identify genetic risk variants underlying inhibitor development in severe hemophilia a patients. Blood 127(23):2924–2933. https://doi.org/10.1182/blood-2015-12-685735
79. Chaix MA, Andelfinger G, Khairy P (2016) Genetic testing in congenital heart disease: a clinical approach. World J Cardiol 8(2):180–191. https://doi.org/10.4330/wjc.v8.i2.180
80. Rachamadugu SI, Miller KA, Lee IH, Zou YS (2022) Genetic detection of congenital heart disease. Gynecol Obstet Clin Med 2(3):109–123. https://doi.org/10.1016/j.gocm.2022.07.005
81. Hoffman JI, Kaplan S (2002) The incidence of congenital heart disease. J Am Coll Cardiol 39(12):1890–1900. https://doi.org/10.1016/s0735-1097(02)01886-7
82. Yang YQ, Wang J, Liu XY, Chen XZ, Zhang W, Wang XZ, Liu X, Fang WY (2012) Novel GATA4 mutations in patients with congenital ventricular septal defects. Med Sci Monit 18(6):Cr344–Cr350. https://doi.org/10.12659/msm.882877
83. Schott JJ, Benson DW, Basson CT, Pease W, Silberbach GM, Moak JP, Maron BJ, Seidman CE, Seidman JG (1998) Congenital heart disease caused by mutations in the transcription factor NKX2-5. Science 281(5373):108–111. https://doi.org/10.1126/science.281.5373.108
84. Yamagishi H, Srivastava D (2003) Unraveling the genetic and developmental mysteries of 22q11 deletion syndrome. Trends Mol Med 9(9):383–389. https://doi.org/10.1016/S1471-4914(03)00141-2
85. Tennessen JA, Bigham AW, O'Connor TD, Fu W, Kenny EE, Gravel S, McGee S, Do R, Liu X, Jun G, Kang HM, Jordan D, Leal SM, Gabriel S, Rieder MJ, Abecasis G, Altshuler D, Nickerson DA, Boerwinkle E, Sunyaev S, Bustamante CD, Bamshad MJ, Akey JM (2012) Evolution and functional impact of rare coding variation from deep sequencing of human exomes. Science 337(6090):64–69. https://doi.org/10.1126/science.1219240
86. Pasipoularides A (2018) The new era of whole-exome sequencing in congenital heart disease: brand-new insights into rare pathogenic variants. J Thorac Dis 10(Suppl 17):S1923–s1929. https://doi.org/10.21037/jtd.2018.05.56
87. Winlaw GMBaDS (2015) Next generation sequencing in congenital heart disease: gene discovery and clinical application. Next Generat Sequenc Applic 2(1):1000113. https://doi.org/10.4172/jngsa.1000113

88. Liu Y, Chang X, Glessner J, Qu H, Tian L, Li D, Nguyen K, Sleiman PMA, Hakonarson H (2019) Association of rare recurrent copy number variants with congenital heart defects based on next-generation sequencing data from ramily trios. Front Genet 10. https://doi.org/10.3389/fgene.2019.00819
89. Soden SE, Saunders CJ, Willig LK, Farrow EG, Smith LD, Petrikin JE, LePichon JB, Miller NA, Thiffault I, Dinwiddie DL, Twist G, Noll A, Heese BA, Zellmer L, Atherton AM, Abdelmoity AT, Safina N, Nyp SS, Zuccarelli B, Larson IA, Modrcin A, Herd S, Creed M, Ye Z, Yuan X, Brodsky RA, Kingsmore SF (2014) Effectiveness of exome and genome sequencing guided by acuity of illness for diagnosis of neurodevelopmental disorders. Sci Transl Med 6(265):265ra168. https://doi.org/10.1126/scitranslmed.3010076
90. Saunders CJ, Miller NA, Soden SE, Dinwiddie DL, Noll A, Alnadi NA, Andraws N, Patterson ML, Krivohlavek LA, Fellis J, Humphray S, Saffrey P, Kingsbury Z, Weir JC, Betley J, Grocock RJ, Margulies EH, Farrow EG, Artman M, Safina NP, Petrikin JE, Hall KP, Kingsmore SF (2012) Rapid whole-genome sequencing for genetic disease diagnosis in neonatal intensive care units. Sci Transl Med 4(154):154ra135. https://doi.org/10.1126/scitranslmed.3004041
91. Willig LK, Petrikin JE, Smith LD, Saunders CJ, Thiffault I, Miller NA, Soden SE, Cakici JA, Herd SM, Twist G, Noll A, Creed M, Alba PM, Carpenter SL, Clements MA, Fischer RT, Hays JA, Kilbride H, McDonough RJ, Rosterman JL, Tsai SL, Zellmer L, Farrow EG, Kingsmore SF (2015) Whole-genome sequencing for identification of Mendelian disorders in critically ill infants: a retrospective analysis of diagnostic and clinical findings. Lancet Respir Med 3(5):377–387. https://doi.org/10.1016/S2213-2600(15)00139-3
92. Meng L, Pammi M, Saronwala A, Magoulas P, Ghazi AR, Vetrini F, Zhang J, He W, Dharmadhikari AV, Qu C, Ward P, Braxton A, Narayanan S, Ge X, Tokita MJ, Santiago-Sim T, Dai H, Chiang T, Smith H, Azamian MS, Robak L, Bostwick BL, Schaaf CP, Potocki L, Scaglia F, Bacino CA, Hanchard NA, Wangler MF, Scott D, Brown C, Hu J, Belmont JW, Burrage LC, Graham BH, Sutton VR, Craigen WJ, Plon SE, Lupski JR, Beaudet AL, Gibbs RA, Muzny DM, Miller MJ, Wang X, Leduc MS, Xiao R, Liu P, Shaw C, Walkiewicz M, Bi W, Xia F, Lee B, Eng CM, Yang Y, Lalani SR (2017) Use of exome sequencing for infants in intensive care units: ascertainment of severe single-gene disorders and effect on medical management. JAMA Pediatr 171(12):e173438. https://doi.org/10.1001/jamapediatrics.2017.3438
93. Thung DT, Beulen L, Hehir-Kwa J, Faas BH (2015) Implementation of whole genome massively parallel sequencing for noninvasive prenatal testing in laboratories. Expert Rev Mol Diagn 15(1):111–124. https://doi.org/10.1586/14737159.2015.973857
94. Mellis R, Chandler N, Chitty LS (2018) Next-generation sequencing and the impact on prenatal diagnosis. Expert Rev Mol Diagn 18(8):689–699. https://doi.org/10.1080/14737159.2018.1493924
95. Qiao F, Hu P, Xu Z (2019) Application of next-generation sequencing for the diagnosis of fetuses with congenital heart defects. Curr Opin Obstet Gynecol 31(2)
96. Epstein RJ, Lin FP (2017) Cancer and the omics revolution. Aust Fam Physician 46(4): 189–193
97. Hinkson IV, Davidsen TM, Klemm JD, Kerlavage AR, Kibbe WA, Chandramouliswaran I (2017) A comprehensive infrastructure for big data in cancer research: accelerating cancer research and precision medicine. Front Cell Dev Biol 5:83. https://doi.org/10.3389/fcell.2017.00083
98. Griffith OL, Gray JW (2011) Omic approaches to preventing or managing metastatic breast cancer. Breast Cancer Res 13(6): 230. https://doi.org/10.1186/bcr2923
99. Harrod A, Lai C-F, Goldsbrough I, Simmons GM, Oppermans N, Santos DB, Győrffy B, Allsopp RC, Toghill BJ, Balachandran K, Lawson M, Morrow CJ, Surakala M, Carnevalli LS, Zhang P, Guttery DS, Shaw JA, Coombes RC, Buluwela L, Ali S (2022) Genome engineering for estrogen receptor mutations reveals differential responses to anti-estrogens and new prognostic gene signatures for breast cancer. Oncogene 41(44): 4905–4915. https://doi.org/10.1038/s41388-022-02483-8
100. Ross-Innes CS, Stark R, Teschendorff AE, Holmes KA, Ali HR, Dunning MJ, Brown GD, Gojis O, Ellis IO, Green AR, Ali S, Chin SF, Palmieri C, Caldas C, Carroll JS (2012) Differential oestrogen receptor binding is associated with clinical outcome in breast cancer. Nature 481(7381):389–393. https://doi.org/10.1038/nature10730
101. Hodis E, Watson IR, Kryukov GV, Arold ST, Imielinski M, Theurillat JP, Nickerson E, Auclair D, Li L, Place C, Dicara D, Ramos AH, Lawrence MS, Cibulskis K, Sivachenko A, Voet D, Saksena G, Stransky N, Onofrio RC, Winckler W,

Ardlie K, Wagle N, Wargo J, Chong K, Morton DL, Stemke-Hale K, Chen G, Noble M, Meyerson M, Ladbury JE, Davies MA, Gershenwald JE, Wagner SN, Hoon DS, Schadendorf D, Lander ES, Gabriel SB, Getz G, Garraway LA, Chin L (2012) A landscape of driver mutations in melanoma. Cell 150(2):251–263. https://doi.org/10.1016/j.cell.2012.06.024

102. Shen T, Pajaro-Van de Stadt SH, Yeat NC, Lin JC (2015) Clinical applications of next generation sequencing in cancer: from panels, to exomes, to genomes. Front Genet 6:215. https://doi.org/10.3389/fgene.2015.00215

103. Dutton-Regester K, Hayward NK (2012) Whole genome and exome sequencing of melanoma: a step toward personalized targeted therapy. Adv Pharmacol 65:399–435. https://doi.org/10.1016/b978-0-12-397927-8.00013-0

104. Tan PY, Chang CW, Chng KR, Wansa KD, Sung WK, Cheung E (2012) Integration of regulatory networks by NKX3-1 promotes androgen-dependent prostate cancer survival. Mol Cell Biol 32(2):399–414. https://doi.org/10.1128/mcb.05958-11

105. Bhatia-Gaur R, Donjacour AA, Sciavolino PJ, Kim M, Desai N, Young P, Norton CR, Gridley T, Cardiff RD, Cunha GR, Abate-Shen C, Shen MM (1999) Roles for Nkx3.1 in prostate development and cancer. Genes Dev 13(8):966–977. https://doi.org/10.1101/gad.13.8.966

106. Bray F, Ferlay J, Soerjomataram I, Siegel RL, Torre LA, Jemal A (2018) Global cancer statistics 2018: GLOBOCAN estimates of incidence and mortality worldwide for 36 cancers in 185 countries. CA Cancer J Clin 68(6):394–424. https://doi.org/10.3322/caac.21492

107. Bandyopadhyay S, Mitra R, Maulik U, Zhang MQ (2010) Development of the human cancer microRNA network. Silence 1(1):6. https://doi.org/10.1186/1758-907x-1-6

108. Harfe BD (2005) MicroRNAs in vertebrate development. Curr Opin Genet Dev 15(4):410–415. https://doi.org/10.1016/j.gde.2005.06.012

109. Xie L, Yang Z, Li G, Shen L, Xiang X, Liu X, Xu D, Xu L, Chen Y, Tian Z, Chen X (2013) Genome-wide identification of bone metastasis-related microRNAs in lung adenocarcinoma by high-throughput sequencing. PLoS One 8(4):e61212. https://doi.org/10.1371/journal.pone.0061212

110. Daugaard I, Venø MT, Yan Y, Kjeldsen TE, Lamy P, Hager H, Kjems J, Hansen LL (2017) Small RNA sequencing reveals metastasis-related microRNAs in lung adenocarcinoma. Oncotarget 8(16):27047–27061. https://doi.org/10.18632/oncotarget.15968

111. Han YN, Li Y, Xia SQ, Zhang YY, Zheng JH, Li W (2017) PIWI proteins and PIWI-interacting RNA: emerging roles in cancer. Cell Physiol Biochem 44(1):1–20. https://doi.org/10.1159/000484541

112. Liu Y, Dou M, Song X, Dong Y, Liu S, Liu H, Tao J, Li W, Yin X, Xu W (2019) The emerging role of the piRNA/piwi complex in cancer. Mol Cancer 18(1):123. https://doi.org/10.1186/s12943-019-1052-9

113. Kosaka N, Iguchi H, Ochiya T (2010) Circulating microRNA in body fluid: a new potential biomarker for cancer diagnosis and prognosis. Cancer Sci 101(10):2087–2092. https://doi.org/10.1111/j.1349-7006.2010.01650.x

114. Luthra R, Chen H, Roy-Chowdhuri S, Singh RR (2015) Next-generation sequencing in clinical molecular diagnostics of cancer: advantages and challenges. Cancers (Basel) 7(4):2023–2036. https://doi.org/10.3390/cancers7040874

115. Robins H (2011) Detecting and monitoring lymphoma with high-throughput sequencing. Oncotarget 2(4):287–288. https://doi.org/10.18632/oncotarget.270

116. Ferret Y, Caillault A, Sebda S, Duez M, Grardel N, Duployez N, Villenet C, Figeac M, Preudhomme C, Salson M, Giraud M (2016) Multi-loci diagnosis of acute lymphoblastic leukaemia with high-throughput sequencing and bioinformatics analysis. Br J Haematol 173(3):413–420. https://doi.org/10.1111/bjh.13981

117. He J, Wu J, Jiao Y, Wagner-Johnston N, Ambinder RF, Diaz LA Jr, Kinzler KW, Vogelstein B, Papadopoulos N (2011) IgH gene rearrangements as plasma biomarkers in Non-Hodgkin's lymphoma patients. Oncotarget 2(3):178–185. https://doi.org/10.18632/oncotarget.235

118. Alonso CM, Llop M, Sargas C, Pedrola L, Panadero J, Hervás D, Cervera J, Such E, Ibáñez M, Ayala R, Martínez-López J, Onecha E, de Juan I, Palanca S, Martínez-Cuadrón D, Rodríguez-Veiga R, Boluda B, Montesinos P, Sanz G, Sanz MA, Barragán E (2019) Clinical utility of a next-generation sequencing panel for acute myeloid leukemia diagnostics. J Mol Diagn 21(2):228–240. https://doi.org/10.1016/j.jmoldx.2018.09.009

119. Gray PN, Dunlop CL, Elliott AM (2015) Not all next generation sequencing diagnostics are created equal: understanding the nuances of solid tumor assay design for somatic mutation detection. Cancers (Basel) 7(3):1313–1332. https://doi.org/10.3390/cancers7030837
120. Desai K, Hooker G, Gilbert K, Cropper C, Metcalf R, Kachroo S (2021) Real-world trends in costs of next generation sequencing (NGS) testing in U.S. setting. J Clin Oncol 39(15_suppl):e18824–e18824. https://doi.org/10.1200/JCO.2021.39.15_suppl.e18824
121. Luthra R, Patel KP, Routbort MJ, Broaddus RR, Yau J, Simien C, Chen W, Hatfield DZ, Medeiros LJ, Singh RR (2017) A targeted high-throughput next-generation sequencing panel for clinical screening of mutations, gene amplifications, and fusions in solid tumors. J Mol Diagn 19(2):255–264. https://doi.org/10.1016/j.jmoldx.2016.09.011
122. Bacher U, Shumilov E, Flach J, Porret N, Joncourt R, Wiedemann G, Fiedler M, Novak U, Amstutz U, Pabst T (2018) Challenges in the introduction of next-generation sequencing (NGS) for diagnostics of myeloid malignancies into clinical routine use. Blood Cancer J 8(11):113. https://doi.org/10.1038/s41408-018-0148-6
123. Gaide O, Emerson RO, Jiang X, Gulati N, Nizza S, Desmarais C, Robins H, Krueger JG, Clark RA, Kupper TS (2015) Common clonal origin of central and resident memory T cells following skin immunization. Nat Med 21(6):647–653. https://doi.org/10.1038/nm.3860
124. Matos TR, de Rie MA, Teunissen MBM (2017) Research techniques made simple: high-throughput sequencing of the T-cell receptor. J Invest Dermatol 137(6):e131–e138. https://doi.org/10.1016/j.jid.2017.04.001
125. Kirsch IR, Watanabe R, O'Malley JT, Williamson DW, Scott LL, Elco CP, Teague JE, Gehad A, Lowry EL, LeBoeuf NR, Krueger JG, Robins HS, Kupper TS, Clark RA (2015) TCR sequencing facilitates diagnosis and identifies mature T cells as the cell of origin in CTCL. Sci Transl Med 7(308):308ra158. https://doi.org/10.1126/scitranslmed.aaa9122
126. Vandeputte M (2021) The journey from next-generation sequencing to personalized medicine? Biochemist 43(6):4–8. https://doi.org/10.1042/bio_2021_192
127. Ley TJ, Mardis ER, Ding L, Fulton B, McLellan MD, Chen K, Dooling D, Dunford-Shore BH, McGrath S, Hickenbotham M, Cook L, Abbott R, Larson DE, Koboldt DC, Pohl C, Smith S, Hawkins A, Abbott S, Locke D, Hillier LW, Miner T, Fulton L, Magrini V, Wylie T, Glasscock J, Conyers J, Sander N, Shi X, Osborne JR, Minx P, Gordon D, Chinwalla A, Zhao Y, Ries RE, Payton JE, Westervelt P, Tomasson MH, Watson M, Baty J, Ivanovich J, Heath S, Shannon WD, Nagarajan R, Walter MJ, Link DC, Graubert TA, DiPersio JF, Wilson RK (2008) DNA sequencing of a cytogenetically normal acute myeloid leukaemia genome. Nature 456(7218):66–72. https://doi.org/10.1038/nature07485
128. Beltrand J, Busiah K, Vaivre-Douret L, Fauret AL, Berdugo M, Cavé H, Polak M (2020) Neonatal diabetes mellitus. Front Pediatr 8. https://doi.org/10.3389/fped.2020.540718
129. Lightbody G, Haberland V, Browne F, Taggart L, Zheng H, Parkes E, Blayney JK (2019) Review of applications of high-throughput sequencing in personalized medicine: barriers and facilitators of future progress in research and clinical application. Brief Bioinform 20(5):1795–1811. https://doi.org/10.1093/bib/bby051

Correction to: The Study of Small RNA Sequencing from Biological Samples (miRNA-seq)

Weam Othman Elbezanti, Kishore B. Challagundla, and Manoj K. Pandey

Correction to:
Chapter 10 in: Valerie J. Carabetta and Olaitan Akintunde (eds.),
High Throughput Gene Screening: Methods and Protocols,
Methods in Molecular Biology, vol. 2866,
https://doi.org/10.1007/978-1-0716-4192-7_10

The chapter "The Study of Small RNA Sequencing from Biological Samples (miRNA-seq)", written by Weam Othman Elbezanti, Kishore B. Challagundla, and Manoj K. Pandey, was originally published with incorrect affiliation of Professor Weam Othman Elbezanti. The correct affiliation of the author was Department of Surgery, Cooper University Hospital, Camden, New Jersey, USA. The affiliation details have been updated in the chapter.

The updated version of this chapter can be found at https://doi.org/10.1007/978-1-0716-4192-7_10

Valerie J. Carabetta and Olaitan Akintunde (eds.), *High Throughput Gene Screening: Methods and Protocols*, Methods in Molecular Biology, vol. 2866, https://doi.org/10.1007/978-1-0716-4192-7_17,

Index

A

APOBEC 74, 83, 93
Assays for transposase-accessible chromatin using sequencing (ATAC-seq) 17, 111–120, 166, 167, 272

B

Bacteria 41, 62, 68, 189–191, 193, 197, 199–202, 231–242, 278, 294, 295
Bacterial genomes 31–42, 119
Bar-coded primer 65
Bismark 77, 87, 95
Bisulphite sequencing (BS-seq) 74, 75, 84, 91, 92, 94, 272

C

Cancers 14–16, 51, 59, 60, 112, 159, 176, 224, 265–282, 288, 297, 301–306
Capture 14, 15, 18, 46, 49, 52, 93, 125, 130, 160, 162, 165, 209, 221–224, 231–242, 255, 270
Cell lysis 102, 103, 108, 115, 116, 232, 233, 235
Chromatin 12, 17, 18, 99–101, 103, 104, 107, 108, 111, 112, 160, 165–167, 245–247, 249, 259, 272, 273, 278, 301
Chromatin accessibility 101, 112, 114, 166, 167, 272, 273
Chromosome conformation 18, 231–242, 247
Complementary DNA (cDNA) 4, 15, 16, 125–137, 139, 140, 142–155, 157, 160, 162, 164, 165, 167, 168, 176, 177, 179–187, 216, 218, 220, 221, 226, 252, 256, 257, 269

D

Data analysis 16, 22, 46, 50–51, 78, 150, 169, 170, 183, 259
Deep sequencing 231–242, 247, 249, 294
DNA methylation 73, 76, 87, 92, 112, 247, 249, 272, 273
DNA sequencing (DNA-seq) 3–6, 9, 12, 14–16, 21, 31–33, 208, 212, 249, 266, 267, 279, 290

E

Enzymatic methyl-sequencing (EM-seq) 73–96

F

Formaldehyde modification 210
Formalin fixed paraffin embedded tissue 281

G

Gene expressions 16, 59, 99, 100, 111–113, 125, 159–170, 175, 189, 190, 193, 198, 199, 201, 202, 231, 268, 269, 277, 279, 295, 301, 303
Genome 3, 31, 45, 75, 99, 112, 159, 190, 249, 266, 287
Genome organization 18, 231–242, 246
Gut microbiome 59–69, 279, 281

H

Hfq 190–192, 195–199
Hi-C 18, 231–242, 247
High-throughput sequencing (HTS) 5, 15, 87, 112, 269
Hybridization capture 46

I

Infections 112–115, 117, 118, 120, 199, 201, 269, 281, 293, 294, 296, 297

L

Library generation 115–117
Library preparation 12, 19, 33, 42, 46, 48, 74, 75, 78, 84, 92, 93, 105–107, 125–127, 145–147, 164, 168, 177, 208, 210, 220, 234, 239, 249, 258
Long-read assembler 34, 35

M

Machine learning 190, 196–197, 202
Massively parallel sequencing (MPS) 265–282
Micrococcal nuclease (MNase) 17, 100–104, 108, 247

Valerie J. Carabetta and Olaitan Akintunde (eds.), *High Throughput Gene Screening: Methods and Protocols*, Methods in Molecular Biology, vol. 2866, https://doi.org/10.1007/978-1-0716-4192-7,

Micrococcal nuclease digestion and sequencing (MNase-seq) 109, 112
MicroRNAs (miRNAs) 16, 175–179, 181–184, 186, 187, 193, 303
MNase digestion 100–105, 108

N

Nextflow 78, 84, 85, 93, 95
Next-generation sequencing (NGS) 3–22, 32, 45–53, 59–69, 73, 100, 103, 105, 107–109, 176, 177, 179–183, 185–187, 207, 208, 212, 213, 223, 245–260, 266–269, 274, 276–282, 287–307
Normalization 61, 66, 114, 116, 118, 185, 249, 250
Nucleic acid stabilizer 62
Nucleosome occupancy 112

O

Oncogenesis 265–282, 305
Oxford Nanopore 13, 31–42, 290

P

Pathogenicity predictions 46, 50, 51
Personalized medicine 287–307
Poly(A) selection 126, 127, 129–136

Q

qPCR 115–118, 217, 223, 260

R

Regulatory networks 189, 190, 195, 202
Regulatory RNAs 196, 202
Ribosomal depletion 153
RIL-seq 190–202
RNA-induced silencing complex (RISC) 176
RNA-sequencing (RNA-seq) 15, 16, 125–157, 159, 160, 196, 208, 215, 218–220, 223, 268, 297, 298

S

Second generation sequencing 266
SeqMonk 78, 87–91
Sequential extraction 208, 210–212, 224
Short-read sequencing 50, 59–69, 168
Single-cell 14–16, 159–170, 272
Single cell RNA-seq (scRNA-seq) 16, 159–170, 269, 272, 277
Single-molecule sequencing (SMS) 12–13, 270, 282, 290
Small RNAs (sRNAs) 189–202, 303
16S rRNA sequencing 66
SYBR stain 178, 183

T

Targeted enrichment 47, 49–50
Ten-eleven translocation methylcytosine dioxygenase 2 (TET2) 74, 81, 82, 93, 94
T4-phage beta-glucosyltransferase (T4-BGT) 74, 93
Transcriptomes 15, 160, 162, 165, 195, 198, 220, 268, 269, 272, 274, 276, 291, 296, 303
Transposition 115, 116, 118, 119, 165, 167

V

Virus infection 111–120

W

Whole-genome bisulfite sequencing (WGBS) 17, 73–76, 87, 89, 93, 95

MIX
Papier aus verantwortungsvollen Quellen
Paper from responsible sources
FSC® C105338

If you have any concerns about our products,
you can contact us on
ProductSafety@springernature.com

In case Publisher is established outside the EU,
the EU authorized representative is:
Springer Nature Customer Service Center GmbH
Europaplatz 3, 69115 Heidelberg, Germany

Printed by Libri Plureos GmbH
in Hamburg, Germany